FLYING KNOW IT ALL

To

Bow, Mos, and Ray

FLYING KNOW IT ALL

A–Z simply explaining
piloting, navigation, instruments,
airliners, business jets, GA, drones,
military aircraft, missiles, MAD,
AF447, and little-known facts

by

Christopher Bartlett

Published December 2025

ISBN 978-0-9560723-9-9

Publisher
OpenHatch Books
"Opening the hatch
of say a spaceship
on a new world."

Gillingham, Dorset, UK
https://chrisbart.com

PREFACE

It will take time to realize just how many "gems" there are in this book.

While it can be handy for reference, it is outstanding as a tool for creating a treasure trove of in-depth knowledge.

It differs from conventional aviation reference books by adding a human and sometimes controversial touch, thereby making it more engaging and stimulating to browse.

The 640-page printed version is a great present for those whose work involves aviation, including the military, not forgetting the young who can also benefit from learning some background facts which few of them have been told about these days.

Note the *Table of Contents* is at the end of the book in the printed version, and at the beginning in the Kindle where one only has to click on an entry to go straight to it.

An asterisk * indicates the event (crash etc.) is in our classic book

Air Crashes

and

Miracle Landings

FYI

https://chrisbart.com

NOTE!

TABLE OF CONTENTS
AT THE END

~ A ~

ACARS (Aircraft Communications, Addressing, and Reporting System)

ACARS, pronounced AY-CARS, started off in 1978 as a crude system for reporting events, such as the closing and opening of the cabin doors, back to the airline. It paid for itself by preventing crews falsifying their time sheets.

Since then, ACARS has become ever more sophisticated, reporting in-flight technical information and abnormalities not only to the airline but also to the engine maker. The airline can therefore have the appropriate parts ready when the aircraft lands, and, if the problem is deemed too serious, another aircraft can be prepped. The engine maker has even on occasion called the airline while the aircraft is still airborne to recommend a change of regime to protect the engine.

Being a means of digital communication, ACARS is also used for many other purposes, such as requesting updated weather reports, air traffic control clearances, and to communicate with Dispatch. However, in the US and over the Atlantic, the separate **CPDL (Controller–Pilot Data Link)** is now increasingly being used for clearances.

ACARS featured in three major disasters:

1. **9/11 The Warning Sent by United Airlines to Flight UA93*.**
   [* indicates an account in *Air Crashes and Miracle Landings*]
   After three of the hijacked airliners had struck New York's Twin Towers and the Pentagon, United Airlines used ACARS to warn the pilots of the delayed fourth hijacked aircraft (UA93) not to allow access to the cockpit.
   Unfortunately, it was not worded strongly enough to prevent that happening, not that cockpit doors were at the time reinforced enough to prevent determined entry.

Having had time to learn the fate of the other airliners, the passengers on UA93 confronted the hijackers with the struggle ending in the aircraft crashing in open country.
See **9/11: Two Unarmed F-16s Scrambled to Down 4th Airliner**.

2. **Air France AF447**
ACARS first came to public attention in 2009 when Air France Flight AF447* en route from Rio de Janeiro to Paris disappeared over the South Atlantic with no MAYDAY call . The only clues apart for a small amount of wreckage were ACARS messages until the voice and data recorders were recovered from the bottom of the ocean two years later.
See **AIR FRANCE AF447 DROPS FROM 37,924 FT INTO SEA***

3. **Greatest-Ever Aviation Mystery**
MH370*, Malaysian Airlines Boeing 777 flight which mysteriously disappeared on a flight between Kuala Lumpur and Beijing in 2014, never to be found. Clever analysis of ACARS "handshakes" showed it ended up thousands of miles away off Perth, Australia, in the South Indian Ocean in the opposite direction to that it was meant to be going.
See **MH370 "Greatest Aviation Mystery Ever."**

ACAS (Airborne Collision Avoidance System)

See **TCAS**, which is really the same thing.

Accelerate–Stop Distance **on Runway**

Overall distance required to accelerate from commencement of the takeoff run to takeoff decision speed (V_1), abort the takeoff, and then brake safely to a halt, with some runway remaining. In this test, the brakes become exceedingly hot, and the overheat protection plugs in the wheels may melt to release the air (nitrogen) to prevent the tires from exploding.

See **Taking Off (Critical V-Speeds).**

Accident Models/Accidentology

Some of the best academic work on accidents has been done by academics working in fields outside aviation such as nuclear power where the fate of millions of people might be at stake.

There are dozens, if not hundreds, of books on the subject, which can seem so simple at some levels and so complex at others. Though not for the fainthearted, Wiegmann and Shappell's, *A Human Error Approach to Aviation Accident Analysis*, published by Ashgate, explains how various academics have treated the subject.

See **Air Crashes, Swiss Cheese Accident Model**, and **Normal Accident**.

ACE (Agile Combat Employment): Dispersed "Austere" Airbases

The US Air Force has mainly been combatting second- (or, rather, third-tier) adversaries posing little threat to its many overseas airbases. However, a peer or near-peer adversary could render such bases inoperable, or at the very least unable to service aircraft, simply by attacking them with swarms of drones or hypersonic missiles, some of which would certainly overcome even the most sophisticated defenses.

Consequently, from 2017 onwards the US Air Force began to contemplate a new *modus operandi* called ACE (Agile Combat Employment). This consists of establishing multiple "austere" airbases on US islands and at partner allies' airfields, the idea being that multiple bases would dilute any attack, with some bases only suffering limited damage or even not attacked at all.

Ideally, the US Air Force would pre-position equipment, but where not possible, airlift dispositions would be put in place to fly it in. Backup and maintenance personnel would have to be proficient in multiple specialties rather than in just one as is usually the case now.

It is taking time to implement partly because getting budgetary approval from Congress for the extra equipment is a lengthy process. Even so, short takeoff and landing (STOL) aircraft not requiring obvious runways are facilitating its realization, though hangar-type facilities are still required for some maintenance and servicing, especially of stealth aircraft.

The US Army can already engage in ACE-like operations, covertly moving rocket launchers about as the Germans did in forests with the V-2 in World War II.

Islands such as Guam, seen by the US military as fixed **Aircraft Carriers**, could even today be used to launch Precision Strike Missiles (PrSMs) able to target moving objects such as ships in the South China Sea. In fact, arrangements have been made for a company to create facilities in Guam for the servicing of off-base **Standoff** operations. Furthermore, improving relations between the US and the Philippines could provide extensive areas for similar ACE deployments there.

ACE is not a new concept. At the height of the Cold War, straight sections of the German autobahns were hardened for conversion to takeoff/landing strips for NATO aircraft with the sides widened and barriers instantly removable. The road to the airport in Singapore is one such example—and many other countries are doing likewise.

While ACE largely covers a *modus operandi* for full-size aircraft, in late 2023, the US military revealed a plan called **Replicator** that would involve the manufacture of thousands of drones in the belief that deploying so many drones with the US's flair for innovation would more than compensate for China's numerical advantages.

Following the destruction in 2025 of many expensive Russian bombers at major airbases inside Russia by Ukraine using drones launched from "civilian" vehicles brought in surreptitiously and parked nearby, many governments believe that their aircraft are similarly vulnerable.

With publicly available satellite photos showing rows of expensive military aircraft parked out in the open at bases in the US such aircraft would be vulnerable there too. The problem is that they often need long runways, and an ACE bases could be located and attacked. Expensive sites with storage under thick concrete roofs would be necessary.

Aces Learnt from Mistakes, Value of Confidential Reporting

In World War II, some German fighter aces claimed the experience of being shot down several times—and, of course, surviving—was what made them so good. The same was no doubt true for Battle of Britain pilots flying over home soil and likely to be picked up by their own people, and even in some cases *shot down in the morning and up again in the afternoon.*

Incidentally, in World War II, US aces were sent back to the US to train new pilots, unlike the German ones who were kept on station, scoring more but less likely to survive.

All this shows how one learns from mistakes and the great value of schemes whereby pilots can report incidents anonymously for the benefit of all.

See **ASRS (Aviation Safety Reporting System)**, a scheme run by NASA in the US for the confidential reporting of incidents. This corresponds to **CHIRP** in the UK and **EUCARE** in Europe.

There is also the system called **OFDM (Operational Flight Data Monitoring)** that automatically reports to an outside organization flight deviations from the norm, such as coming in too high above the glide path. The outside organization analyses the data and can notably pick up on dangerous trends.

Elsewhere, we have mentioned how artificial intelligence (AI) incorporated in autopilots could learn how mistakes by pilots are made and work out how crashes could be avoided. Apparently, such AI systems can take the knowledge gained on one type of airliner and apply it to another.

Acronyms

Every new project envisaged by the military in the US seems to have an acronym, partly because the names are such mouthfuls. Even though acronyms are usually explained at the beginning of articles and speeches, anyone not working in that field can soon forget what they mean and lose the thread.

In the headings, we put the most common acronyms first with the full meaning in brackets afterwards. Otherwise, it is the other way round, with the acronym second.

Acronyms with Number for Repeated Letter [Military]

In military usage, a number (2, 3 …) in an acronym indicates a repeated letter such as in the quite common C2 (for CC) which *can* stand for Command and Control (CC). There is C3 for Command, Control, and Communications, and even C4 in the often used C4ISR standing for Command, Control, Communications, Computers, Intelligence, Surveillance, and Reconnaissance!

The US Air Force alone has thousands of acronyms, and, in fact, one can have a handy acronym for anything in the military.

Active Sidesticks

Airbus is considering introducing active sidesticks where they are electro-mechanically linked so they move together, and the pilot can see what their colleague is doing.

The Gulfstream G500 and G600 business jets now have active sidesticks of which the company is very proud.

"Adam" PA Code for Missing Child, "Bravo" PA Code to Scare People

Codes allow organizations to convey messages over public address systems without causing undue alarm. A code has the further advantage of being concise and something that staff can be trained for, thus enabling them to spring into action immediately.

A well-known one that may be heard at an airport in the USA is "Adam," meaning a child is missing. It is derived from the name of the six-year-old son of a host of Fox TV's *America's Most Wanted* who was abducted from a Sears department store in Florida in 1981 and sadly subsequently found dead.

One code, "Bravo," is an exception to the rule of not being intended to alarm people in that it is intended to do just the opposite, and make them cower or even scatter enabling, say, an individual with an explosive device to be identified. With Bravo having become too well known, other scary codes are being conjured up to replace it, but remember that on a cruise ship Bravo can still mean there is a fire on board.

Of course, the **Chimes** and dings heard by passengers on airliners are codes, usually without sinister meanings, such as that a flight attendant should answer the intercom or that the aircraft has reached 10,000 feet, and they can leave their seats by the exits and start their tasks.

As meanings differ from airline to airline, we cannot cite them usefully. Furthermore, many airlines will not publicize them because of the security implications, although Qantas did once without including those relating to security.

Codes on board the aircraft can even be earmarked for particular passengers. One such being, "Is there a friend of Bill on board?" intended for anyone belonging to Alcoholics Anonymous (AA). (On joining AA, one is always told about how it was founded by "Bill.")

Adaptive Cycle Engine (ACE) / Variable Cycle Engine (VCE)

Traditionally, aircraft engines have been designed to perform optimally under specific regimes, be it takeoff, subsonic, supersonic, or hypersonic (over Mach 5) flight.

Fighter aircraft engines pose a particular conundrum, for they must, on the one hand, be able to maneuver with maximum thrust in actual combat but, on the other hand, have the range to reach their combat area with sufficient **Loiter/Dwell Time**.

An ACE/VCE engine able to adapt to the different regimes would solve the problem. If the sixth-generation combat aircraft that the US envisages are to have the desired range, they would need ACE/VCE engines, which still must be perfected. The latest US fighters are often slower than the older ones because being designed to engage at considerable distances from the enemy, not such a premium is put on speed, with range considered more important.

On the civilian aviation side, ACE/VCE would really be essential for economical supersonic airliners, which need great thrust at slow speed for takeoff and the ability to fly at various speeds sub-sonically and, finally, supersonically. The supersonic Anglo-French Concorde airliner had to use after burners to take off.

Of late, there have been great advances in engine design with Lockheed Martin now talking about a **Turbine-Based Combined Cycle (TBCC) Engine** for the ~ **Lockheed SR-72 "Son of Blackbird**."

ADF (Automatic Direction Finder)

Equipment that determines the direction of a radio transmitter. Can be tuned to an **NDB (Nondirectional Beacon)** or even to an AM radio station, and therefore it is not limited to line-of-sight broadcasting stations, as would be the case with FM.

Can be set in two modes:

1. Needle pointing to transmitter (relative to the nose of aircraft).

2. Needle pointing relative to magnetic north.

ADIRU (Air Data Inertial Reference Unit)

An important unit on the modern airliner (beginning with the A320 and 777). It integrates two key sources of data for not only the instrument displays but also the autopilot and associated systems needed to fly the aircraft.

The ADIRU usually has two distinct partitions with independent processors:

1. The Air Data partition that provides critical air data including airspeed, **Mach**, angle of attack, air temperature, and barometric altitude.

2. The Inertial Reference partition that provides attitude, heading, ground speed, and current position data used for navigation. It also measures body accelerations and G for flight control.

Units providing air data to the ADIRU, such as the angle of attack sensors and pitot tubes, are relatively simple but crucial.

Those providing the inertial reference data are highly sophisticated, consisting nowadays of **Ring Laser Gyroscopes** and linear accelerometers. Compared with the traditional mechanical "spinning-top" gyroscopes, these modern laser gyroscopes are infinitely more robust and reliable. Incidentally, they do not resist toppling.

~ Initiation (Alignment) of the ADIRU or IRS

The pilots must initialize their ADIRU or if, as on the Boeing 737, they do not have one, their stand-alone **Inertial Reference System (IRS)**, every time they are powered up, so the system knows where they are and in which direction the aircraft is pointing. This is called alignment and must be done—usually before starting off at the airport—without the aircraft being buffeted by high winds or shaken by goods or passengers being loaded.

To carry out the initiation the pilots enter the latitude and longitude coordinates for their geographic coordinates into the system and wait while it determines the aircraft heading by sensing the roll and pitch of the aircraft as it moves in conjunction with the rotation of the Earth. Alignment is fastest at the Equator where the Earth's angular rotation is greatest, taking only 5 minutes there, 10 minutes at latitude 60 degrees (Oslo, St. Petersburg, Whitehorse [Canada]), and 17 minutes at extreme latitudes, but, of course, not at the poles.

It has built-in compensation to account for the rotation of the Earth—otherwise after 12 hours it would think the aircraft is upside down!

In the past there have been rare instances of problems caused by malfunctioning of the ADIRU, oddly enough with two or three occurring off Perth Australia, leading to suggestions they could have been caused by emissions from some military installation. The most notable case was Qantas Flight 72*. However, Australian investigators could find no evidence of such a cause.

Over long distances, inertial reference positioning can drift by as much as 10 nautical miles in contrast to **GPS**, which, all being well, retains its accuracy. However, one never knows if GPS positioning will be available, with it possible for the system to be disabled, say, in time of war, or hacked by terrorists. Though less accurate, and prone to drifting, inertial navigation is here to stay as a trusty backup. Its inherent reliability is why, for instance, it is used in **Missiles**.

See **Quantum Inertial Reference Systems** to see how one day inertial reference systems could become even more accurate or rather less subject to drift over long distances.

Administration or Agency? [US Usage]

Journalists not brought up in the US should note that the *A* at the end of the acronyms for well-known US organizations might well stand for administration and not agency.

Administrations:

1. FAA (Federal Aviation Administration)
2. NASA (National Aeronautics and Space Administration)
3. TSA (Transportation Security Administration)

Agencies:

1. CIA (Central Intelligence Agency)
2. FEMA (Federal Emergency Management Agency)
3. NSA (National Security Agency)

A *Department* is what is called a *Ministry* in other countries. Note that TSA and FEMA are both part of the Department of Homeland Security (DHS), created in the aftermath of **9/11**.

ADS (Automatic Dependent Surveillance)

"Automatic dependent surveillance" is easier to understand if one bears in mind that *dependent* means absolutely everything depends on information *transmitted* by the aircraft (including geographical position determined by GPS and possibly cross-checked with inertial navigation), rather than reflection of a radar beam as in military radar (see "Primary radar" under **Radar**).

There are two ways in which ADS is used: ADS-B, where B stands for *broadcast, and ADS-C, where C stands for contract.* They are explained in detail below.

ADS-B (Automatic Dependent Surveillance–BROADCAST)

See **NextGen**.

This is the type of ADS most widely used, with the other, ADS-C, mainly limited to oceanic areas with poor or no VHF radio coverage.

ADS-B is transmitted permanently by radio and can be picked up by aircraft up to 100 nautical miles away or more and, of course, by controllers on the ground. In fact, anyone on the ground with a suitable receiver can pick it up and many indeed do, passing details on to aircraft tracking systems such as Flightradar24.

Those transponders automatically broadcast a full range of information far exceeding that "squawked" by the traditional one.

And notably:

1. Height, ground speed, and airspeed.

2. Exact position (determined by GPS and cross-checked with other navigational instruments such as inertial navigation).

3. Climb or sink rate, rate and direction of turn, or absence thereof.

4. Latest control inputs, thus indicating what the aircraft is about to do. Thus, the controller does not have to wait to see what is happening.

Of course, information traditionally given out by airliner transponders, such as identity, squawk code, etc., is included.

This data is picked up by both air traffic control and other aircraft within a radius of up to, say, 150 miles.

Systems such as **TCAS (Traffic Warning and Collision Avoidance Systems)** can thus be even more effective since they will know what action the pilots are about to undertake even before the aircraft reacts, something that might take crucial seconds for the controller to pick up on. Furthermore, it can reduce the risk of any collision scenario developing in the first place.

Of course, the controllers will also have their traditional **Radar** monitors, both secondary and primary.

Advantages are:

1. *Greater accuracy* allows aircraft to be closer together and means aircraft would not have to be held back so much as now when there is congestion.

2. *Works in the absence of radar coverage*, such as in isolated areas, or in the lees of mountains.

3. *Often possible to preclude potential conflicts.*

4. *Also functions on the ground*, even indicating when an aircraft is just moving off or a runway incursion is about to occur. It would have prevented the worst ever multi-aircraft aviation disaster, when two Boeing 747s collided in Tenerife in 1977.

5. *Other features*, such as weather depiction and advisories, can be incorporated.

ADS-C (Automatic Dependent Surveillance–CONTRACT)

ADS-C is mostly transmitted over the ACARS network via satellite and unlike ADS-B can operate anywhere, and notably over oceans. ADS-C can operate either through **CPDL (Controller Pilot Data Link)** communications or separately on its own.

Whereas ADS-B continuously interacts with ground controllers and other aircraft, ADS-C only works by arrangement with one or more specific Air Traffic Services Unit(s) (ATSU), hence the use of the word Contract.

As said, it only connects intermittently, say, at 15-minute intervals, although this is being reduced to a couple of minutes or so to allow aircraft to be closer together in some busy places.

On receiving a logon request, the Air Traffic Air Traffic Services Unit (ATSU) must know which contract applies.

It can be a:

1. Periodic Contract
   at regular intervals.

2. Demand Contract
   when contact is specially needed, which includes an emergency mode.

3. Event Contract
   when something specific happens such as passing a given waypoint.

The system includes **CPDL (Controller Pilot Data Link)** communications.

Although not needed in every case, ADS-C can, if required, send a vast amount of data.

The aircraft in the **AF447 ONLY A MINUTE TO SAVE THEM** crash we describe further down was equipped with ADS-C but because of a change in arrangements prior to departure was not logged on. Had it been working, the wreckage might well have been found much sooner.

AERCAP

World's largest lessor.

AEW (Airborne Early Warning)

The same as **AWACS (Airborne Early Warning and Control System)** but without the ability to control battles.

AESA (Active Electronically Scanned Array) Radar

These radars are a great feature of the very latest combat aircraft. They do not require the traditional mechanically rotating antenna and have many advantages. They are also used on the latest AWACS aircraft, which therefore no longer have the prominent large-diameter mushroom pod for a rotating antenna stuck on top.

The AESA antenna is made up of an array of hundreds of tiny elements (Transmit/Receive Modules [TRM]) making the system very reliable since failure of one will not disable the system as each is independently linked to the computer.

Its beam is "steered" electronically by controlling the phase of the radio waves transmitted and received by each of the multiple radiating elements in the antenna. This digitally controlled scanning nature of the AESA allows it to quickly scan any direction in comparison to a mechanically scanned radar whose "view" is constrained by the direction it is facing and how quickly its motors can turn it.

AESAs not only have the above advantages and outstanding performance but are also less susceptible to electronic jamming since they can vary the frequency used thus preventing the jammer broadcasting the one used to confuse it.

The varying and intermittent nature of their transmissions also means they do not have an obvious signature.

Africa's New Hub Airport, Ethiopian Airlines

With an initial completion date set for 2029, it is expected that the airport will accommodate 60 million passengers annually. Future expansions will increase capacity to 110 million passengers per year.

Initially, it will have two parallel runways and will be built at Bishoftu, about 40 km (25 miles) southeast of the capital Addis Ababa, and stretch across 35 square km (13.5 square miles).

Though Africa with its many airlines is seen as somewhat dangerous as regards flying, Ethiopian Airlines has a very good reputation as a well-run airline and will make the new airport a significant hub.

After Burner

Sometimes aircraft such as fighters or the supersonic Concorde airliner may at times need more thrust than the engine or engines can nominally supply. This can be achieved by injecting fuel into a combustion zone (called the after burner) situated after the final turbine with the conflagration providing additional thrust.

However, this is very inefficient, resulting in much greater fuel consumption. A fighter might only need this for 30 seconds or a minute in a combat situation, but even this, if they have been on station for long, may risk them not having enough fuel to get home.

Concorde only needed the after burners to compensate for extra drag on takeoff and when going through the sound barrier. On takeoff the extra drag arose when the aircraft rotated and the large delta wing attained a sharp angle to the airflow.

See **Adaptive Cycle Engine (ACE)/Variable Cycle Engine (VCE).**

Age of Aircraft Is Measured in Cycles

Though the greater the chronological age of the aircraft the greater the number of takeoffs and landings (called cycles), the number of cycles can differ greatly according to the routes flown. An ultra-long-haul aircraft might only have one cycle per day, whereas an aircraft used on short-haul flights might have as many as five or six.

In considering the age of a metallic airliner, it is the number of cycles that is regarded as most important. This is because when an aircraft climbs to cruising height the pressurized part and notably the cabin expands, with the reverse happening when it descends back to land. The metal making up the fuselage stretches and shrinks each time, and if this is repeated many times it can fail, especially at points where stress is concentrated.

Though metals and alloys initially seem to bend with no problem, they can behave oddly with repetitive bending. The phenomenon can be simply demonstrated by opening a paperclip and then repeatedly bending the wire until after a minute or so it suddenly stiffens, and then with further bending finally snaps. The same can happen on aircraft, particularly at places where stress might by design or unintentionally, say, due to repairs with one row of rivets instead of two, be concentrated.

Examples of such crashes are:

1. The world's first jetliner, the Comet, where in 1954* the fuselage failed.

2. Japan Airlines 747 Flight JL123 in 1985* where the rear bulkhead of the Boeing 747 failed due to a repair with one row of rivets instead of two.

3. Aloha Airlines 737 Flight 243 in 1988* where part of cabin roof blew off leaving passengers alive but sitting in the open. Repetitive short hops between the Hawaiian Islands combined with age and the Hawaiian hot damp air had left the fuselage weakened. See **Decompression (Loss of Cabin Pressure)**.

Nowadays, with some airliners such as the Boeing 787 and Airbus A350 having fuselages made of **Composites**, the situation is somewhat different in that they are not affected by repetitive stress in the same way. The problem there is detecting any hidden weakness since they are not electrically conductive.

Interestingly, the hull of a submarine is similarly subject to aging according to the number of cycles (1 deep dive and return to surface = 1 cycle). At one time the Russians were making their submarine hulls from titanium only to find that even that super metal could become brittle and fail with too many cycles. However, modern nuclear-powered submarines do not need to surface so often, and they are outmoded over decades before repetitive-stress-fatigue becomes a factor.

Age of Pilot

The maximum permitted age for commercial airline pilots varies from country to country. With older people who have looked after their health—as airline pilots normally do—there has been a move to allow pilots in the US and elsewhere to work to 65, and even 67, rather than 60. In the US, pilots' unions and even the airlines have opposed recent suggestions that the maximum age of pilots be raised, even slightly.

This is something of a contentious issue at the airlines because of the differing financial situations of the pilots concerned. With so much salary-wise at airlines based on seniority rather than ability, junior pilots do not want to see captains staying on as seat-blockers. Senior pilots with advantageous pension plans are quite happy to retire early, but those not so blessed, perhaps having joined the airline later in their careers, may want to continue.

In France, the government has been thinking of forcing through legislation to raise the age limit from 60 to 65 in the face of the threat of strike action by some pilots as some of them are so highly paid (after so many previous strikes) that they can afford to retire early in comfort.

While the major airlines may be pretty good at weeding out those whose faculties have deteriorated with age, this may not be true for business and general aviation. It is worth noting the following article published in *Aviation Week* on June 24, 2008, by Patrick R. Veillette, PhD, entitled "Tombstone Mentality." He consulted 100 business aviation colleagues to learn their concerns about near misses with light aircraft, contaminated runways, and so on. As regards age, he says the following:

> The aging pilot population within certain segments of the business aviation industry gave serious concern to many. Colleagues who frequently fly with the over-sixty group notice deterioration in important sensory, perceptual, cognitive, and motor skills that are important in piloting. They have noticed changes to an aging pilot's ability to learn, memorize, tolerance to fatigue, sleeping habits, physical changes, and so forth.

He adds that in the domain of general aviation, the policy of avoiding two old pilots being paired together is not properly followed, a mistake the major airlines would be unlikely to make.

It should be noted that a good number of airline pilots at age 60 in the US are off work for health reasons. The ability to continue flying to age 65 or 67 probably depends very much on the individual.

In the US it is Congress and not the FAA that fixes the maximum age of pilots.

AGL (Above Ground Level)

Abbreviation used to indicate the height above ground level, as opposed to sea level. An example would be 5,000 feet AGL. On aviation charts, the height of obstacles above ground level is shown in brackets below their height above mean sea level.

"Ahead of the Aircraft" and "Behind the Aircraft" [Meaning]

Being "ahead of the aircraft" is a term used—especially when pilots are about to land—to describe the situation where they have everything well in hand and are not rushing, where they are said to be "ahead of the aircraft."

Conversely, if they are rushing, playing catch-up, and liable to make mistakes, they are said to be "behind the aircraft."

Thankfully, largely due to great advances in technology such as GPS coupled with charts of the terrain and **TCAS**, hitting mountains and midair collisions are now unlikely. However, problems tend to happen especially where a controller "kindly" allows the pilots to take a shortcut resulting in them getting behind the aircraft in their haste to adapt to the new circumstances.

The crash of American Airlines Flight 965 at Cali, Columbia, near Christmas in 1995*, shows how pilots on becoming rushed on being offered an unplanned straight-in landing requiring last-minute changes can get confused, no matter how able they are and how perfect their records.

Another example is the collision between a Lockheed Super Constellation and Douglas DC-8 over New York City in 1960 when the DC-8 was allowed to take a shortcut*.

AI Changing Aviation

Artificial intelligence can help fly the aircraft and also analyze millions of incidents, picking out trends. It can even use knowledge gained from the way pilots resolved a situation on one model of aircraft and apply it to a similar incident on a different one.

AI Changing Warfare

A hyper-intelligent Korean four- or five-year old child was once asked how the Vietnam War could be ended and said both sides should be given as many weapons as possible so they killed each other off and could fight no more. AI might come to the same unwelcome conclusion.

One feature of AI in warfare is that everything could be speeded up with terrible consequences, if not the end of the world we know, due to no time for second thoughts. The great advantage of a manned bomber as opposed to an ICBM is that it can easily be recalled and used for negotiating.

At the moment, people working on AI in warfare say they are trying to let it only go so far, with the most dreadful decisions subject to human control.

Drones equipped with AI can, for instance, seek out and be guided to targets very effectively.

AIDS (Aircraft Integrated Data System)

Air Crash Survival Greater When Shock Dissipated

Modern motor cars are designed so that the front crumples and absorbs the shock in a head-on collision. The fuselage of an aircraft can absorb considerable shock if it hits first, and even break apart into sections thereby further attenuating the deceleration in stages. Thus, the most horrendous-seeming crash with the aircraft breaking up may have more survivors than one that appears benign with it virtually intact.

An example of a seemingly benign crash but with no survivors was that of a British European Airways (BEA) tri-engine Trident in 1972*, which stalled and belly flopped onto a field three minutes after taking off from London's Heathrow Airport because someone in the cockpit mistakenly retracted the droops (like slats) on the leading edges of the wings.

Though its fuselage appeared relatively intact, and there was no fire, only one occupant survived, only to soon expire in hospital. This extreme mortality rate was because there was little under the cabin floor to absorb the pancaking shock. Had it come down nose-first with the front crumpling and perhaps breaking off, there might well have been some survivors at the rear.

Contrast that with the "miracle DC-10 crash-landing" at Sioux City Airport in 1989*, where coming in with no hydraulics and the controls therefore not working, the pilots were maneuvering by varying engine power alone. A wing touched the ground at the edge of the runway causing the aircraft to flip, spin, and break up into five pieces. Even so, there were 185 survivors. See ~ **Captains Haines and Fitch, UAL Flight 232, July 19, 1989*** and **Safest Seat**.

Ailerons

Linked airfoils at the trailing edge of the wings that control the banking of the aircraft—one goes up while the one on the opposite wing goes down. Usually there is an inner and outer one, with both deployed at low speeds, and only the inner ones at high airspeeds since with less leverage they put less stress on the wing roots.

In low airspeed situations where the ailerons may not suffice, the air brake on the wing being pushed down can be flipped up to help push it down.

Air China (CA) Mainland China's Flag Carrier

The IATA code for Air China based in Beijing is CA because AC was allocated to Air Canada. Note that the code for Taiwan's China Airlines is CI.

Other Chinese airlines flying internationally include Cathay Pacific CX, China Cargo Airlines CK, China Eastern Airlines MU, Hainan airlines HU, and Shanghai Airlines FM.

Air Crash Risk Now 40 Times Less than in 1960s and 1970s

Considering the enormous increase in the number of flights, there would be a major air crash several times a week, were the incidence the same as in the 1960s and 1970s. Now even a whole year or decade can go by without even one in the US.

The reasons for this great improvement are many and include:

1. Lessons learned.
2. Confidential reporting programs such as those mentioned under **Aces Learnt from Mistakes**.
3. **SMS (Safety Management Systems).**
4. Advances in onboard technology including **GPWS (Ground Proximity Warning System)** warning pilots if they are in danger of hitting the ground and **TCAS (Traffic Collision Avoidance System)** warning pilots if they are in danger of colliding with another aircraft and, if necessary, ordering appropriate avoidance action. Usually, one goes up and the other down. Also, improved **Weather Radar** so pilots can avoid the most dangerous storms, and so on.
5. Advances in technology on the ground such as windshear detectors at airports, radar, and other means of surveillance enabling air traffic controllers to monitor in real time the movement of aircraft on runways and taxiways at airports. Automatic warning to controllers of conflict if something is on the runway when are aircraft landing. [There was such a warning at Tokyo's Haneda Airport when a Japan Airlines Airbus A350 landed on top of small plane with six occupants going to help victims of an earthquake. However, the controller did not notice it from one minute earlier. All on the airliner evacuated safely without anyone bringing luggage!]
6. Advances on the ground and in the air such as **ADS-B (Automatic Dependent Surveillance**) helping with **Air Traffic Control (ATC)**.

7. **CRM (Crew Resource Management).**
This used to be called Cockpit Resource Management but changed to "Crew" because cabin crew, ground crew, and maintenance personnel all have a part to play. The idea behind resource management is that people work as a team with even junior people such as first officers having their views considered, unlike in the old days where the captain was like God.

Air Cushion Seats

Some premium cabin seats now use air cushions, which have advantages over foam in that they can be rendered soft when the seat is upright and hard when horizontal, besides saving a considerable amount of weight.

Air Force One

"Air Force One" used to be an inspiring term for Americans, partly because of the dramas, including movies, built around it, and partly because of the awe and respect associated with the presidency.

Even though the call sign Air Force One was first routinely used by controllers in the 1950s to designate any aircraft happening to be carrying the president of the United States, the Boeing 707 used by President Kennedy was the first dedicated aircraft to be so designated by the public—or rather the media.

In fact, there are now two Air Force Ones to ensure one is always available.

For some time, two modified Boeing 747-200s have been used as Air Force Ones. They are equipped with a shower, accommodation for the president and their family, a conference room, and quarters for the press, officials, and Secret Service. Naturally, high-tech defensive mechanisms and secure communication facilities are key features. These make them complex and very expensive.

The next Air Force Ones, ordered by President Trump, are going to be modified Boeing 747-8s, with the work taking much time and problems arising between Boeing and a subcontractor.

Boeing thought it could save money by converting two 747-8s on hand that were to go to a Russian airline that had gone bankrupt. In fact, the new more stringent FAA regulations for wiring and the extra wiring required for the president's aircraft with its sophisticated communications and other equipment meant building two new ones would be much cheaper. Boeing regrets having accepted this contract with its fixed prices.

See **Ramp Freeze, Airport Brought to Standstill for VVIP**.

In a nuclear war, it is more likely the US president would be on one of the ~ **Doomsday Planes** which have even better communications and protection against **EMP (Electromagnetic Pulse)**.

~ "Marine One," "Navy One," and "Executive One"

If the president is aboard a Marine craft such as the helicopters often seen taking off from the White House lawn, the call sign is Marine One; if it is a Navy craft it is Navy One; and if it is a civilian craft, it is Executive One.

~ Doomsday Plane (Boeing E4B)

While the US president might have to stay airborne in a crisis, there is another aircraft, the so-called *Doomsday Plane*, or **National Airborne Operations Center (NAOC)**, that similarly must be able to remain airborne in a crisis so that commanders can liaise with the president, secretary of defense, and commanders on the ground, at sea, and even under water in submarines.

Its communications facilities are far more comprehensive than those on Air Force One and include a very-low-frequency antenna that can be trailed five miles behind for communicating with submarines, and a hump on top of the fuselage for super-high-frequency and Milstar military satellite communications. There are four such aircraft based at Offutt Air Force Base just south of Omaha, with one always ready to fly at an instant's notice with one engine kept running permanently. These are getting old and need replacement.

In fact, they are being replaced by four Boeing 747-8 aircraft acquired from Korean Air by Sierra Nevada Corporation who have been awarded a $13 billion contract by the Pentagon to modify them.

9/11 (Nine-Eleven) is the only occasion the Doomsday Plane has been deployed exercising its mission for real.

The Russians have just announced a more modern Ilyushin-96-400M "Doomsday Plane" to replace the Il-80 Maxdome ones.

AIR FRANCE AF447 DROPS FROM 37,924 FT INTO SEA*

One would not normally harp on pilots' personal circumstances when writing about a fatal air crash, but do so here for a proper understanding. The details mentioned really concern an unfortunate *combination* of circumstances rather than anything truly egregious.

Air France Flight AF447, an Airbus A330 widebody, disappeared over the South Atlantic on June 1, 2009 with the cause a complete mystery. There had he been no **MAYDAY** calls or mention of trouble and only routine **ACARS** messages needed for maintenance.

Some debris including the large vertical stabilizer were found floating on the ocean off Brazil.

MEDIA CONTRIBUTED BY THE MARINHA DO BRASIL

Despite assistance from a nuclear submarine, the search failed to find the wreckage on the seabed before the batteries in the **Pingers** in the voice and data recorders ran out after a month. However, clever review and analysis of the search area by an American company enabled it to be located two years later, and the **Voice and Flight Data Recorders** recovered. To the great relief of Airbus and operators, the recorders showed there had been nothing intrinsically wrong with the A330 itself and revealed what had happened.

Basic Information Regarding Flight

The long flight between Paris and Rio de Janeiro necessitated three pilots, a captain and two first officers (FOs), with them supposedly taking turns to rest.

Of the two first officers (FOs), the most senior had had much more route flying experience, but had been promoted to Air France management. As such, he no longer flew regularly and, in fact, was only on the flight to keep his pilot's license current, something he had already achieved by flying the outward leg three days earlier from Paris to Rio de Janeiro.

The other first officer, whom we shall refer to as the junior FO, would fly the leg back to Paris. In practice, this would normally only mean the few minutes taking off from Rio and landing back in Paris with the autopilot and flight management system doing most of the work.

It has been alleged that many at Air France considered the easy overnight flight back to Paris was a good opportunity to recover from fun things at Rio.

As Air France did not have protocols mandating who should be assumed to be in charge in the captain's absence, the pilot flying (PF) would be the first to deal with any issues, and in this case, the junior FO so nominated by the captain before departing for his rest.

After the aircraft's evening departure from Rio de Janeiro, the senior FO was the first to take his rest, leaving the captain and junior FO at the controls and chatting in the knowledge everything on the **cockpit voice recorder (CVR)** would be overwritten after two hours.

Almost three and a half hours into the flight, with the A330 out of radar range over the South Atlantic, the captain decided it was time for him to "skedaddle" for his rest and called the senior FO back from the rest station situated adjacent to the cockpit to take his place. But first let us look at the underlying human factors.

Human Factors

1. **Female Companions**
   With the prospect of a three-night layover in Rio, the captain and the junior FO had brought along female companions. For the captain, in the middle of a divorce, it was an opera-singing off-duty flight attendant. For the junior FO, it was simply his wife.
   While there was nothing particularly wrong with this in normal circumstances, had the captain not had his special companion in the cabin he might have stayed 20 minutes longer in the cockpit before going for his rest and still been there at the crucial moment the autopilot disengaged. It most certainly would have meant he would have been in the crew rest station adjacent to the cockpit and easily summoned, allowing the senior FO to concentrate on what the junior FO was doing when the aircraft was about to stall.

2. **Captain's Lack of Sleep**
   The fact that the captain had told the junior FO he had only had an hour's sleep and "It was not enough" only became public knowledge in the course of the parallel judicial enquiry. The BEA investigators said they had omitted it from the **Cockpit Voice Recording (CVR)** transcript because it was private! Not only would this sleep deficit have impaired his performance, it might have led him to leave the cockpit that little bit earlier. Of course, the captain might have been bragging about his sexual prowess and not have been really so tired, and anyway no one would be in best condition in the middle of the night.

3. **Fraught Relationship between the Two FOs**

   Their bad relationship is quite evident from the tone of the original French version of the cockpit voice recording where the junior FO in speaking to the captain referred to the senior FO—away resting—snidely as "him" rather than by his first name or even family name.

   Incidentally, the captain and junior FO would have cemented their relationship in the course of the Rio helicopter tour they are said to have made with their female companions on the day of their departure. An outspoken aviation commentator with good aviation contacts in France told this author he had heard about the two pilots "hating each other's guts" on a visit there. According to the UK's **PPRuNe (Professional Pilots Rumour Network)** the captain and junior FO had started off as cabin crew and therefore felt a certain affinity leaving the management senior FO an outsider.

   Anyway, their poor relationship *might* have meant **the** junior FO stubbornly ignored the senior FO's repeated **"Go back down!" instructions at the most critical time** when they were going to stall.
   Had he followed that instruction the aircraft would not have stalled and it would have be a non-event.

TIMELINE OF THE DISASTER

> Sixteen minutes after the captain's departure, the aircraft encountered a cloud of hyper-cooled ice crystals, causing the **Pitot Tubes** to ice up.
> With them no longer sending credible airspeed data to the computer, the autopilot disengaged at 00.00 (0 min. 0 sec.).

–00:19 [19 seconds before autopilot disengagement.]
There is the sound of the impact of ice crystals.

> *Neither pilot alluded to the noise. The investigators noted that at the time pilots had little knowledge regarding the phenomenon. Nevertheless, the senior FO (pilot monitoring) took it upon himself to reduce the airspeed down toward Mach 0.8 and switched on the engine anti-icing, no doubt in response to the indication of ice shown by the instruments.*

–00:02 to 00:00:
A cavalry charge—a repeated trilling sound—warning the autopilot is about to disengage.

00:00 AUTOPILOT DISENGAGES

00:01 Junior FO:
"I have the controls."
00:03 Senior FO:
"All right."
00:05
Two synthetic voice stall warnings:

> *The French investigators (BEA) call them "Stall 1" warnings, no doubt because they were warning of the risk of a stall, and said these first ones occurred perhaps because the junior FO being "startled" pulled back too sharply on his side stick.*

"Stall … Stall."
00:06 Senior FO:
"What is that?"

> *Despite the stall warnings, the junior FO was still for some reason keeping the nose up.*
> *Due to the blockage of the pitot tubes the computer had erroneously indicated a slight drop in height, which the junior*

FO may have wanted to correct, but this does not explain why he continued to climb so steeply.

00:08

Two more SV stall warnings, the last one truncated.

"Stall, S—"

00:10 Junior FO:

"We haven't got a good display … of speed."

00:11 Senior FO:

"We've lost the speeds …"

00:17 Senior FO:

[Reading from the ECAM (Electronic Centralized Aircraft Monitor)]

"ALTERNATE LAW, PROTECTIONS LO …?"

The ECAM was telling them that they were in Alternate Law without the normal protections preventing them endangering the aircraft.

What the CVR transcribers could not decipher was the word "LOST," so how could the junior FO understand, especially as he was at the same time trying to say something about the thrust levers? Over such a vital point the senior FO should have ensured his colleague understood.

00:19 to 00:23 Senior FO:

"Wait, we're losing … Wing anti-ice. Watch your speed. Watch your speed."

In saying "speed," it is quite likely the senior FO was referring to the rate of climb not airspeed, the indication of which they had lost.

00:23 Junior FO:

"Okay, okay, okay. I'm going back down."

00:26 Senior FO:

"Go back down."

00:27 Senior FO:

"According to that we're going up. According to all three you're going up, so go back down."

00:30 Junior FO:

"Okay."

00:31 Senior FO:

"You're at … Go back down."

00:37 Junior FO:

"We're in a … We're in a climb."

00:45 Senior FO:

"Blast! Where is he?" [French expletive difficult to render in English]

It seems the senior FO had already attempted to summon the captain by banging on the partition between the cockpit and crew rest station, as well as by using the call button.

00:46

"Stall … Stall … Stall … Stall … Stall …"

Almost-continuous synthetic voice stall warnings start. The French investigators call them "Stall 2" warnings because they represent an incipient or actual stall as opposed to the earlier ones denoting the risk of one.

01:01 Senior FO:

"Blast! Where is he?"

The junior FO made slight nose-down inputs alternately to the right and to the left.

The rate of climb that had reached 7,000 feet a minute had dropped to 700 feet a minute, with the roll varying between 12 degrees to the right and 10 degrees to the left, suggesting the junior FO was struggling just to keep the wings level and not concerned with the pitch.

The airspeed shown on the left side increased suddenly to 215 knots (Mach 0.68), while the speed displayed on the left primary flight display (PFD) remained invalid for 29 seconds.

"Stall 2" warnings continue.

01:05 They reach a maximum height of 37,924 feet at which point the aircraft stalls completely.

1:20 Senior FO:

"Do you understand what's happening or not?"

01:28 Junior FO:

"I don't have control of the airplane anymore now."

01:30 Junior FO:

"I don't have control of the airplane at all."

"Stall 2" warnings continue.

00:01:37 Return of the captain

Sound of cockpit door opening.

The captain's return to the cockpit coincided with them dropping roughly through the altitude (35,000 feet) it had been when he had departed a little earlier.
He was not to know they had just executed a senseless steep climb without added power which meant the stall warnings were credible.

Captain:

"Er, what are you [doing]?"

According to Bill Palmer, author of a very illuminating book, "Understanding AF447," *analyzing the crash from a technical angle, the captain probably came to the cockpit from the cabin because of the strange behavior of the aircraft, not because he had heard the senior FO's summons, hence the question, "What are you doing?"*

01:39 Senior FO:

"What's happening? I don't know, I don't know what's happening."

01:42 Junior FO:

"We're losing control of the aircraft there."

01:42 Senior FO:

"We lost all control of the aircraft. We don't understand anything. We've tried everything." [Not true.]

01:53 Junior FO:

"I have a problem; it's that I don't have vertical speed indication."

The aircraft was dropping so fast that the display could not show the actual sink rate, though there was a colored mark indicating it was downwards and "off-the-dial."

01:59 Junior FO:

"I have the impression that we have some crazy speed, no? What do you think?"

The cacophony generated by the aircraft mushing downwards with the air howling as it splayed off the wings and fuselage may have made the junior FO believe they were traveling too fast and needed the air brakes.

Senior FO:

"No, above all don't extend [the air brakes]."

02:09 Senior FO:

"What do you think about it? What do you think? What do we need to do?"

02:10 Captain:

"I don't know, it's going down."

[Impossible to translate the colloquial "there" that the captain added and the junior FO also starts using.
Some exchanges omitted.]

02:22 Senior FO (with stall alarm still sounding):

"You're climbing."

02:23 Senior FO (stall alarm still sounding):

"You're going down, down, down."

Captain:

"* * * [Going down]"

02:25 Junior FO:

"Am I going down now?"

> *It's confusing because they talk about going down without specifying whether it is anything to do with the pitch.*

02:27 Senior FO:

"Go down."

Captain:

"No, you climb there."

02:34 Junior FO:

"(It's) Okay, we're in TOGA."

> *The junior FO thought it was "okay" because normally at low altitude where he had been trained to handle stalls, being in* ***TOGA (Takeoff/Go-Around)*** *mode would normally get them out of trouble. However, this was not true in thin air at high altitude.*

02:37 Junior FO:

"On alti, what do we have here?"

> **This, sadly, was almost the last chance the captain had to save them as there was not only the need to recover from the stall but also the need to stop the 10,000+ ft/min precipitous rate of descent.**

02:39

Stall warnings continue.

Captain:

"* * *, it's impossible."

> *The junior FO again asked about altitude, without getting a reply, and in the following 20 seconds the captain's main*

concern was to tell him to get the wings level, rather than get the nose down.

The synthetic voice stall warnings continued.

03:13 Junior FO:

"We're there, we're there, we're passing level 100 (10,000 feet)."

03:20 Junior FO:

"But I've Been In Maxi Nose-Up For A While."

*This remark finally makes the captain realize—**TOO LATE**—what the junior FO has been doing wrong all along.*

03:38 Captain:

"No, no, no, don't climb."

03:40 Senior FO:

"So, give me the controls; the controls to me; the controls to me."

03:41 Junior FO:

"What is …
How come we're continuing to go right down now?"

Adding to the confusion was the fact that the creators of the Airbus software had silenced the stall warnings below 60 knots.

This meant that the stall warnings restarted when they did something right and speeded up (above 60 knots). Also, with the aircraft at an angle the air was striking the pitot tubes at an angle giving misleading speeds.

03:35 Junior FO:

"Go ahead. You have the controls. We're still in TOGA, eh."

00:03:50

On taking control, the senior FO failed to lock out the junior FO's sidestick by pressing on the red priority button on his own, thus allowing dual inputs with the computer averaging them. As a result the junior FO cancelled out many of the senior FO's actions. Anyway, it was far too late to avoid crashing into the sea.

04:00 Captain:

"Watch out, you're pitching up there."

04:02 Junior FO:

"Well, we need to; we are at 4,000 feet."

04:11 Synthetic voice "sink rate" warning:

"Pull up!" "Pull up!" "Pull up!"
"Pull up!" "Pull up!" "Pull up!"
"Pull up!" "Pull up!" "Pull up!"

4.21 The A330 strikes the ocean with

Vertical Speed -10,912 ft/min (-124 mph, -200 km/h)
Ground Speed 152 knots
Pitch Angle 16.2 degrees nose-up
and Roll Angle 5.3 degrees left

The 152-knot ground speed suggests that at **TOGA** at low altitude they were close to extricating themselves from the stall but, of course, not from the -10,912 ft/min. precipitous descent.

Captain had Hardly Any Time to Save Them

Only three minutes elapsed from the return of the captain to the aircraft's impact with the sea. This meant they dropped some 35,000 feet in that time, or at over -10,000 feet per minute, as shown by the flight data recorder.

The captain would have needed to take the following steps to save them **upon realizing they had stalled,** which he failed to do until perhaps right at the end when the junior FO said he had been **continuously pulling back on his sidestick**.

1 20 seconds
to take the senior FO's place in his seat or tell the junior FO what to do in the hope he would follow his instructions.

2 Another 20 seconds
to get the nose down—perhaps having to reduce engine thrust and possibly wait for the trim to adjust before being able to do so.

3 Another 15 seconds, for the aircraft to gain enough airspeed to recover and begin to fly properly.

4 Another 50 seconds, making a total of 105 seconds , to fully break the vertiginous descent without overstressing the aircraft, which might have required almost 8,000 feet.

The time required to save the situation once had the captain correctly identified the problem was 105 seconds, while the time he had before actually hitting the water was only 164 seconds.

This only left him around 59 seconds or just about a minute to correctly work out what was wrong and save them. Even supposing our estimates of the time required are too high he can hardly had more the a minute and a half to come to the correct conclusion and save them. Had he arrived just a little earlier he might have become noticed the aural stall warnings which only made sense if he knew about the climb.

A minute is not much time **even** for someone in daytime and in tip-top condition to correctly analyze the situation without being properly informed.

For someone in the middle of the night when mental performance is at its lowest level, let alone someone suffering from lack of sleep, it would be extremely difficult.

The Role of the Senior FO

Though the junior first officer made fatal mistakes, almost certainly due to not having been trained to handle stalls at high altitude, the senior first officer's performance left much to be desired.

The execrable relationship between him and the junior FO and the lack of an Air France protocol saying the more experience pilot should automatically take command in a crisis, must have made him think only the captain could handle him and even take over. This led to him diverting his attention away from what junior FO and the aircraft were doing to summoning him at the most critical moment when the aircraft was about to stall.

When the senior FO towards the end formally took over the controls he failed to continually press his priority button to lock out his colleague who was cancelling what he was doing. This suggests his formally taking over early on would have been similarly meaningless.

One aviation commentator has claimed that the fact the senior FO had his seat set right back all the time indicated a lack of seriousness with him only being on the flight to keep his pilot's license current. The author thinks it was very likely the senior FO expected the captain would want to take his place and therefore kept his seat back.

To give senior FO his due, he had prior to the fatal stall repeatedly told the junior FO to go down, with the junior FO only appearing to comply but not.

On the **Crew Resource Management (CRM)** front, there was much left to be desired, particularly where the senior FO did not ensure his colleague understood him saying they were "in **Alternate Law** and protections were lost." Not only was what he said almost incomprehensible, the junior FO was simultaneously talking about something else.

Though the senior FO was far more educated and more experienced than the junior FO, he had specialized in meteorology at the prestigious aviation academy he attended. Had he specialized in the theory of flight he might have taken the stall warnings following the steep climb more seriously.

Transferred to management, he had perhaps lost the assuredness of a regular line pilot.

Other Factors

Other factors were:

1. Air France's poor choice of pitot tubes (possibly to favor a French manufacturer) that led to the autopilot

disengaging due to icing in the first place. Better ones were being progressively installed on the company's A330s but not quite yet on that particular aircraft. It should be noted other pilots had been able to continue flying safely despite icing of those pitot tubes.

2. As said, lack of training for coping with stalls and flying Airbus aircraft in **Alternate Law 2** at high altitude, was surely a major factor with the junior FO being quite sure he was "okay" in **TOGA (Takeoff Go-Around)** mode.

3. Also, had the **ECAM (Electronic Centralized Aircraft Monitor)** indicated "Airspeeds Unreliable," as the reason for the **autopilot** disengaging, they would have had a better understanding of what was happening, perhaps been less unnerved, and more importantly more likely to follow the simple procedure they had been taught for dealing with that, as had other A-330 crews where the pitot tubes had iced up.

4. The stall warnings should not have been deactivated below 60 knots giving the impression that when they speeded up above 60 knots they were doing something wrong. However, when that was happening it was mostly too late for them to recover anyway.

5. Had there been **Active Sidesticks** with the stick on the senior FO's side moving in unison with that of the junior FO, the senior FO, and later on the captain, might had been able to see that the junior FO had been constantly pulling back. However, prior to the fatal stall the senior FO had been aware they were climbing and had several times told the Junior FO to go down with no success.

6. In our entry **Flight Time—Pilots Required for Long-Haul and Ultra-Long-Haul,** we say that whatever the circumstances there should always be a captain as well as a first officer present in the cockpit.
   Not only was there only one captain rostered for AF447, Air France did not have a protocol for specifying who—presumably the most experienced officer—should

automatically take command and possibly take over in a crisis.

The Junior First Officer

No one knows why the junior FO kept on climbing when he first took the controls, though it is possible he was having so much difficulty keeping the wings level he was not paying attention to the aircraft's pitch. At the beginning, an indicated drop in airspeed had erroneously caused the instruments to show a very slight drop in height but not nearly enough to warrant a continuous steep climb.

The junior FO did not really follow the senior FO's repeated instruction to **GO DOWN.** Had he done so the aircraft would not have stalled in the first place and the "other factors" mentioned would not have come into play.

However later, and too late, in the presence of the captain, he said he had been in "maxi nose up for quite a while," so he knew what he was doing, believing that they were "Okay in TOGA." Perhaps it was that that belief that killed all on board, and was because he had not been trained by Air France to handle stalls at very high altitude.

The fact that that the senior FO did not make sure he understood "protections" were lost in alternate law certainly did not help him.

Choice of Pitot Tubes

Apart from the lack of training for flying manually at high altitude, those at Air France responsible for the purchase of the pitot tubes were at the root of the disaster.

Of course, they could never have imagined any pilots could behave as those did. There had been other cases of those pitot tubes icing up with the pilots managing to fly on perfectly safely. Anyway, conscious of the problems those e pitot tubes, those on that aircraft were about to be changed.

Lessons Learned by All Airlines

Other airlines could not believe their pilots would ever do anything similar," but nevertheless gave them extra training to cope with flying manually at high altitude and extricating themselves from a stall there, however improbable that should be. The disaster also drew the attention of aviation authorities worldwide to **Loss of Control In-Flight (LOC-I)**.

Air France itself seems to have learned much, and it has taken multiple measures to become now a very safe airline.

Reference material from official BEA investigation:

See
https://bea.aero/fileadmin/uploads/tx_elyextendttnews/presentation.rapport.final.05juillet2012.en_04.pdf
for Final Abbreviated BEA Report in English,
and
https://bea.aero/fileadmin/uploads/tx_elyextendttnews/annexe.03_07.pdf
for Annex to BEA report showing the many details obtained from the AF447 flight data recorder including altitude, sink rate, and sidestick positions. It is in French and not so easy to follow, but even seeing the altitude as they climb to the stall, then drop down to the sea, and finally futilely pull back on the sidestick at the last moment makes it all too real. There may also be an English version.

See also **Manual Flying**.

See also the excellent book *Understanding AF447* by Bill Palmer looking at the crash from a technical perspective.

Air Marshals

Although many countries' airlines, and notably El Al, do have security people on board their airliners, the US has taken the idea of having incognito armed marshals on board the most seriously. The exact figure is classified but following the events of **9/11** the number of air marshals apparently jumped from 33 to between 3,000 and 4,000. Following that, hiring stopped but restarted as attrition depleted their numbers.

On July 16, 2015, the Homeland Security Subcommittee on Transportation questioned the incoming head of the Transportation Security Administration, Rod Allison, about the air marshal program, saying that though they supported it, there were complaints about air marshals. These included passengers being angry at being bumped for the benefit of marshals with no explanation. Republican Earl "Buddy" Carter described a flight where six marshals sat in first class, and then more marshals from a canceled flight joined them.

They thought that with reinforced cockpit doors, there was less need for a marshal, and more focus ought to be put on flights representing the greatest risk.

The presence of air marshals may have dissuaded some individuals with bad intentions, but there do not seem to be many, if any, cases where they have saved the day.

The US Air Marshals Service trains a few pilots and aircrew to handle guns and deal with hijackers, with those pilots having permission to keep the weapon in the cockpit.

Air Rage

The term *air rage* conjures up images of drunken individuals causing disruption. However, there was a case in India where enraged passengers acted in unison. So frustrated were they that the aircraft was being kept for ages in a holding pattern, they all rushed toward the cockpit to confront the pilots, thereby shifting the center of gravity so far forward that the aircraft was in danger of crashing.

Burly military types who happen to be on board can help subdue an unruly passenger. Subjugation must not get so out of hand that the individual succumbs—due to strangulation—as indeed happened in one case.

Some flight attendants have undergone rigorous training through the FAA for dealing with and protecting themselves from passengers posing a threat, and it would not be pleasant to be on the receiving end of some of the "nasty" maneuvers they have been taught.

While a panicking passenger trying to open a door to get out may instill fear in nearby passengers, they should not worry, unless the aircraft is extremely low, and even then, the pressure differential forcing the plug doors shut means they are usually impossible to open.

A crazed passenger trying to take control of the aircraft is another matter. On a BA flight to Nairobi in December 2000* a crazed man wanting to *save* the aircraft entered the cockpit and tried to take over the controls. The force he exerted on the control column was enough for the autopilot to disengage, and the aircraft dived so precipitously it was in danger of breaking apart even before hitting the ground. Fortunately, he was overpowered seconds before the situation became irretrievable.

The reinforced and locked cockpit doors installed after **9/11** to keep out hijackers have virtually eliminated the possibility of such a passenger gaining entry. Unfortunately, they once prevented a Germanwings captain on a toilet break from regaining access to the cockpit when his suicidal copilot was intentionally crashing the aircraft*.

Air Traffic Clearance

Permission from ATC notably, but not exclusively, to taxi, take off, land, climb, descend, or enter controlled airspace. The clearance includes instructions on what to do, and, of course, must be repeated back by the pilot to confirm they have heard and understood correctly. The worst-ever air disaster, in which two Boeing 747s collided in fog on the runway at Tenerife*, had the first officer still reading back the clearance for the actions after takeoff as the captain began his catastrophic takeoff roll.

Air Traffic Control (ATC)

Even in 1950, air traffic control (ATC) in the US was very primitive, and in many areas, it was virtually nonexistent. After all, the country was enormous, and airliners were so few that apart from areas where they would be funneled (bunch up), there was plenty of room for them to avoid each other when flying under visual flight rules following the principle of "see and be seen."

Furthermore, at the time, the controllers did not contact the aircraft directly but usually through the intermediary of the airline's dispatchers, though they did in the vicinity of large cities or the airport of departure or arrival. Consequently, in vast uncontrolled areas with no radar coverage, pilots would report to their companies when passing over the waypoints on their route.

In 1956, Transworld Airlines Lockheed Super Constellation Flight 2 bound for Kansas City with 64 passengers and 6 crew on board collided with United Airlines Douglas DC-7 Flight 718 bound for Chicago, with 53 passengers and 5 crew. Both aircraft were flying under "visual flight rules near cloud" and were getting close together because they were treating their passengers to a sight of the Grand Canyon en route.

Four years later there was another major tragedy, that time with two airliners under air traffic control colliding over New York City. This led to the introduction of speed limits for aircraft approaching airports and a rethink of the whole ATC system in the US.

Air Traffic Control: Facilities

Different sets of controllers handle the various stages of a flight. In the US they would be referred to as follows:

1. **Ground control**
   Handling taxiing at the airport.

2. **Tower**
   Handling takeoffs and landings.

3. **Departure control**
   Handling period between control by the tower and

handover to the air route traffic control, called a *center* in the US.

4. **Center**
   Handling traffic using air routes in the region. For instance, there will be Chicago Center, New York Center, and so on. The aircraft is handed from center to center until it wants to land, at which point it descends and is handed over to approach control.

5. **Approach control**
   Handling the aircraft until handover to the tower for the actual landing.

Australia is working on a special highly digitalized air traffic control system for drones for it is estimated drone flights there will rise to 60 million per year by 2043. This would include package delivery drones, air taxis, and other UAVs. At present, the figure is about 1.5 million drone flights a year.

Air traffic control in many countries is conceptually antiquated, making it difficult to cope with increasing traffic. However, new technologies (see **ADS-B**) should allow aircraft to fly closer together and on more direct routes compared with the traditional airways between radio beacons.

Air Traffic Control: Pilot Interaction, Phraseology

Commercial flying has become almost unbelievably safe these days thanks to advances in technology and the lessons learned. However, miscommunication between pilots and air traffic control, already largely responsible for the worst multi aircraft disaster ever at Tenerife*, remains a problem.

2024 started with a major incident at Tokyo's Haneda Airport in which a light aircraft carrying rescuers going to the aid of people caught in an earthquake mistakenly entered a runway just as a Japan Airlines A350 was touching down there with proper clearance.

All on the light aircraft were killed on the spot, while the A350 that had caught fire continued down the runway, coming to a stop some two-thirds of the way down. All on board evacuated safely, showing that while an aircraft with a Carbon Fiber Composite Polymer (CFCP) fuselage can burn, in the short-term it may resist flames from burning fuel better than one made of aluminum.

Though the final report into the Haneda accident is not out, this was surely due to miscommunication with ATC. Could it have been the controllers were speaking English as they should at an international airport to which the light aircraft pilots were unaccustomed?

As said, in recent years in the US, there have been worrying near misses on runways and taxiways, some of which were most likely due to miscommunication between pilots and ATC.

Training pilots realistically to handle communications with air traffic control is essential, and here artificial intelligence could play a role in the future. For in 2024, the President of Embry-Riddle Aeronautical University had an op-ed in *Aviation Week* saying, "AI Can Help Flight Students Reach New Heights." It seems it can enable students to practice without a tutor present using a system recognizing and analyzing what they are saying and replying like a proper controller handling other flights as well. It will even be able to pick up on mistakes such as use of words that should not be used lest they be misinterpreted and lead to a disaster.

The FAA's AIM (see **How to Find Information about Aviation**) has a section on **Air Traffic Control (ATC)** and a chapter called "Radio Communications Phraseology and Techniques" covering this complicated subject—too complex to cover here and well worth consulting.

However, we can make some general observations, namely that language should be kept simple so controllers in other countries are not confused. (For instance, see **Charlie Bravo (CB)—Meaning Cumulonimbus**.)

Another important point is that pilots should always include their call sign in communications even though it can be obvious they are continuing a communication.

Air Traffic Control: ATC as Career in US

There is an interesting website https://123atc.com/
"For and about Air Traffic Controllers in the United States,"
where they say you can:

1. View details about all FAA air traffic control facilities in the US.

2. View air traffic controller salary information.

3. View a database of all airline call signs authorized in the US.

4. View a glossary of terms used in air traffic control.

5. Learn how to apply to become an air traffic controller.

6. View the FAA payroll calendar.

It cites the advantages of a career as an air traffic controller, including it being among the highest-paid jobs obtainable without a college degree (though a degree is helpful), good benefits and job security, and the unique ability to retire with a good pension at age 50 or after 20 years of service

Besides that, one is free once off the clock, meaning no phone calls, emails, or work at home. Finally, the government pays you while training.

Of course, to be accepted for training and then qualify as an air traffic controller one must have certain unique qualities and be able to work under pressure.

Though there are the advantages cited above, being a controller is a very demanding job, and despite the mandatory intermittent breaks while on duty, it can take a great toll on one, mentally and physically. Many controllers consider leaving after age 40, and that includes Europe.

See **Gander**.

Air Traffic Flow Management (ATFM)

A largely computer-based system adjusting departure and arrival times so aircraft arrive at their destination at a time when they can land immediately, rather than many arriving simultaneously and having to burn up fuel while having to hold.

Though the great virtue of AFTM was seen as saving fuel, it is now being realized that if done in a sophisticated way with computers integrated with the **CPDL (Computer-Pilot Data Link)** it can greatly reduce the burden on air traffic controllers and thus improve safety.

It can mean aircraft having to wait quite a long time on the ground for permission to take off when their destination airport is saturated with traffic and perhaps aircraft have to be spaced out due to weather conditions.

Air Transport Association (ATA) [US] Now A4A

Washington-based lobbying group for the airline industry. Name now changed to **Airlines for America (A4A).**

Allegedly, the ATA recently tried to divert passenger outrage at long delays by blaming corporate jets, when, in fact, though corporate jets have equal rights to the airlines as regards taking off and so on, on a first come, first-served basis, they usually use different airports.

Underlying this dispute is a larger one, namely that according to the FAA's own estimates, GA (General Aviation) planes, which include both corporate jets and weekend flyers, account for 16 percent of the air traffic control system's overhead but contribute only 3 percent of the earmarked fees to run it. At major US airports, corporate jets pay next to nothing compared with airliners.

AIRBUS

When the idea of a European twin-engine wide-body aircraft was first seriously floated in the late 1960s, there was much confusion and lack of coherence both in Europe and the US regarding which airliners should be developed.

Worthy British airliners, such as the Trident and VC10, were commercial failures largely because they were ultimately designed to meet the supposed requirements of British airlines.

It was a bitter pill to swallow for the British to later learn that the *original* spec for the Trident coincided exactly with what airlines in the US finally decided they needed only for it to be downsized, or "emasculated" as one engineer said, to suit British European Airways (BEA), which only flew short routes to European destinations and being in a temporary downturn thought it would be unable to fill the seats.

Conversely, the VC10 was over spec'd for the benefit of British Overseas Airways Corporation (BOAC), which in the dying days of the British Empire flew long routes via airports with poor runways and hot climates where engines perform badly. This resulted in an excellent aircraft but one that though popular with passengers had a reputation for being fuel hungry, killing its prospects on key routes such as those across the Atlantic where even a 2 percent extra fuel burn was significant.

The bitter pill in that case was that those developing countries were soon to upgrade their main airports rendering the over spec totally unnecessary.

On the other hand, US manufacturers made the mistake of trying to satisfy everyone, including the relatively few that needed the three engines mandatory for flying over wide expanses of water. They compounded that mistake by building two tri-jets, the DC-10 and the Lockheed L-1011, when there was only room in the market for one, thus giving the Europeans an opening. Had that not been the case, the launch of the Airbus A300, the world's first twin-aisle, twin-engine wide-body, would have been even more difficult than it was.

According to John Newhouse's classic and well-informed book *The Sporty Game*, 1966 was a key year. This was not only because of the agreement between Boeing and Pan American regarding the 747 but also because it was the year Frank Kolk of American Airlines sent Boeing, Douglas, and Lockheed the specifications of a new airliner, much smaller than the 747, that would best fit his company's forecast growth in passenger traffic. He called it the Jumbo Twin, a double-aisle, wide-body aircraft capable of carrying as many as 250 passengers 2,100 miles at subsonic speed.

Lockheed and McDonnell Douglas were thereby encouraged to pursue their wide-body ventures—but, for the reasons mentioned above, with three engines rather than two. This allowed the Europeans to go ahead with their twin-engine "Airbus" along the lines proposed by Kolk. Frenchman Roger Béteille, who was the mastermind of the Airbus project at the time, allegedly said Kolk's list of requirements constituted a significant input to the Airbus project.

Hard though it may be to believe, the Airbus was not getting the lion's share of French government funds allotted to aviation. That was going to the supersonic Concorde, much favored by President de Gaulle as a means of enhancing France's prestige. This impeded the Airbus project, though Airbus staffers claim the technology developed for the Concorde contributed to subsequent Airbus projects, such as the A320.

Perversely, the Concorde project had the result of making the Americans—not wanting to be left behind—divert technical talent from the Boeing 747 and other projects to work on what they called the **SST (Supersonic Transport)** that supposedly would greatly outdo the Concorde.

Meanwhile, the Germans resolutely refused to support the Concorde project on the grounds it was unlikely to succeed commercially. On the other hand, they supported an Airbus-type project but mainly financially since their aircraft manufacturing base was then so limited—their best engineers and designers having been spirited away to the US under **Operation Paperclip** at the end of World War II, with others ending up in the Soviet Union.

Britain's Hesitant Role

Before and during the conception of the Airbus A300, the British government hoped that Hawker Siddeley would collaborate with US companies. However, when Hawker Siddeley found Boeing wanted them as a mere contractor, shouldering the risk but not sharing in any success, it decided to throw in its lot with Airbus in the knowledge that the father of the project, Roger Béteille, had great admiration for its expertise in wing design and work on the Trident.

Hawker Siddeley then invested its own funds in tooling for the project, with a subsequent loan from the Germans, who had later increased their share when the British government pulled out, fearing it would never recoup its investment. Incidentally, Rolls-Royce, whom the British government had saved from bankruptcy brought about by the costs incurred in developing the RB-211 engine for the Lockheed TriStar, was able to provide the Rolls-Royce RB-207 engine at little extra research cost, thus making the project a lot easier to realize financially.

In 1977 Hawker Siddeley became part of British Aerospace (BAE), which in January 1979 took up a 20 percent share in Airbus Industrie.

Much later, on October 13, 2006, having decided that its future lay in working with the US, BAE relinquished this share, with some argument, as the value of its shares had fallen in the light of revelations regarding the problems associated with the production of the A380 superjumbo.

The wing-making facility near Bristol in the UK was bought by GKN in 2008, so that aspect was still in British hands. In 2015, Airbus, comparing its margins of about 6 percent on commercial aircraft with those of more than 12 percent in GKN's aerospace division, threatened to seek some supplies for the A320 elsewhere unless GKN offered keener prices.

A320 Paves the Way to Success and Overconfidence

The A300 was innovative in that it was the first wide-body, twin-aisle, *twin*-engine airliner. However, sales were sluggish at the beginning, with the early ones being arm-twisted ones with the French and German governments insisting on their flag carriers purchasing it, and it was only in later years that sales picked up, making a good total in aggregate. See **Airbus A300**.

The technically innovative A320 fly-by-wire single-aisle narrow-body short- to medium-range airliner is what placed Airbus in the major league, with many orders even before the first commercial flight. It went on to be, and still is, a great commercial success.

While there is no space here to go into the twists and turns of the Airbus saga, the fact that Airbus performed as well as it did up until the development of the Airbus superjumbo, the A380, is something of a wonder. That is considering the inherent inefficiencies of an organization with international partners and two countries (France and Germany), vying for power via prima donna personalities, all the time making sure the other could not become the key decision maker.

Airbus subsequently created the A340/A330 family with the airframe able to be fitted with either four engines or just two. At the beginning it went all out promoting the four-engine A340 but was blindsided when the **ETOPS** regulations (that specify how far an airliner can fly from an airport it can divert to) were modified to allow flights over oceans just in time to allow the Boeing 777 to be the great success it has been.

Nevertheless, the twin-engine version, the A330, has continued to prove relatively successful.

The A380 Gamble

Those, including the author, witnessing the maiden flight of the A380 at Toulouse in southern France in April 2005 did not realize that much of the voluptuous maiden's underwear was missing.

This only became publicly known much later when Airbus informed launch customers that its aircraft would be delayed, mainly due to problems with the wiring, which could not be connected. The consequent fall in the value of Airbus shares just when British Aerospace was exercising its option to sell them back angered BAE. There were also accusations of insider trading by senior people at Airbus in the know.

Nicola Clark, then an *International Herald Tribune* reporter and aviation and transportation journalist based in Paris, whose articles also appeared in the *New York Times*, has studied the affair in depth and written a detailed, prize-winning article entitled "A Humbled Airbus Learns Hard Lessons." Published as early as December 14, 2006, the article told how when some 200 German technicians came to Toulouse with hundreds of kilometers of precut electric wire to thread painstakingly through the rear and front fuselage sections made in Hamburg, they found the lengths were often insufficient for them to reach the corresponding connectors in adjacent sections.

The technicians pointed this out, but management did not take their views seriously, perhaps because problems were to be expected for so complex an aircraft. According to Clark, obfuscation, and failure to deal with the problem early on, proved extremely costly.

Almost unbelievably, the problem had arisen because the Germans insisted on using an out-of-date American 2D program for the wiring layout, while the French were using a better 3D program. Clark suggests that it was not just a question of German pride and avoiding the costly nuisance of changing to another program but also the fact that the less efficient 2D program required more workers and would boost employment there. She was not able to get an answer on this point from union representatives.

Airbus said the problem had simply arisen from use of incompatible software, which probably glosses over the fact that the complexity of the paths the wires followed meant that the German 2D program was underestimating the length of wire needed for it to reach the specified point (connector) on the adjacent section of the fuselage.

After paying airlines compensation for delays and later for taking aircraft out of service for replacement of some wing parts that used a brittle aluminum alloy, the A380 has been relatively free from problems.

Airbus has been disappointed by sales of the A380, with US carriers not interested, partly because of the nature of their networks. Yet, for Emirates, which ordered or committed itself to ordering 140 as of March 2016, it has proved a moneymaker. This is largely due to the nature of its network and the fact that working with the A380 on such a large scale means it can be more efficient.

Emirates made it the centerpiece of its operation with luxury suites and a first-class section with showers. These only allowed passengers to use a limited amount of water as the authorities would not allow the water used to be cleaned and reused in the toilets.

However, in April 2016 Airbus told its suppliers to reduce deliveries of parts in line with a production rate of only 1.7 per month, even though lower fuel prices were supposed to give the aircraft a boost. Unfortunately, the A380 was too large to fill regularly on many routes, and even on those where it could, the airlines generally prefer to use several smaller aircraft to provide the frequency that profitable business and first-class passengers seek.

Though the A380 proved popular with passengers, with four engines it was not only expensive to run and maintain but also expensive to buy—something made worse by the fact that there was no second-hand market. Orders from by far its biggest operator, Emirates, that had been keeping it on life support dwindled. In early 2019 Emirates converted most orders to other aircraft including the excellent A350. With that, Airbus decided to terminate production by 2021. It was an impossible situation with each aircraft sold at a loss. However, there are claims that the lessons learned in developing the technically superb A380 enabled Airbus to make the A350 a great success.

At the height of the Covid-19 pandemic, airlines with A380s took them out of service with many stored in deserts and some saying they regretted ever purchasing them. But now, with passenger demand rebounding more than expected, airlines are changing their tune and bringing them back into service. Emirates says it is filling every seat on its multiple A380 flights to Heathrow. With delays to the Boeing 777X, maybe Emirates wishes the A380 was still in production. It would even be happy to see an A380neo, with more advanced engines and other improvements. The problem for Airbus is that only Emirates with its diverse market could fill such a large aircraft all year round. Interestingly, British Airways flying out of slot-restricted Heathrow Airport has brought all of its A380s back into service.

Airbus, now owned by EADS, was doing very well with sales of the A320, A330, and A350, though cash flow was impacted by having to wait before aircraft already constructed could be delivered, due to the inability of suppliers to fix the interiors on time. Also, deliveries of the A320neo, with the innovative geared turbofan engine made by Pratt & Whitney, had been delayed while the engine maker sorted out problems. However, long term those may not be significant, with the engine beating expectations.

At the time of writing, Airbus is doing particularly well with its various iterations of the narrow-body A320 with versions even being used for long-haul flights. The stretched A321 series is doing well too, though the ultra-long-range version with an extra "rear center" fuel tank fully fitted in the fuselage is taking time to be certified with EASA insisting on replacement of the composite fairings on the underside with metal ones more resistant to sliding along the ground in the albeit unlikely event of a wheels-up landing or one where the undercarriage collapses.

The A220 (Originally Produced by Canada's Bombardier)

When Delta placed a large order for a mid-size airliner made by the Canadian company Bombardier, Boeing objected, although it didn't make a similar-sized airliner. Boeing claimed it was unfair competition due to the "unfair" subsidies the project had received and managed to get crippling duties imposed. Airbus stepped in and managed to take over the project, making the aircraft at its plant in the US to avoid duties. The airliner is now called the A220 and is very popular with passengers, though Airbus has yet to make a profit, partly because the relatively small Bombardier did not have the clout to negotiate good prices from suppliers that Airbus had to continue to use.

Boeing in response unsuccessfully tried to come to a similar arrangement with Embraer of Brazil, but that fell through.

Of late, Airbus has been looking at the possibilities for hydrogen-powered eco-friendly aircraft.

It is reported that Airbus hopes to no longer rely on model launch aid from governments on extremely favorable terms repayable on success but not on loss, but like Boeing on juicy military contracts.

~ Airbus A220 (A220) 2016/250+**

[** Entered service, not first flight.]
Narrow-body airliner originally called the C-Series designed by Canada's Bombardier Aerospace with subsidies from the government of Quebec. As explained above, this was a project Airbus took over, completing the aircraft at its facilities in the US thus circumventing the tariffs.

The aircraft is now proving very popular.

~ Airbus A300 (A300) 1974/561

The Airbus A300 was the first wide-body twin-engine airliner, and although sales were slow at the beginning, the aircraft eventually proved a great success because the increasing reliability of jet engines had made the twin-engine design viable as well as economical.

In the early 1970s it filled a gap on the third rung of the wide-body-airliner market. On the top rung sat the "queen of the sky," the four-engine Boeing 747 jumbo. On the second rung, with not enough room for both, jostled the McDonnell Douglas DC-10 and the Lockheed TriStar L-1011. Though the TriStar was clearly the more technically advanced of the two wide-body trijets and certainly the safer, it failed to fully establish itself in the market largely due to the late delivery of the innovative Rolls-Royce engines. McDonnell Douglas had tragically cut some corners to get its DC-10 in first and tie in several key US carriers.

After its maiden flight in 1972, the A300 entered commercial service a year and a half later in 1974. Many of the early purchases were arm-twisted ones, with the French and German governments urging their national carriers to purchase it. Only from 1978 onwards did sales really pick up, and although they never soared, the aircraft had advanced features, making it a steady seller. A total of 561 units had been produced when production ceased in 2007.

An early decision to opt for US General Electric CFM engines instead of a Rolls-Royce engine still under development made it possible to put the A300 into production so quickly, besides facilitating sales to the US. Pratt & Whitney engines later became an option.

This decision to forgo the Rolls-Royce engine infuriated the British government, though ironically much of the success of the A300 was due to the wings, which were designed and manufactured by the British company Hawker Siddeley with the help of engineers "lured" from Germany. The British had been hoping Hawker Siddeley would tie up with Boeing instead on such work, but Boeing merely wanted to use Hawker Siddeley as a subcontractor, shouldering considerable risk but with little prospect of sharing any success. Had Boeing offered to make the British company more of an equal partner, the A300 and subsequent Airbus aircraft might never have been the successes they were and indeed are.

~ Airbus A320 (A320) Family 1988/11,600+

The A320 family consists of narrow-body short- to medium-range twin-engine jets carrying just over 200 passengers. There is the smaller A319, and even smaller A318, and the larger A321. They are direct competitors of the Boeing 737, which at times they have outsold.

The most computerized airliner when first produced proved too sophisticated for some users, who could not fully master the computerized controls.

Just after it entered service, an Air France captain demonstrating a brand-new A320 at a tiny air show engaged in a risky maneuver to show the crowd how the computer would allow the A320 to seem to hang on the edge of a stall with its nose preening, and then dramatically climb away. Coming in from Basel on the Franco-Swiss border only five minutes' flying time away, with not enough time to stabilize their approach, the pilots found the spectators were not located next to the paved runway as expected. Instead, the spectators were alongside a shortish general aviation grass airstrip at an angle of 40 degrees to the left of it. The strip was not only half the length of the paved runway, but the overrun space at the end was also only half as much.

Having had to adjust course at the last minute, the captain had commenced his stunt much too low, with the bank for the last-minute turn perhaps having contributed to their loss of height. Cleverly hanging there over the strip at the edge of the stall, the pilots finally saw the tall trees ahead. Too late they tried to initiate a go-around, but with the engines idling and needing eight seconds to spool up, and the aircraft already struggling at the limit of its operating envelope, the aircraft had no reserve of energy. After what seemed a desultory effort to heave itself up over the trees, it sank into them, with spectators shocked by the sight of a billowing cloud of smoke suggesting a terrible disaster*.

With many children on board, it could have been a disaster, but amazingly, 133 of the 136 people on board managed to escape—perhaps because they had no carry-on luggage, plus many were young and physically able, and as first-time flyers had paid attention to the safety briefing. The three that perished were a disabled boy, a young girl who had been unable to undo her seat belt, and a valiant woman who had come to assist her.

TV news bulletins around the world played and replayed the dramatic footage.

In France, the accident became a contentious affair. The captain in command was blamed and fired, but for years he continued to claim it was not his fault and that the French authorities had tampered with the recordings in the flight recorders to save the reputation of the aircraft. He could, for example, justifiably point out he had not been briefed properly about the high trees and location of the spectators. But effecting such a hazardous maneuver with passengers on a not yet stabilized aircraft at low height with a last-minute change of course taking them over terrain they had not expected to fly over was irresponsible to say the least.

Boeing then mocked Airbus over the incident, telling visitors to the company that the A320's computer had *done great* to perfectly land it on the trees—with Boeing continuing to use its old 1960s technology on its 737, ending with the MAX disasters.

Despite this inauspicious start, the A320 has come to be one of the most successful and reliable airliners, and the A321 will extend this money-spinner's life until technology evolves enough for a completely new design to be worthwhile. One engine option is the radically new Pratt & Whitney geared turbofan, which is claimed to be especially quiet and economical. Though as efficient as expected, there have been teething problems with the engine.

The A320 competes directly with the Boeing 737, and it was Airbus's decision to produce more fuel-efficient new engine (neo) versions that led to Boeing rushing the ill-fated 737 MAX into production rather than taking time to develop, and get approved, a contemporary new design.

There is also the A321 model, about 7 meters longer, which is able to take more passengers but with a slightly shorter range, though a more efficient new engine option (neo) extends this.

There has been talk of Airbus using a new system to make a thin more aerodynamic composite wing to give it a further edge.

~ Airbus A321XLR

The single-aisle Airbus A321XLR does not look much different from its siblings, but it will be able to fly 4,700 nautical miles (or about 5,400 regular miles, or more than the distance from New York to Athens) without stopping. That is around 1,000 miles more than most narrow-body models, though the 737 MAX may give it some run for its money. It will be able to fly for up to 10 hours, about double the duration most airlines use for a single-aisle aircraft. Smaller airlines such as those from Malta and Latin America are showing interest, with Icelandair seeming to be the first to launch flights.

~ Airbus A330 1994/1700 and Airbus A340 1993/1,600

In the early 1990s, Airbus designed a wide-body airliner with twin-engine and four-engine variants, which, since they both had many structural parts in common, would in theory be cheaper to produce.

Initially, Airbus put the four-engine variant to the fore because the regulations as they then stood meant that it could fly many oceanic routes two or more hours' flying time from diversion airports. At first this seemed justified with Singapore Airlines even using it on the world's longest scheduled nonstop flight—from Newark Liberty International Airport to Singapore's Changi Airport, some 9,534 miles (15,343 km), taking about 18.5 hours.

Unfortunately for Airbus, the ETOPS regulations governing twin-engine airliners then changed, and it became possible for approved twin-engine jetliners flown by approved operators to fly routes taking them as much as two or more hours from airports to which they could divert. As a result, the more expensive to buy and maintain four-engine A340 lost its appeal.

Boeing shortly afterwards introduced the twin-engine 777, which right from the start was ETOPS approved for flights far from land. It was a runaway success, with sales not only dashing the hopes of Airbus selling the A340 but even eating into those of Boeing's own jumbo 747-400.

Though the four-engine A340 never lived up to expectations, the twin-engine variant, the A330, began selling steadily. There is now the "new engine option," called the A330neo, which has more efficient engines and a choice between the novel Pratt & Whitney geared turbofan and the CFM engines.

An unforeseen advantage of having a twin-engine aircraft originally designed to have four engines as an option is that the strengthened wing and vacant emplacements (where the other two engines would have been) made it an ideal platform for an air-refueling tanker. It consequently initially won the contract to supply the US Air Force with a US company doing the expensive finishing work. However, intensive lobbying by Boeing resulted in new bids having to be submitted with the spec tailored so that Boeing would inevitably win. Airbus withdrew and Boeing won with the 767 platform—perhaps much to the regret of the US Air Force in view of the many problems it is having with it, including debris left inside the fuel tanks.

~ Airbus A350 (A350) 900 2015/500+

Having expended so much energy and resources on development of the A380 superjumbo, a chastened Airbus initially proposed a jazzed-up A330 to compete directly with Boeing's 787 Dreamliner. This idea was met with derision by prospective clients, who made it clear that they wanted an innovative design at least on a par with the Boeing 787. Airbus gave in and came up with a new design using composite materials, like the 787, to be called the A350.

Benefiting perhaps from lessons learned from the problems that Boeing had encountered with the Dreamliner, the A350 was delivered without too much pain and is proving to be an excellent aircraft.

Singapore Airlines restarted its ultra-long-haul flight from Singapore to New York with a long-range version.

Qantas has chosen the A350-1000 version for its Sunrise Project nonstop flights from Sydney to London and New York. To meet the latter requirement Airbus is merely having to slightly increase the maximum takeoff weight and add an extra center fuel tank. Flights were to start in 2023, but that was delayed due to Covid-19, and EASA insisting on more metallic protection under the extra fuel tank to lessen the risk of fire in the event of a belly landing, putting the launch date back to early 2027. At one time it was assumed that the pandemic would mean the project being abandoned, but Covid-19 might be an argument for nonstop flights with less intermingling with passengers from other countries.

The 12 Airbus A350-1000s ordered for Project Sunrise will have 6 First Class Suites, 52 Business Suites, 40 Premium Economy Seats, and 140 Economy Seats. There will be a "Wellbeing Zone" available to all, where passengers can stretch their legs, exercise, and stand around. Initial service will be Sydney to London, followed by Sydney to New York with flight times of 18-20 hours.

Qatar Airways was a launch customer for the A350, even showing it off at air shows. However, when one of the A350s was found to have loose paint near the tail when being repainted for a special livery, the relationship between the airline and Airbus went increasingly downhill. Airbus claimed it was not a safety issue, but Qatar Airways said it was, with its regulatory authority grounding a fair number as unfit to fly. Qatar Airways demanded compensation for the grounding, with Airbus saying the airline had not needed them during the Covid-19 downturn, and that it was just clever maneuvering, and furthermore Qatar Airways was harming the reputation of the aircraft—some other airlines had found problems with the paint but did not withdraw any from service on such grounds.

It ended up in court in London, with Airbus refusing to supply other aircraft Qatar Airways had ordered as it was a "difficult" customer—that was probably true for in the past the airline had on occasion refused to accept aircraft, say, for poor-quality internal fittings and finish.

Qatar then ordered aircraft from Boeing but risked finding itself in a weakened negotiating position having only one supplier. Finally, in early 2023 the two parties agreed to settle their differences "amicably" rather than in court. This meant that Qatar Airways could get delivery of the urgently needed airliners on order rather than going to the back of the queue.

One point not mentioned is that the temperature aircraft are exposed to while standing outside at Qatar's Doha Airport are extreme, perhaps worse than at Dubai. Geographically, the Gulf super connector hubs are well placed as staging posts for traffic between Europe and the East. However, apart from the high temperatures, which to some degree can be mitigated by having many flights taking off at night or early morning, there is the additional problem of air-blown sand.

~ Airbus A350 (A350) 1000 2018/592+

The 1000 version is larger and has much larger diameter Rolls-Royce engines, which unlike the smaller ones on the 900 had problems when used by Middle East carriers with their hot temperatures and sandy conditions. At the Dubai airshow in 2023 Emirates did not follow through on an expected order for the A350-1000 with Emirates always-outspoken president, Sir Tim Clark, telling reporters that he wasn't happy with the performance of the larger A350-000's engines. Airbus was given a consolation order for some A350-900s.

Clark said he was not happy with the "time on wing" Qatar Airways was finding with its A350-1000s, which would mean A350-1000s frequently could not be used for revenue service.

However, with Qatar seeming able to use them profitably there was probably another major factor in Emirates' decision, namely that Rolls-Royce had just before decided to change its contract terms to ensure it always made a profit, and was taking a hard line with Emirates regarding compensation for downtime—usually Emirates could negotiate very favorable terms in view of the large orders it could place.

Nevertheless, Rolls-Royce cannot afford to reject such a major client and is working hard to solve the problems, which will result in savings elsewhere.

~ Airbus A380 2007/271

Passengers love Airbus's truly double-deck superjumbo A380.

Emirates Airlines, with its strategically placed Middle East hub, had over 100 in its fleet, serving over 35 destinations, even with showers in some of the A380's first-class sections.

Entry of the superjumbo into service was delayed, partly due to a discrepancy in software programs used for electric wiring at two different production locations—either would have worked well.

The launch customers were Singapore Airlines, Emirates, and Qantas. All three airlines have put the emphasis on space, comfort, and facilities for their premium passengers rather than on maximizing the total number of seats. The aircraft was proving extremely popular with passengers.

Airbus conceived the A380 at a time when the Boeing 747 was the "Queen of The Sky" and had proved very profitable for its maker. However, the landscape was changing, with wide-body twinjets, such as the 777, carrying many more passengers than first envisaged, and due to changes in the ETOPS regulations, they were allowed to fly over oceans. There then arrived the very efficient composite Boeing 787 and the Airbus A350 with great range, which enabled passengers to fly point to point rather than via hubs.

Another problem with the A380 was the investment required for an airport to be able to handle it—special air bridges, wider taxiways, and so on. Interestingly, Boeing's 777X has folding wingtips enabling it to have a greater wingspan like the A380 but still use conventional facilities.

The A380 often needs the space that two aircraft would require. This meant the A380 could only operate to a limited number of destinations—five airports in the UK and only 14 in the much larger US.

Another problem was that the A380 did not fit the US market where the route structure was different, involving many large airports with space for multiple runways unlike, say, the UK with one key one with insufficient slots.

Production of the A380 ended in 2021 and a number of airlines put many in storage during the Covid-19 passenger downturn with the companies implying they might not return. However, as of the end of 2022 there was renewed interest in them with many returning to service and Emirates possibly sorry they are not still in production. This great resurgence of the A380 is partly due to the delayed Boeing 777X widebody.

In 2025 the A380 became very much in favor with major airlines such as Lufthansa, Singapore Airlines, British Airways, Quantas, Etihad, and Korean Air apart from Air France putting it to very good use. In the case of Air France, the prohibitive cost of modifying the tired old-fashioned cabins to please premium passengers made them decide to withdraw them.

Aircraft Carrier

The point of aircraft carriers is to project power to places one could not otherwise reach. They are exorbitantly expensive to construct and operate, since besides their large crews they need ships, submarines, and aircraft just to protect them.

The gigantic US carriers risk proving a hostage to fortune, with the potential loss of their large crews escalating to a more deadly confrontation. In this respect, Britain's two newly built but not nuclear-powered aircraft carriers will perhaps serve as stalking horses as their sinking and the sacrifice of their crews might not lead to so dangerous an outcome.

Although the US has perhaps a dozen giant carriers, only two or three or so can be immediately deployed in time of peace. Two or more can be completely out of action due to maintenance and upgrading of systems. Unlike in wartime, others may be in port but not immediately ready with crews on board for deployment. Also, unlike China's carriers, the US ones operate in multiple theaters.

Interestingly, the large US carriers now have the protrusions at the sides welded on, rather than bolted for easy removal for passage through the Panama Canal. The carriers are nuclear powered and can take the roundabout route at full speed, and having them welded means they are both stronger and cheaper.

The latest US carriers of the Gerald Ford class have improved nuclear power plants able to provide the greatly increased surge electrical power needed for the new electromagnetic launch systems, laser defense weapons, and to allow for the even greater electrical requirements of future systems. They are also designed with equipment to facilitate manipulation of munitions, equipment, parts, and even aircraft, and so require smaller crews. The major problem with the relatively numerous older carriers is going to be their inability to meet the ever-increasing electrical demand.

"Airborne" Aircraft Carriers?

Like the battleship in World War II, the golden age of the giant *waterborne* aircraft carrier might partly be over with some *airborne* carriers taking their place. By this we mean aircraft or even drones from which aircraft, hypersonic missiles, and drones can be launched.

In addition to being less vulnerable, the relatively high (air) speed of an airborne carrier could be a plus when launching supersonic and hypersonic aircraft and drones equipped with ramjets or scramjets that only work when moving fast.

Instead of having expensive destroyers and submarines protecting them, an "airborne carrier" could have, say, 50 or more cheap drones doing that, with the carrier weaving among them to further confuse an enemy. These sacrificial drones could intentionally emit heat and noise, not to mention electronic countermeasures. Also, such carriers could launch from **Standoff** positions, and they would not be putting thousands of sailors and aircrew at risk.

Aircraft Dispatcher [US], Flight Operations Officer [Europe, Africa]

Also called flight dispatcher in Asia and the Middle East, and flight operations officer in many other places.

The role is perhaps taken most seriously in the US, with aircraft dispatchers sharing 50/50 responsibility with the captain of the aircraft, with either being able to refuse permission for the aircraft to depart. They must be certified, and for safety reasons they are walled off from the commercial management of the airline.

The major US airlines may have as many as 500 dispatchers who work remotely at the operations centers of the airline with no physical meeting up with the pilots. They must be very capable, being responsible (in addition to the captain) for ensuring there is the required amount of fuel and selecting the optimum flight path considering the weather, en route winds, and restrictions, not forgetting the dangers represented by conflict zones where the aircraft could be shot down! In this, they may be helped by computers using artificial intelligence (AI).

Aircraft Emergency Recovery Systems

See **Whole Aircraft Recovery Parachute System.**

Aircraft Registration Codes

The code prefixes on aircraft denoting their country of registration are not immediately obvious except in a few cases, such as G for Great Britain, I for Italy, D for Germany, RA for Russian Federation, and JA for Japan. US-registered aircraft start with the letter N, while those registered in China begin with B, with further letters or an additional number being subcategories for Hong Kong, Macau, and Taiwan.

See http://en.wikipedia.org/wiki/Aircraft_registration.

Air-Launched Ballistic Missile (ALBM)

Ballistic missile launched from a standoff bomber.

See **Missiles**.

Air-Launched Cruise Missile (ALCM)

Cruise missile launched from a standoff bomber. Look-down radar-equipped fighters can destroy them, but as, say, the B-52 can launch 20 at a time from a standoff position, it is difficult to achieve 100 percent defense, especially as there can be multiple bombers.

Airfoil [US], Aerofoil [UK]

A wing, propeller blade, or similar body passing through air to provide lift, change of direction, or, in the case of a propeller, forward force.

Airframe

Essentially, the aircraft structure minus the engines. A well-maintained airframe can be updated, and see its value increased, by installing the latest avionics, making it easier and more pleasant to fly, and even more fuel efficient.

Much of the airframe on the B-52 that entered service in 1955 is very rugged and with it to be used in standoff scenarios is still going strong.

Airframer

Though the media refer to Airbus and Boeing as aircraft manufacturers, a more appropriate word increasingly used in the aviation business is *airframer,* to distinguish them from engine makers. Airframers make most of their profit when they sell an aircraft, whereas engine makers may hardly make any profit at that point and would sometimes be satisfied with no profit, since the profits come from service, maintenance, and parts.

Airline Codes

These are IATA codes used in flight numbers and so on. There are two codes used to designate airlines: the two-letter IATA code, such as BA for British Airways, which is put before the flight number in timetables and airport indicator boards; and the three-letter ICAO code. Because of the shortage of two-letter codes, some regional carriers are allowed to have codes used by a regional carrier in another part of the world when there is no chance of them getting confused.

To help avoid confusion and to make it easier for the recipient to know the call is destined for them, air traffic controllers add another name when calling aircraft. Some of these are a hangover from the old days. For instance, British Airways aircraft are given the prefix *Speedbird.* When Captain Chesley Sullenberger ditched his Airbus A320 in the Hudson River after his engines failed due to a bird strike, his call sign was *Cactus 1549* because that was the designation kept when American Airlines merged with American West Airlines, which flew to arid places with cacti.

Airline Deregulation Act [US]

The Airline Deregulation Act, passed by the US Congress in 1978, gradually released the grip of the Civil Aeronautics Board (CAB) on air travel in the US. Up until then airfares had been regulated to allow the airlines a return of 12 percent on flights that were 55 percent full. Getting approval to fly a new route could take years, with the possibility that it would finally be refused by the CAB. Internationally, IATA would fix fares at high levels, with the result that air travel was very much for the rich or people traveling for business.

After 1978 there followed a period, with the introduction of larger aircraft able to transport the masses, where charter flights were smashing prices, and flag carriers were selling seats officially at full price and unofficially through bucket shops at large discounts.

Airliner Nomenclature [Airbus A3xx] [Boeing 7x7]

The first Airbus was the A300, and it was simply called that because it was expected to have 300 passengers. The name was retained even when the actual number of passengers was somewhat fewer.

Subsequent models were a play on A3xx.

In the case of Boeing, the company simply chose to start with the number 7 for jetliners, because that was the one used to denote a jet aircraft as opposed to, say, a piston one. The 7 was added because it sounded better and did not give the impression it was the first and only one.

Subsequent models were a play on 7x7.

Airlines for America (A4A), Formerly ATA

Powerful trade organization for US airlines and instrument for lobbying Congress. On October 27, 2015, Delta Air Lines Inc. (DL) elected to leave A4A at the end of April 2016; shortly afterwards, A4A terminated its membership with immediate effect. In 2020, Delta rejoined.

The sums spent on lobbying are considerable, and furthermore the airlines shower individual members of Congress with favors regarding the booking of flights and treatment.

Airmanship

Sums up the way pilots handle their tasks, exercising above all good judgment as well as dexterity.

See **Famous Feats of Airmanship**.

Airport Surface Collision Prevention Aids [US]

In recent years the number of incidents on the ground at airports has seemed to be increasing. They have frequently been "near misses" that could have been disasters.

To deal with this the FAA has started rolling out systems to assist controllers. Some are quite simple and merely act as a memory aid.

1. Runway Incursion Device (RID).
   This consists of a device with buttons that controllers press to indicate when ramps, taxiways, or runways are in use.
   When a controller issues a clearance the runway or taxiway in question is selected by pressing its button, which stays lit and colored red until that runway or taxiway is marked clear by pressing another button.
   It can be set to give aural alerts to warn the controller to check a taxiway or runway before issuing a clearance.
   For instance, it can show when runways are closed or occupied by surface vehicles.
   Tailored for the particular airport, it can handle up to eight runways, taxiways, or ramps.
   The controllers will have several linked devices at their disposal.
   It should be in place at 76 airports by 2027.

2. Surface Awareness Initiative (SAI).
   Uses ADS-B data to monitor surface traffic including ground vehicles at airports lacking a surface surveillance tool.

3. Approach Runway Verification (ARV).
   It can alert controllers should an approaching aircraft be lined up to land on the wrong runway, on a taxiway, or even at the wrong airport!
   This would have been useful in the narrowly averted disaster at San Francisco when an Air Canada aircraft landing at night mistook a taxiway for the runway and almost landed on top of four airliners waiting there.

Airports

Airports vary in size from gigantic ones covering an area larger than New York City through tiny ones such as London City Airport squeezed onto London's docklands to tiny fields.

For perhaps one of the most comprehensive lists of the worlds airports country by country see https://www.flightradar24.com/data/airports.

It shows 1,501 airports for the US, which is a little surprising considering Steven Varsano founder of **The Jet Business** has pointed out that the US has by far the most business jets because there are 5,000 airfields they can use there. It is not surprising if one realizes this flightRadar24 list is only for "proper" airports able to handle commercial traffic.

~ Busiest Airports

The significance of this can sometimes be confusing as some cities with no space for an enormous airport have several medium-size ones serving the general area. An example would be London, with its six airports. Some lists rank airports according to the number of passengers including arrivals.

As these are always changing and can be found on the Internet, we do not give the actual figures.

In the US

Some such as the three below are difficult to understand until one realizes they function very much as hubs and are not uniquely serving their catchment area.

1. Hartsfield–Jackson Atlanta International Airport.
   This is the busiest airport essentially because it is a strategically located hub within two hours' flying time of anywhere in the mainland US, and it is the main one for Delta Air Lines.
2. Dallas Fort Worth International Airport.
3. Denver International Airport.

Los Angeles, Chicago, and New York's JFK are not far behind.

Interestingly, though in the US Northeast, New York City, for example, retains its status as a premium business class traffic hub; some say business is moving south, and Dallas, Houston, and Miami are the modern day business hubs.

Outside the US

The busiest airports include:

Dubai International Airport, Beijing Capital International Airport, London's Heathrow Airport (with just two runways and not open overnight), Tokyo Haneda International Airport, iGA Istanbul Airport, and Indira Gandhi International Airport.

Furthermore, Dubai will be the world's leading airport in the first three months of 2025 (Q1) in terms of available seats. It has 15.1 million departing seats, surpassing the next busiest airport, Hartsfield–Jackson Atlanta International Airport, by 2 percent. These are international flights; Atlantic holds the record for domestic flights, and the number of flights will pick up in the summer.

~ Largest Airports According to Area

1. Dubai International Airport is being replaced by a very much larger airport called Al Maktoum International Airport after the name of the ruler of Dubai, which will very likely be the largest in the world with five parallel runways and able to handle 260 million passengers. It will cover 301 square miles.
2. Denver International Airport covers 53 square miles.
3. Istanbul Airport opened in 2018, covers 30 square miles, and is strategically placed between Europe and Asia. It does not suffer the extremely high temperatures and sandy conditions of the Middle East hubs that degrade engine performance and subject them to greater wear.

~ Most Dangerous Airports

Lists are published regarding the world's most dangerous airports, but they are deceptive as regards the danger they represent for passengers. This is because usually only the best pilots and specially trained and approved pilots are allowed to fly into them. Also, pilots will be extra careful, and as one knows accidents tend to happen when easing off.

As regards major airports regularly flown into by full-size airliners, Hong Kong's Kai Tak was perhaps the one most cited because of the mountains and the lack of flat open space on the approach.

The traditional straight-in approach was impossible, and aircraft had to come in towards a mountain at an angle to the runway, at the last minute make a sharp turn at a spot marked by a chequerboard, and come down to the runway with passengers looking into the windows of apartment buildings. Over the years there were several mishaps but not many, all things considered. However, with the increasing traffic a new airport was opened on an island and reclaimed land quite some distance from built-up areas. Kai Tak is no more.

Airports may be dangerous to land on, or sometimes worse, to take off from for several reasons:

1. **Nearby mountains or hills**, or even high obstacles. These could be struck on landing due to navigational error or failure to observe Minimum Safe Altitude (MSA) at various points. They could make a takeoff or **Go-Around** difficult, even necessitating a steep climb on takeoff.

2. **High elevation**, or high temperature, where the engines perform less efficiently.
   Some such airports have extra-long runways to cope with that, meaning they are not particularly dangerous. An example is Denver International Airport at an elevation of 5,434 ft/ 656 m with **Runway** 16R/34L, with a length of 16,000 ft (4.88 km), and the longest public-use **runway** in North America and the seventh longest in the world. Five other runways are 12,000 ft/3,658 m in length.

3. **Short runways.**

4. **Weather conditions**
   can make any airport dangerous, though if they are so bad it will normally be closed. Also, pilots have certain minima and, say, if at 500 ft they cannot see the runway properly, the landing must be aborted.
   Of course, some airports, for example in North America

where there can be ice, snow, and poor visibility, can be more prone to bad weather than others.

5. **ILS (Instrument Landing System)** out of order. An airport that otherwise is very safe can become dangerous if the ILS is out of order and the pilot is not used to landing without them, as happened when a Korean Asiana Airlines 777 was landing at San Francisco in 2013*.

The most dangerous airport in the world is said to be Lukla near Mount Everest in Nepal shown in the photo below. It has all the dangerous features mentioned from 1 to 5 above.

Following a number of accidents including one where a pilot tried to land in bad visibility and just below the runway killing all on board apart from himself, the Civil Aviation Authority of Nepal set high standards for pilots. To land there they must have completed 100 short-takeoff-and-landing flights, have at least one year of such experience in Nepal and have perfectly executed ten flights into Lukla with a certified instructor.

[Sam Chui]

As one can see from **Sam Chui's** photo, the runway is very short, 1,729 ft, and surrounded by high mountains. At the top of the 12 percent slope behind the aircraft is a solid wall and mountainside meaning an overrun on landing is impossible, as is a go-around after touchdown. Aircraft take off downhill in the other direction to be faced with a sheer drop at the not so far end.

Only a certain type of aircraft, such as the one shown, can operate there. They must have powerful engines to cope with taking off from a short runway in the high-altitude thin air and handle well at high altitudes.

Airport Codes: Three-Letter IATA, Four-Letter ICAO Codes

The derivation of many three-letter IATA codes used for timetables and ticketing, such as LHR for London Heathrow, is obvious, though some are awfully hard to fathom.

Furthermore, there are the four-letter ICAO airport codes used on navigational charts, by air traffic control, and for flight planning. Though these codes are not guessable, there is a certain logic in that the first letter represents a region such as "E" for Northern Europe and "L" for Southern Europe. The second letter represents a country with sometimes a certain logic there, too.

The first two letters for Great Britain are "EG." The four-letter code for London's Heathrow is "EGLL"; London Gatwick is "EGKK." Large countries such as the US, Russia, and China have the region identifier to themselves with areas within them sometimes having the second letter.

Airships

Prior to World War II, the German Zeppelin airships had had much success with many trips to the US with passengers, and the round-the-world trip in 1929 (see **Lady Grace Drummond-Hay, "Around the World by Zeppelin"**).

However, the Hindenburg disaster that occurred on May 6, 1937, at Lakehurst, New Jersey, United States, signaled the end of the hydrogen-filled passenger-carrying rigid airship. There were 97 people on board (36 passengers and 61 crewmen), with 35 (13 passengers and 22 crewmen) losing their lives in the fiery inferno. There was also a fatality on the ground.

Rather like the "Miracle on the Hudson" the worldwide publicity with terrible photos was so great that the impact of the disaster was more than it might have been. There were many suggestions as to the cause, including sabotage, though in the end static electricity was thought to be the most likely.

From time to time there are news stories of plans for giant helium-filled airships, sometimes for transporting heavy items such as generators to isolated sites such as mines. For that there is the problem of how to keep the craft down when unloaded. One idea was to have water as ballast; another was to compress much of the helium.

Airspace Classification, A, B, C, D, E, F, G

Airspace is classified with various applications applicable. Perhaps the most notable one is Class B (Bravo) airspace near major airports where aircraft must have permission to enter lest they collide with an airliner taking off or landing. Because airliners mostly fly under the direction of an air traffic controller, they *de facto* have permission to enter and do not have to request it.

Most countries follow the **ICAO** classification, with some variations. The classes range from A to G, with G having no restrictions other than common aviation order ones.

In the United States it is roughly as follows:

Class A

In the US, airspace from 18,000 feet **MSL** (above mean sea level) up to and including FL600, including the airspace overlying the waters within 12 nautical miles of the coast. Aircraft must operate under IFR (Instrument Flight Rules) unless otherwise authorized. Note that in the US it is particularly high because parts of the country have very high mountains. For other countries it is much lower.

Class B (Bravo Airspace)

The area taken up by Class B (**Bravo Airspace**) around the busiest major airports such as Los Angeles, New York, and Chicago varies with height, being narrower at the bottom where aircraft will only be so low when close to the runway and very much wider at the top where aircraft can be high up when far from the airport. It looks somewhat like an inverted tiered wedding cake and usually extends from the ground to 10,000 feet MSL. In the US, prior air traffic control clearance is required to operate there, and all aircraft thus cleared receive separation services. VFR operations, "clear of clouds," is allowed. Sometimes general aviation aircraft stray into this to the fury of controllers fearful of a disaster, with some GA pilots proving awkward.

Class C and D

These are for smaller airports. Every pilot must establish two-way radio communications with the ATC facility providing air traffic services prior to entering the airspace and thereafter maintain those communications while within the airspace.

For Class C, around airports such as John Wayne in Southern California, San Jose and Oakland in Northern California, and Midway in Chicago, VFR aircraft are only separated from IFR aircraft within the airspace.

For Class D, around airports such as Santa Monica, Palo Alto, and San Carlos in California, DuPage near Chicago, and Teterboro near New York City, no separation services are provided to VFR aircraft.

Class E

Generally, if the airspace is not Class A, Class B, Class C, or Class D, and is controlled airspace, it is Class E airspace. Class E airspace is always below Class A airspace (18,000+ feet MSL).

Class F

Often omitted.

Class G

Do what you like subject to normal aviation norms. Somewhat like driving your car off-road in a farmer's field.

Airspeed, Knots, Mach

The speed of aircraft through the air as opposed to over the ground. As the aircraft gets higher the air becomes less dense, and the *indicated airspeed* determined by the pitot tubes lessens even though the aircraft is striking the air faster.

Near the ground, pilots are concerned with their airspeed in knots (kt) or kilometers per hour (km/h), but at cruising height at high altitude they worry about their airspeed relative to the speed of sound or their Mach number, which they can set their airspeed indicator to show.

See **Coffin Corner** regarding flying at very high altitude with little margin between flying too fast and too slow.

Airstair, Parachuting via Airstair from Boeing 727

Airstairs are sets of steps built into aircraft. Notably, they are at the rear of the Boeing 727 and can be deployed without the airport having to provide one. With airports now equipped with jet bridges, allowing passengers to pass directly from the aircraft into the terminal, there is no need for them.

In 2012 the 727's almost unique airstairs were put to good use when scientists wanted to study the effects of a crash on humans. They placed highly sophisticated human dummies with sensors able to measure stress on individual bones in a 727 that was purposely crashed in a desert in Mexico (because the FAA would not agree to purposely crashing an aircraft in the US) after the pilot had parachuted away via the airstair.

The scientists found that those sitting at the front in first class would all have been killed, those in the middle over the wing would probably have mostly survived, and those at the rear to have certainly done so. The brace position was to a certain degree better, while the dummy with no belt slid forward under the seats in front like a sledge. As already mentioned, all crashes are in some way unique, so you cannot say that those at the front will always be worse off. Nevertheless, the four sole survivors in the world's worst-ever single-aircraft crash—Japan Airlines Flight 123*—were in the tail section, which broke off.

See **D. B. Cooper** who notoriously parachuted out of a 727 with a large sum of extorted money and was never found.

AirTag: Lost Luggage

People put Apple AirTags in their checked in luggage in the hope that it will help them find it should it not appear on the carrousel at the airport when they arrive. Until recently they have been angered to know where their lost item is only to find the airline insisting on relying on only their own system of tags.

Though some airlines, including Delta, United, British Airways, Lufthansa, Quantas, and Virgin Atlantic have set up internal ways at selected airports to link up to these tags, Apple has the "Share Item Location" function on the new iOS 18.2 which should facilitate this.

Incidentally, AirTags do not depend on GPS but link to other iPhones that happen to be in their vicinity.

Airway

Designated routes followed by aircraft. Countries usually have two sets of airways, with different designations according to their height. Some countries follow the US system of designating the (lower) airways as Victor airways and the upper air routes as jet routes (say, 18,000 feet and above).

Airways are usually between VOR beacons, with the pilot following the outbound radial and then the inbound radial of the next one, switching between the two once the latter's signal strength becomes superior, which is not necessarily halfway, since VOR signal strengths vary.

Oceanic routes have a different nomenclature.

With the eventual introduction of GPS-based ADS-B, airways are likely to lose their importance as aircraft take routes that are more direct. However, they have proved their worth and would have to be kept for use in the event of something going wrong with the satellite-based system.

Airworthiness Directive (AD)

These are mandatory orders by authorities, such as the FAA, for rectification of a defect found after certification of the aircraft model. They may be immediate, in which case the aircraft may not fly until the rectification has been performed, or mandatory within a certain time or number of flying hours, or even the next time the aircraft is due for routine maintenance.

They are greatly feared by airlines, since such directives could ground a large part of their fleet at peak periods, and hence they will try to find ways to soften the stance of the authorities, as in the case of the DC-10 cargo hold door mechanism, where there was eventually a "gentleman's agreement" with the FAA, and later a disastrous crash abroad*.

Aisle

Narrow-body aircraft have a single aisle, with passengers often sitting three abreast on either side.

While two aisles would improve comfort, the main thinking behind this is that two aisles would decrease turnaround times, since people can board and disembark much more quickly with two aisles. With a single aisle it only takes one person to bring boarding—or worse still an emergency evacuation—to a halt.

Aisle or Window Seat? Psychology

An aisle seat is good if you want the freedom to move around and a window seat fine if you want to be undisturbed and perhaps less likely to be infected by sick passengers passing by.

Interestingly, according to *The Independent*, some psychologists say there might be another reason for the choice of seat: People who select window seats are "selfish and controlling" and those who prefer aisle seats are more likely to be introverts. Though there might be some truth in this it may only be incidental and should not be taken as a given fact.

See **Weight and Dimensions of Passengers**.

ALARP (As Low as Reasonably Practicable) [Risk in Military Context]

In military operations as opposed to civil aviation, one cannot seek zero risk. Instead, one must aim to keep it *as low as reasonably possible*, hence the acronym ALARP. It came to public notice in the UK during the inquest into the lives of military personnel lost in the Nimrod (AWACS) disaster in Afghanistan in 2006.

Alcohol Limits for Pilots

Airlines have strict regulations concerning the consumption of alcohol by pilots before flying. Some say none within 8 hours and others none within 24 hours of flying. In the UK, the legal limit for airline pilots is 20 micrograms of alcohol per 100 milliliters of breath, while for drivers (motorists) it is 35 micrograms of alcohol per 100 milliliters of breath.

Occasionally, pilots have been arrested way over the limit, but these incidents do seem to be exceptions. Airlines have programs to try to identify crew members with alcohol problems and even help them rather than let the problem stay hidden and untreated.

In fact, fatigue appears to represent a far greater safety risk as regards impact on a pilot's performance.

Airspeed Indicator (ASI)

A key instrument indicating airspeed. In modern cockpits, airspeed is often shown in an easy-to-comprehend tape (band) alongside the artificial horizon, as well as by the traditional dedicated airspeed indicator.

Alerts

Alerts are visual and audible (or the two combined) warnings that something is wrong. Pilots can be inundated by them all sounding at the same time, resulting in confusion, their being startled, and engaging in some unwarranted action.

On the other hand, some powerful warnings such as the **GPWS (Ground Proximity Warning System)** warning that one is about to hit the ground can mask the low-speed warning when crash landing or ditching as happened when Sully ditched in the Hudson River*—it should have been switched off, but that was at the end of the checklist intended for engine failure at relatively high altitude with plenty of time. (Sully came in too slowly and could not flare.)

Pilots are liable to ignore alerts or even turn the warning systems off in the event of repeated spurious alerts.

Experiments unconnected with aviation have shown that when one is concentrating visually on something, one filters out sounds—for instance, you can be concentrating on your computer and not hear the bath water overflowing! This might partly explain why the aural warnings that the aircraft was *stalling* in the Air France **AF447 ONLY A MINUTE TO SAVE THEM** crash into the South Atlantic in 2009* were ignored.

Algorithm

A sophisticated-sounding word, not easy to define—and that is why it is so handy. It is the series of instructions computers use to achieve an end (do something), solve a problem, or simulate a set of circumstances. While the computer programs on an airliner can be considered as a vast collection of algorithms, the term is usually used restrictively, such as saying, "Company X is developing better algorithms for predicting fan blade longevity in its new engine."

All-New Aircraft Design

The cost of creating an all-new aircraft and getting it certified is enormous and makers, especially now, prefer tweaking existing designs with better engines and more advanced wings. The last thing an airframer wants is to create a new model only to find it in no time overtaken by a significant advance in technology.

Boeing, although under a lot of pressure to create a "797" midmarket aircraft, seems to have cooled to the idea with its CEO saying it might be necessary to wait until the 2030s for a completely new aircraft. In fact, it has much on its plate, and there is no commercial reason for it to fill every gap in its lineup. Its decision to proceed with the ill-fated 737 MAX based on a 1960s design made a lot of commercial sense in that at the time it was having problems with the 787 and much else requiring technical resources. The problem with the MAX was that the MCAS system fatally pushing the nose down was initially designed for use at high airspeeds with other protections, and it was subsequently used for lower speeds with only one mechanical attitude sensor that could easily fail.

The MAX program could finally prove very profitable. However, the dent in Boeing's reputation and that of the FAA means that getting designs approved is no longer automatic for Boeing and now incurs very costly long delays.

Allergies: Great Downside Risk for Airlines

Passengers with life-threatening allergies tell of how they are "badly" treated even by staff of the best airlines. They do not realize the great potential downside for the airline if they have to divert, and particularly so if it is a small airline without an abundance of aircraft to replace the one diverted, not to mention the difficulties encountered by other passengers if it is an ultra-long-haul flight.

If someone should suffer a seeming life-threatening allergic reaction, the aircraft may have to divert to some inhospitable place and the incident may cost the airline a fortune not only in hotels and airport charges but also due to the aircraft being out of position and the airline having to cancel other flights or use other aircraft.

Airlines do try to avoid serving nuts but sometimes find it impossible to stop passengers eating the ones they have brought on board. That said, the nuts problem might have been overblown for tests carried out in a room where people with a peanut allergy were placed near someone consuming nuts showed that traces are not easily airborne and that they were not easily affected.

This suggests that that many of the cases of people with a nut allergy allegedly being badly affected on aircraft were more likely due to traces left on the tables and video screens from previous flights with nuts and nut products often being sticky. Anyone with such an allergy should wipe down all surfaces with wipes or, better still, get someone traveling with them to do it for them.

Recently, experts from Imperial College London and Aviation Medical concluded that research studies, including aircraft simulations, show no evidence to support airborne transmission of nut allergens as a likely phenomenon, and that announcements requesting nut bans are not therefore supported, and may instill a false sense of security.

Incidentally, a passenger with a nut allergy in 2023 felt obliged to buy every packet of nuts passengers had—perhaps 30 or 40 in all—at a cost of several pounds each, showing how serious SHE believed the risk. Of course, she might have been wanting to make a point.

Food and drink are even more difficult to be sure about, whether it be the onboard catering or something the passenger consumes before the flight as happened to the young girl with a severe sesame seed allergy who despite the use of two EpiPens sadly died after a flight from Heathrow to Nice—she had consumed a preflight baguette with sesame seeds sold at the airport in a wrapping that did not indicate it contained them.

Sometimes—much to the consternation of the individual feeling hard done by—cabin crew will play it safe and give hardly anything other than fruit to someone with a severe allergy on long-haul and ultra-long-haul flights if special food for them has not been loaded.

EpiPens can be lifesaving but may not be enough on a long-haul flight as medical treatment is often necessary as well. The person can seem to recover only to have another attack later. The advice, even on the ground, about when EpiPens should be used has changed, with emergency room doctors saying they should be used early rather than later when airways risk getting blocked.

It is essential EpiPens are used correctly. It is said that sometimes they are not applied deeply enough into the muscle. And, regardless of what is in the instructions, the needle should be left in place in the muscle for even 30 seconds after the delivery of the dose to help make sure it is fully absorbed.

Alliance (Aviation Alliance)

Airlines cooperate in alliances honoring each other's air miles, sharing facilities such as airport lounges and so on.

There are three major alliances:

1. **Star Alliance**
   Aegean Airlines, Air Canada, Air China, Air India, Air New Zealand, All Nippon Airways, Asian Airways, Austrian Airlines, Avianca, Brussels Airlines, Copa Airlines, Croatia Airlines, Ethiopian Airlines, EVA Air, LOT Polish Airlines, Shenzhen Airlines, Singapore Airlines, South African Airways, Swiss International Airlines, TAP Air Portugal, Thai Airways International, Turkish Airlines, United Airlines.

2. **One World**
   American Airlines, Alaska Airlines, British Airways, Cathay Pacific, Finnair, Iberia Airlines, Japan Airlines, Malaysia Airlines, Qantas, Qatar Airways, Royal Air Maroc, S7 Airlines, Sri Lankan Airlines.

3. **SkyTeam**
   Aerolineas Argentinas, Aeroméxico, Air Europa, Air France, Al Italia, China Airlines (Taiwan), China Eastern Airlines, Czech Airlines, Delta Airlines, Garuda Indonesia, ITA Airways, Kenya Airways, KLM Royal Dutch Airlines, Korean Air, Lufthansa, Middle East Airlines, Saudia, Scandinavian Airlines (SAS), Tarom, Vietnam Airlines, Virgin Atlantic, XiamenAir. (Aeroflot was formerly listed as a member.)

There are also three very much smaller alliances (Vanilla Alliance, U-Fly Alliance, and Value Alliance). Emirates is not part of any of these alliances but has an arrangement with a few airlines including Japan Airlines and easyJet whereby points gained on its own Skywards program can be gained or used mutually.

Choosing an Alliance

If several carriers fly the route(s) you frequently use for work, consider choosing one in an alliance with airlines serving routes you would like to go on for leisure.

Alpha α (Angle of Attack) [Airbus]

Alpha (α) refers to the **Angle of Attack** (in other words, the angle of the chord of the wings to the airflow). If the angle is too steep and the airspeed insufficient, the aircraft will stall. To cope with wind shear and errors brought about by pilot inattention, where the aircraft might fatally lose lift near the ground, Airbuses using the autothrottle system have a protection mode called "alpha floor" that automatically applies takeoff/go-around power should the angle of attack exceed a certain limit (15 degrees in the case of the Airbus 320). They also have something called "alpha protect" on the speed tape indicating where protections come into play.

Alternate Law as Opposed to Normal Law [Airbus Protections]

This, in the aviation context, is something the public had never heard of until the crash of Air France Flight **AF447 ONLY A MINUTE TO SAVE THEM** into the South Atlantic in 2009*. It specifically pertains to the fly-by-wire system in Airbus aircraft.

~ Normal Law

In *normal law*, that is except in exceptional circumstances, the Airbus computers provide "protections" that prevent the aircraft from flying dangerously. The protections include preventing the pilots pitching the aircraft up or down to an unsafe degree, banking too much to the left or right, subjecting the aircraft to g-forces that would overstress it, unsafe angles of attack (angle at which oncoming air encounters the wings), and combinations of angle of attack and lack of airspeed that could result in a stall, and over speeding that could damage the aircraft. With them functioning, the pilot can in theory do no wrong!

~ Alternate Law

However, for the computer to be able to do all the above it must know what is happening, and notably the airspeed measured by sensors in the pitot tubes set alongside the fuselage.

In the case of **AF447** these had iced up, and the autopilot disengaged with a warning that it was doing so, followed by a warning on the **ECAM (Electronic Centralized Aircraft Monitor)** that they were in Alternate Law and "protections"—that would prevent the pilot doing something putting the aircraft in danger—were lost.

There are two versions, 1 and 2, of Alternate Law, with Alternate Law 2 being the most extreme with the sidestick acting directly on the airfoils, making the stick hypersensitive as in **AF447**.

Altimeter

The altimeter is the key instrument indicating the height of the aircraft. There are two basic types, *the radio altimeter* and the *barometric altimeter*, with each having specific advantages.

~ Radio Altimeter

The radio altimeter uses radio waves bounced back off the ground to give the height by noting the time taken for them to come back. Its great advantage is that it gives true height above ground level (AGL) at that moment in time, which is particularly useful to know when landing and for the synthesized vocal alerts.

Ground proximity warning systems (**EGPWS**) and terrain avoidance warning systems (**TAWS**) that have prevented many accidents are based on the radio altimeter data.

Radio altimeters can fluctuate wildly when the aircraft is near the ground and passing over a clump of trees. To avoid spurious warnings when not landing or near the ground, radio altimeter warnings are often programmed with, say, a 6-second delay which has sometimes led to pilots receiving a warning too late.

Early radio altimeters with the radio wave going straight down perpendicularly to the aircraft were dangerously inaccurate when an aircraft was banking steeply.

This happened to Eastern Airlines Flight 401 in 1972, when pilots trying to work out whether an "undercarriage down" indicator lightbulb was faulty were circling in pitch darkness over the Everglades in Florida and flew too close to the ground when the autopilot that they thought was maintaining their height had unknown to them disconnected as regards height. As they were banking with the radio altimeter sensor facing sideways, they got the ground proximity warning too late and flew into the crocodile-infested swamp*.

Nowadays, radio altimeters have antennas covering an 80-degree arc and can cope with bank angles up to 40 degrees simply by only considering the signal bounced back first.

~ Barometric Altimeter

Though technically primitive as merely depending on the pressure of the air through which aircraft is flying to determine height, it has the advantage that it can show three separate things depending on the pressure setting:

1. **True height above mean sea level (QNE)**
   These elevations are shown on charts and stored in the aircraft computer memory so when it is flying near the ground the pilots can be sure it will not hit something on the ground.
   Aviation charts have **MSAs (Minimum Safe Altitudes)** and show this height for terrain, hills, and mountains.
   Airports have MSAs for the vicinity and the various stages of the approaches.

2. **Flight level (QNH) ensuring aircraft keep apart**
   Above a certain height called the **Transition Altitude (TA)** where there is no danger of the aircraft hitting the ground, they are assigned **Flight Levels** by air traffic control.
   These are in hundreds of feet so that 20,000 feet would

be Flight Level 200.
Above the transition height where there is no danger of aircraft hitting the ground, avoiding them hitting each other is what matters and with all aircraft having the same pressure setting of 1013 hPa or 29.92 inHg (inches of mercury), this can be simply achieved.
Thus the controllers know exactly how high the aircraft are *relative to each other*.
In the UK the transition height applies to aircraft over 3,000 feet, whereas in the US and Canada with their high mountains it is 18,000 feet.

3. **Height above a spot on the ground (QFE)**
This might be the threshold of the runway.
However, apart from small airfields in the UK, this is little used nowadays for the pilots and the computers know the height of the airport and resetting the altimeter would be a distraction and possibly increase the chance of errors. Besides, the radio altimeter is constantly giving the true height of the aircraft above the ground below.
Pilots can always request the QFE setting for the airport if not given.

See **Q Codes (QNE, QFE, and QNH)**.

Angle of Attack (AoA) (α)

Angle of the wings to the airflow. Should it become too great with the airspeed too slow the aircraft will stall. Airbus calls it **Alpha (α)**.

Angle of Attack Sensors

These are simple vanes on the side of the fuselage that move according to the airflow like weathervanes but being mechanical and outside the aircraft they are liable to be damaged by impact with a bird or other object for instance.

In a modern fly-by-wire aircraft the computer would normally take readings from three sensors, and if one disagreed it would ignore it on the assumption it was faulty. However, in a freak accident, pilots performing an acceptance test flight for an Airbus A320 carried out the stall test much too low and crashed. Workers had hosed down the aircraft without protecting the angle of attack sensors, allowing the high-pressure jet of water to penetrate inside the mechanism and remain. When the aircraft reached high altitude, the water froze, locking the sensors at the same angle and making the computer think they were functioning correctly and not automatically preventing the stall. As there were no fare-paying passengers, it did not receive much attention in the media*.

In the case of the two 737 MAX crashes leading to its grounding, Boeing had unbelievably relied on a single sensor.

Anhedral

Civilian aircraft usually have wings that are dihedral in that they slope upwards as they project outwards from the fuselage. This gives the aircraft inherent stability since it tends to automatically right itself after banking, since the lower wing has more lift than the upper one.

On the other hand, the wings of military aircraft, such as fighters, are often anhedral (sloping downwards as they project outwards from the fuselage) to make them inherently unstable and hence more maneuverable. Computerized controls and stabilizers provide apparent stability.

Anti-Access/Area-Denial (A2/AD)

A March 2023 edition of the UK's *Economist* magazine had a long piece about China invading Taiwan and pointed out that China's deployment of A2/AD weapons such as the DF-26 missile, able to strike deep into the Pacific, and new hypersonic missiles that are harder to intercept means the Chinese can fend off American ships and aircraft.

The "DF" in the Chinese missile names stands for "Dong Feng" meaning "East Wind."

See **Acronyms** to understand A2.

Antimissile Technology

Shoulder-launched SAMs (surface-to-air missiles) can be bought on the black market for less than $10,000 and are said to be in the hands of some 30 terrorist organizations. Civilian aircraft on an approach to an airport are particularly vulnerable, as are long-haul flights taking off heavily laden with fuel and unable to climb away rapidly.

El Al has its airliners equipped with antimissile systems, as does **Air Force One**.

AOPA (Aircraft Owners and Pilots Association)

Maryland-based American nonprofit political organization that advocates for general aviation. AOPA's membership consists mainly of general aviation pilots in the United States.

It also has sections in other countries including the UK and Australia.

Approach

The final phase of a flight, requiring intense concentration, as the aircraft comes into land.

Approach Control

Air traffic control unit responsible for handling aircraft as they approach the airport for landing (after they have been handed off to it by the center responsible for traffic in its region). Approach Control, in turn, hands them off to the tower for the actual landing. See **Air Traffic Control (ATC)**.

Apron

The paved area at an airport used for parking aircraft.

Approval and Certification

With the noble intention of guaranteeing safety, everything on an airliner must be approved and certified by the appropriate authority such as the FAA. This can be a time-consuming process. The downside is that when a system is found to have weaknesses, improvements cannot be immediately introduced or are often forgone because the certification process is too costly.

Aptitude to Be a Pilot

There are aptitude tests that can tell whether you have some of the abilities needed to be a commercial pilot. Many would be like those in standard intelligence tests measuring problem solving, identifying patterns, and so on.

Some abilities such as intelligence are innate and cannot be significantly improved with training.

In the US, there was a time when many pilots came from the military where selection was rigorous and those not up to par weeded out. However, one needed to have a college degree to be an officer and pilot.

Incidentally, joining the military was one way to ultimately become a pilot at a major US airline, provided one did not get killed in action.

Though being a pilot seems wonderful—prestige, very good salary (later in career), socially exciting life, and so on, many good pilots such as "Miracle on the Hudson, Sully," became one out of an early passion for flying.

In the US, the FAA does not demand great academic qualifications to be certified as a pilot, but in practice having a college degree may be essential to get accepted by one of the major airlines such as Delta, United, or American Airlines. A college degree can put you at the top of the applicants when there could be thousands. A college degree does show commitment and the ability to concentrate. It also gives one more to fall back on should flying no longer be possible for medical reasons.

In the UK, good GCSEs in, say, English, math, and physics, and A-levels are good to have.

If thinking of a career as a pilot it is a good idea to get someone to take you up in a light aircraft to see if it is really for you.

The first thing to consider is your health, and before spending money on lessons, get checked out to see whether you could easily pass, for instance, the regular FAA medical exam every six months or a year.

One does not need to be a math genius to be a pilot, but an ability to understand technical matters is important and often follows from a natural curiosity regarding how things work. Aircraft and their systems are complex, and problems are mostly dealt with using checklists, though some—the so-called memory items—have to be dealt with instantly with no time to search the QRH (Quick Reference Handbook).

Many qualities are required including the ability to think on one's feet, the ability to keep calm under pressure including under heavy workload, confidence borne of ability, **Resilience**, spatial awareness, ability to work as part of a team, communicate well, neither arrogant nor subservient, and so on.

One problem is that predicting how someone will react in a real crisis is difficult. When military psychologists classed people in a group according to who would be the most heroic in actual combat, many of those judged least likely to be heroic proved to be the heroes in actual combat. Nevertheless, pilots responsible for crashes often had a history of becoming disorganized under pressure.

See **Borderline Pilots, Psychometric Evaluation of Pilots**, and **Mental Health of Pilots**.

APU (Auxiliary Power Unit)

A small power-generating turbine, usually situated at the extreme rear end of aircraft. Used for generating electricity for onboard equipment and air conditioning when aircraft are on the ground and not connected to the airport's electricity supply. Also used to supply bleed air to start the main engines.

The NTSB report into the ditching of Captain Chesley B. Sullenberger's Airbus in New York's Hudson River* after a bird strike concluded that his switching on the APU was a key factor in the safe outcome by providing enough electricity for the computers to fly the aircraft optimally and not stall.

Area 51 near Groom Lake

This area in the desert north of Las Vegas is shrouded in mystery, with it being used for secretly testing secret aircraft such as the U-2, X-15, Oxcart, SR-71, F-117, among others. It was also used to test US aircraft against Soviet fighters that had been obtained secretly, thus finding their weak points.

Rumors abounded that **UFOs**, now called **UAPs (Unidentified Anomalous Phenomena)**, were being studied due to sightings of the secret aircraft tested there. However, just recently has it been revealed that these rumors were actually largely the result of disinformation campaigns to cover up what was being done there. These campaigns, that began years ago, consisted of briefing selected personnel on tests related to UFOs and then having them swear to the utmost secrecy, thus making them believe it was true and a mystery they could not look into.

For a history see, for example:
https://area51.fandom.com/wiki/Area_51#Description.

Also see **"Janet" Call Sign for Secret Airline Flying to Area 51**.

Artificial Intelligence (AI), Military Uses

Military pilots with Aegis-type systems and multiple sensors are inundated with information. Much of it is for nonhostile aircraft as far as 100 miles away. Already, AI is being used to filter the information and interpret it, but now we are coming to sixth-generation aircraft, drones, and missiles.

People working on next-generation fighters and bombers are planning to integrate AI to take much of the mental load off pilots, to some extent flying the aircraft and managing the task of searching sites on the ground while the **pilot** looks out for attack from the air.

See **AI Changing Warfare**.

Artificial Intelligence (AI), Civil Uses

AI is beginning to feature in many civilian roles including the preparation of flight plans where finding the most fuel-efficient route possible can save a great deal of money. Elsewhere, we have mentioned autopilots able to learn from mistakes made on different models of aircraft in a similar situation.

Of course, artificial intelligence could be useful for training pilots.

Ashtray in Toilets

Although smoking is forbidden in airliner toilets, and smoke detectors indicate anyone doing so, provided the perpetrator has not covered them up, the FAA mandates that airliner toilets must have ashtrays. This is to avoid someone fearing being found smoking putting their unextinguished cigarette in the trash bin, which could result in a serious fire because of the amount of paper in the bin.

ASK (Available Seat Kilometers)

Number of seats available multiplied by number of kilometers flown (usually by the whole fleet). Seats may not be available for several reasons, including the need to enable one or more members of the crew to rest, though the larger long-haul aircraft have dedicated crew rest areas. CASK (cost available seat kilometers) is a measure used to compare airlines' costs but needs to be used with care, as the stage length can have a great impact.

ASN (Aviation Safety Network) Database [List of Incidents]

The ASN Safety Database, updated daily, contains descriptions of incidents involving airliners, military transport category aircraft, and corporate jet aircraft safety occurring since 1919.

Airliners considered are those capable of carrying at least 12 passengers.

The incidents, many of which are very minor, can be shown for a given year with a link to the provisional or final accident investigation report. They can also be shown for a given country.

There is also a page with links to other databases including one called WikiBase to which anyone can add incidents for all sorts of craft including military, general aviation, and helicopters. That page also has links to databases relating to eVTOL and drones.

Aspect Ratio

The ratio of a wing's span to its mean chord. Equal to the square of the wingspan divided by the area of the wing.

High aspect ratio wings that are long and narrow have less induced drag, higher lift to drag ratio, and are more stable. Low aspect ratio wings are more maneuverable and are easier to build.

ASRS (Aviation Safety Reporting System)

Scheme run by NASA of the US for the confidential reporting of incidents. It is like the UK's CHIRP or Europe's EUCARE.

"Assumed Temperature" [Boeing] "Flex Takeoff" [Airbus]

With engine power settings automatically calculated by the aircraft computer for optimum takeoff, it is sometimes desirable to "cheat" to use lower thrust and put less strain on the engines. This saves maintenance and money but can only be done where runway length, altitude, and temperature permit.

Airbus calls this procedure "**Flex Takeoff**" while Boeing calls it "**Assumed Temperature**." The pilots simply enter a higher than real ambient air temperature into the flight director so the thrust is lower than it would normally be because the engine temperature cannot exceed a certain limit.

Reducing the power setting can be done in other ways.

Astro-Inertial Guidance Systems

The **U-2** relied on pure inertial navigation systems to ensure it flew the reconnaissance route desired. However, for the SR-71 with its longer missions inertial position error growth posed a problem. Consequently, an astro-inertial navigation system (ANS) using celestial observations to correct these errors and reset the device was developed.

These are also used in missiles, often to complement GPS which can be downgraded or spoofed in time of war. See **RUSSIA Sarmat ICBM**.

Asymmetric Flight

Case where total thrust from engines on one side is not equal to that on the other, say, when an engine has failed. Some four-engine aircraft are allowed to take off (without passengers) with just three engines to go for repairs. Usually only specially trained pilots are allowed to do this.

A British Airways aircraft with passengers on board suffered an engine failure shortly after takeoff from Los Angeles bound for London. Instead of landing as they normally would, the pilots decide to continue on three engines. Unfavorable wind, coupled with having to fly lower than usual with only three engines, meant they could not reach London with the legally necessary fuel reserves and had to land not far short at Manchester (UK).

Though passengers were not in any real danger, since the aircraft had already taken off safely on four engines, the FAA argued with the airline and the UK authorities about imposing a maximum $25,000 fine. This raised the complicated question of jurisdiction in the case of international flights, for the British were arguing that, under their rules, continuing all the way on three engines was permissible.

Indeed, four-engine British Airways aircraft do sometimes come back to London from afar on three engines for servicing. The author experienced such a situation in Japan, where the BA 747 he was about to board had an engine problem and later flew back to London on three engines with a special crew but without passengers.

Asymmetric Warfare

Warfare with a great disparity in terms of size, weaponry, and logistical support, usually meaning that the weaker side must avoid squaring off and instead exploit the stronger side's weaknesses with unsettling pinprick attacks in vulnerable spots, better tactics, and subterfuge, perhaps causing them to overreact and alienate civilian populations, or even tire of the whole enterprise.

Although the US has a military unit called AWG (Asymmetric Warfare Group) set up in the context of the War on Terror advising on such warfare, some in the US try to exclude operations against *terrorists* from this definition, since their inclusion would appear to ennoble them.

ATA (Air Transport Association of America)

Trade association representing and promoting the interests of the major US airlines.

ATC Clearance

Before taking off, an aircraft must obtain an ATC clearance regarding what they are to do after takeoff. For example, "Climb to such and such a height and proceed to x."

ATIS (Automated Terminal Information Service)

Continuous (recorded) broadcast giving meteorological and other conditions at an airport. This includes problems concerning runways, taxiways, and equipment. Frequently consulted by pilots.

Attitude Indicator (AI), Artificial Horizon (AH)

Formerly called the artificial horizon.

Cockpit instrument showing the attitude of the aircraft in the longitudinal axis (pitching up and down) and lateral axis (banking to left and right).

On most Western airliners, the attitude indicator shows the aircraft fixed and the horizon outside tipping to the left or right according to the degree of bank. If the degree of bank is slight, what is happening is obvious, but if it is extreme, it is less so.

Attitude indicators on Russian-built aircraft generally show banking in the opposite way where the horizon is fixed and the image of the aircraft moves to show how it is relative to it, which is more intuitive. Pitch is displayed in the same way in both systems.

Atom Bomb

Very often the yield of nuclear weapons is referred to as a multiple of the destructive force of those dropped on Hiroshima (equivalent to about 15 kilotons of TNT) and Nagasaki (equivalent to about 21 kilotons of TNT) in August 1945.

Despite the lower yield of the Hiroshima bomb, the number of *immediate* deaths (69,000) was greater than the around 37,000 caused by the more powerful plutonium bomb at Nagasaki. This was largely due to the terrain at Hiroshima being flat as opposed to being valley-like at Nagasaki, with the bomb at the former being released over the center of town and over the outskirts in the case of the latter on account of bad weather.

Overall, long-term deaths due to injuries and radiation (leukemia, etc.) were double the immediate ones.

On both the American side and the Japanese side there has been much obfuscation regarding the dropping of the two bombs on Japan and the Japanese reaction, including the fate of the people either surviving in Hiroshima or returning to the devastated city.

See **Thermonuclear Weapons**.

ATPL (Airline Transport Pilot License)

The highest qualification for a commercial pilot. See **CPL**.

ATPCS (Automatic Takeoff Power Control System)

Should an engine on a twin-engine turboprop airliner fail on takeoff there would not only be little or no power from it, but the blades of the propeller slowly windmilling would produce considerable drag. This could be disastrous since the pilots would have little time to **Feather** the prop and increase power on the good engine. The ATPSC system helps manage this.

ATR

Franco-Italian aircraft manufacturer headquartered at Toulouse Blagnac Airport in France. It has a niche market in very basic turboprops selling all over the world due to their simplicity and reasonable cost.

The wings are made at Bordeaux by Airbus and the fuselage by Leonardo in Italy. The engines are from Pratt & Whitney (Canada). The final assembly is by ATR at Toulouse.

Should Embraer proceed with their envisaged advanced turboprop, ATR would face competition and have to produce something more sophisticated.

~ ATR 42 1985/500

Original version of the turboprop assembled by ATR at Toulouse.

~ ATR 72 1989/1,000

Developed from the ATR 42, and as mentioned above it has cornered this niche of the turboprop market for the time being.

Atropine/Belladonna, Nerve Agent Treatment

With fears that extremists, or even a cult as happened in Tokyo, might launch a nerve gas attack, some medical services have been required to keep significant stocks of *atropine*.

Although injection of atropine into the thigh, often with an autoinjector, is not a cure for poisoning with nerve agents such as sarin, it slows their action, giving time for treatment with pralidoxime chloride and the like to have an effect. The problem with atropine is that the high dose required to be effective in giving first aid to those exposed to nerve agents is close to the dose at which nasty side effects such as congestion of the lungs requiring hospital treatment appear.

Nerve gases are heavier than air, so in the event of an incident do not go and lie low in a gutter or cellar. *Instead*, go high up indoors, shut the windows, and turn the heating off to limit the ingress of air through convection.

Cleopatra used a natural form of atropine called belladonna to dilate the pupils of her eyes and make herself more attractive. (Those with glaucoma should only use atropine in life-and-death nerve gas situations, not for seduction.)

Pin prick pupils can be an indication of nerve agent poisoning as in the case of the Novichok affair at Salisbury in England. Novichok, meaning *Newcomer* in Russian, is an extremely potent nerve agent group developed partly because it was difficult to identify, and its component parts were not on the banned list.

ATS (Air Traffic Service)

General term for air traffic control and the services they provide.

ATS (Air Turbine Starter)

A starter for jet engines using air from a ground unit, bleed air from another engine, or from the **APU (Auxiliary Power Unit**. Has the advantage of being much lighter than an electrical starter.

Attitude

The angle of the aircraft on the longitudinal and lateral axes relative to the horizon. In other words, indicates pitch angle and roll angle. The behavior of an aircraft depends on its attitude and, of course, speed and flap settings. The attitude indicator is the artificial horizon. See **PFD (Primary Flight Display).**

Attritable ("Cheap Enough to Sacrifice")

Attritable is a new word used in the US military for weapons including drones (UAVs) that sacrifice reliability and expensive maintenance in favor of low cost, being re-usable for a limited time and even expendable.

Fifth- and sixth-generation bombers and fighters, and especially the stealth versions, are prohibitively expensive and their maintenance extremely costly and time-consuming. This has led the US Air Force to embark on projects such as **Skyborg** to produce attritable drones, that is ones cheap enough to be produced in quantity and not too expensive to lose or sacrifice. While they can be reused, low cost takes precedence over reliability and superlative maintenance.

In a role called ***Loyal Wingman*** they can be required to take a hit in lieu of the manned aircraft they are accompanying. Future air battles might involve a few mother planes and dozens of such drones all linked electronically with integrated artificial intelligence (AI) incorporated.

Unbelievable though it may seem, the term attritable is now being assigned to $33,000,000(!) stealth bomber drones to be designed to operate in conjunction with the upcoming B-21 Raider stealth bomber. Costing half the price, sacrificing them would be better than losing the manned or even unmanned B-21.

Mentioned in this context is the "loyal wingman" *XQ-58A* Valkyrie, built by Kratos designed to work together with a manned fighter and "prepared" to take a hit in its place.

See **Drones**.

ATTOL (Autonomous Taxi, Takeoff, and Landing Program

An Airbus program showing that an airliner can be made to taxi to the runway, take off, and come back to land automatically without the pilot intervening.

Autoland [Airliners]

System permitting aircraft to land in poor or even zero visibility, initially developed and introduced in the UK in the 1960s for the Hawker Siddeley Trijet, though the Aérospatiale Caravelle was the first airliner to be officially certified for Category III autolandings.

Autoland did not catch on at first in the US, not because there was anything inherently wrong with the concept but because near-zero visibility in the northern US was normally associated with blizzards and stormy conditions, in which aircraft could not land anyway. Paradoxically, the pea-soup fogs that incentivized the British to become number one in the autoland field have now largely disappeared, due to coal now rarely being used to heat homes.

Autoland has now been perfected and fitted to many airliners and found to work well but has the great drawback that the proximity of other aircraft can affect the ILS signals on which it depends. This means aircraft must be spaced out (kept twice as far or even farther apart than normal when landing), which in turn means that airports already operating near their limits, such as London's Heathrow, must reduce traffic considerably, and hundreds of flights must be delayed or canceled. New systems using GPS are being developed.

In addition, in the US where the waveband for 5G cellular phone networks is close to that used in airliner radio altimeters needed for regular landings and even more so for autoland systems, the FAA imposed restrictions for certain aircraft.

There have been stories of flight attendants instilling fear in passengers by telling passengers, "The pilot is not able to land here because of the weather. We will have to divert to an airport where he can." This usually simply means the airline had not thought it worthwhile spending money getting the pilot legally qualified for something like Autoland as it was deemed unlikely to be needed on the routes they flew.

Autoland [General Aviation]

There have been inspiring tales of a passenger taking over a general aviation aircraft and landing it successfully following the incapacitation of the pilot. Though rare, passengers always fear their pilot becoming unwell and would welcome a system that could save them in that eventuality.

Advances in technology mean that there are now autoland systems for GA aircraft for emergencies.

One such is Garmin's GA Autoland that can engage automatically if it determines the pilot is unable to fly or when a passenger presses the AUTOLAND button.

Not only does it land the aircraft but continuously reassures the passengers, informing them audibly and visually what it is doing, which will be:

1. Selecting the best airport for landing considering runway length, distance, fuel range and other factors.
2. Liaising with air traffic control.

3. If it is navigating around hazardous terrain, obstacles, and weather.
4. How long before they land.
5. When they are about to land.
6. Landing the aircraft on the runway and bringing it to a stop.
7. Shutting down the engine(s) so the emergency services alerted by air traffic control can take over.

The system is available on select Garmin G3000 flightdeck-equipped aircraft, including the Piper M600 SLS, Cirrus Vision Jet, Dahler TBM 940 and 960, Hondajet Elite II, and Beechcraft Denali.

Interestingly, plans for the Tempest Air Dominance Fighter joint venture between the UK, Italy, Japan, and others include a similar autoland system to save the aircraft in the event of the pilot blacking out during a high-G maneuver.

Automated Takeoff

Computers, as we have said elsewhere, fly aircraft more precisely on a regular basis than humans.

Automatic Flight Control System (AFCS)

Also called IFCS (Integrated Flight Control System).

Autonomous Airliners: No Pilot/One Pilot

Especially in the case of long-haul and ultra-long-haul flights, two pilots are now in the cockpit for hours with virtually nothing to do as the programmed **FMS (Flight Management System)** and autopilot are doing all the work, mostly better than they could. Even now with the flight plan inserted manually or more often on regular scheduled routes loaded already prepared into the flight management system (FMS), the pilots activate it shortly after takeoff and sit back with the autopilot usually remaining engaged until just before landing.

If the aircraft and runway are certified for autonomous landings the aircraft can even land itself smoothly and safely.

Airbus even did a test, with pilots in the cockpit doing nothing but ready to take over, if necessary, with the aircraft automatically taxiing out to the runway, taking off, and after some circuits, landing, and taxiing back to its starting point without any input from the pilots. This demonstrated that even with present technology a completely automated flight from start to finish was feasible.

Of course, as things now stand, pilots need to be present as backup, and to deal with changes and hazardous situations like diversions, turbulence, or emergencies including the autopilot disengaging.

Airframers are looking at the possibility of having just one pilot at the controls during cruise when not much is happening, particularly on long-haul and ultra-long-haul flights to obviate the need for three or even four pilots. There is always the fear that a lone pilot might be suicidal and kill everyone on board. That could be overcome by having some members of the cabin crew being given some basic piloting training and having them sit with the single pilot, which could be good for them mentally.

Autopilot

Originally, the autopilot simply eliminated the need for pilots to hold an aircraft on course manually by making it automatically maintain a certain heading, height, and airspeed controlled by the auto-throttle. Not only was this a godsend on long flights, but it is particularly helpful in busy situations, such as when taking off and climbing out, where constant adjustments would otherwise have to be made to the throttle settings. In fact, the computers can usually fly the aircraft better than the pilots can manually.

The systems have become progressively became more and more sophisticated, so that rather than using the control column or sidestick, many maneuvers are nowadays made by adjusting the flight director, which tells the autopilot or the pilots (via markings on the **PFD (Primary Flight Display)** what is intended.

Finally, to cap it all we now have the FMS (flight management system), which, like HAL, the computer in the movie *2001 A Space Odyssey* knows everything and can control much of the flight, telling the autopilot/flight director and the pilots what to do. Using it, the whole flight can be programmed from departure to arrival, passing over hundreds of waypoints (radio beacons).

Inputting these would be time-consuming and likely to introduce errors. Fortunately, since an airline's aircraft fly the same scheduled routes day after day, they can be found in the database.

Besides having details of the waypoints, the FMS holds all sorts of data regarding terrain, location, forbidden zones, runways, diversion airports, weight of fuel, and fuel that will remain on arriving at destination, and it is able to calculate the optimum airspeed and height.

The controls are in consoles on the deck between the pilots, consisting of a keyboard and a screen with maybe ten keys on each side and keys below. The keys on either side of the screen are "dynamic keys" in that their function (shown alongside them on the screen) varies according to the mode selected by the key underneath. This makes it unnecessary to have hundreds of keys and multiple screens.

As mentioned previously, this makes it possible for the aircraft to take off, fly the route for hours, hold, and land itself virtually without the intervention of the pilots.

Activating the autopilot in tricky situations, such as taking off in a storm, frees up both pilots enabling them to better work out the best course of action.

Interestingly, besides reducing speed, pilots tend to engage the autopilot on encountering turbulence as it can handle it—react more quickly—than they can,

Autorotation (Autogiro/Helicopter)

The movement alone of a rotor-supported craft through the air will make the rotor rotate and provide lift, and, in fact, the autogiro, with no engine driving the rotor, works on this principle.

Though helicopters do have engine-driven rotors, the rotors can free-wheel and rotate independently of the engine, and given sufficient forwards airspeed, keep rotating to allow the craft to come down safely—in other words, not drop like a brick if the engine fails.

Helicopters with media people on board filming a fixed event or incident on the ground need to keep moving ahead slightly to ensure they can come down to earth safely should the engine fail.

Autothrottle/Autothrust

While the autopilot can deftly or aggressively adjust the elevators and ailerons to keep the aircraft on the designed course and altitude, the manipulation of the engines which can be very demanding if done manually is mostly done by a system called the autothrottle [Boeing] or Autothrust [Airbus].

When engaged it adjusts the power of the engines, so the aircraft flies in concordance with the pilots' desires. The sophisticated ones can function in all phases of flight—takeoff, climb out, cruise, descent, and landing—as entered into the flight management system, individual controls. On Boeng aircraft the throttle levers actually move, whereas on Airbus aircraft the stay still, usually in the detent position,

They make life easier, but pilots always need to monitor their airspeed, especially when low and about to land to make sure the autothrottle/autothrust system is working as expected.

There was an accident at San Francisco on July 6, 2013, where an Asiana Boeing 777-200 landing there was initially above the glidepath and then descended below it on short final with the engine thrust in consequence reducing to idle.

Expecting the autothrottle to be keeping them at their reference speed as usual, the pilots did not realize it had disengaged due to their action, and that they were too slow to recover from being too low. Luckily, most of the aircraft got over the airport sea wall with the tail breaking off.

Most occupants managed to evacuate before the fire took hold, while several were thrown out via the gap left by the separation of the tail. Those survived except for one unfortunate high school girl who, lying inert on the ground and covered by foam used for firefighting, was run over by a firetruck because the driver was unable to see it was a person.

AUW (All-Up-Weight)

Total loaded weight of aircraft, including usable fuel, passengers, and freight, not only at takeoff but also at any time during the flight. BOW (basic operating weight) is the weight of the empty aircraft itself plus crew, drinking water, engine oil, hydraulic fluid, unusable fuel, and anything else required at departure, while EW (empty weight) is BOW minus crew.

Avgas (Aviation Gasoline)

The high-octane gasoline used for piston engines and UAVs. Lead—now largely banned for use in motor vehicles—is used to achieve the high-octane number. Also, it differs from mogas (motor gasoline) in that it is made less volatile to prevent air locks at low atmospheric pressures.

Authorities have dragged their feet as regards the elimination of pernicious lead from avgas.

"Aviate! ... Navigate! ... Communicate."

This order of priorities is drummed into pilots in initial training and even later. This is because accidents have been caused by all the pilots being distracted by something, even a failed lightbulb, with no one flying the aircraft.

Some accidents have possibly resulted from pilots being distracted by air traffic controllers, who are now told not to harass pilots declaring an emergency and avoid having them continuously change frequencies.

There was a case in England where the pilots of a Boeing 737 who on feeling vibration and smelling smoke mistakenly shut down the good engine. Just when the captain was about to review with the first officer what they had done, he was interrupted by air traffic control, cancelled the review, and subsequently crashed just short of the runway because they had neither the speed nor the height to restart the good engine.

AVIATION INFORMATION (Some Recommended Sources)

1. *Mentour Pilot*
   https://www.youtube.com/c/MentourPilotaviation
   By a Swedish Boeing 737 training captain. His explanatory videos with emphasis on piloting and safety have been so good he has been able to build up a very wide following with over 2 million subscribers. What is more, his success has not only enabled him to take time off from the major airline he worked for to devote himself to the video offerings but also benefit from the support of a team of great technical people for visual presentation and even research into details that others miss.
   See also his website https://mentourpilot.com
   He is also creating a cell phone app called MENTOUR *APP*.

1. *Blancolirio*
   https://www.youtube.com/@blancolirio/videos
   While this almost "one man" operation does not have the bells and whistles of Mentour Pilot, it provides a great human factor analysis of events whether major or minor such as the problems faced by a pilot executing an emergency landing at a busy airport such as Los Angeles or a helicopter pilot ferrying a sports personality in fog.

2. For profound interpretation of major happenings in aviation there is first and foremost: *LeehamNews*
   https://leehamnews.com.
   Aimed more towards professionals, payment is required for full access, though much can be gleaned free of charge.
   A top feature, entitled *Bjorn's Corner*, has analyses by Bjorn Fehrm, one of the first people to clarify what the MCAS problems with the Boeing 737 MAX were all about.

3. Magazines such as
   Aviation Week
   https://aviationweek.com

and its sister publication *Air Transport Digest*.

Flight Global
https://flightglobal.com

AEROSPACE
published by Royal Aeronautical Society, UK
aerosubs@aerosociety.com (+44 (0) 20 7670 4354) for subscription details.

4. There is also *AIM*.
https://www.faa.gov/air_traffic/publications/atpubs/aim_html/
This is the FAA's *Aeronautical Information Manual*, and *The Official Guide to Basic Flight Information and ATC Procedures*.
Also, by the FAA, there is the
Pilot's Handbook of Aeronautical Knowledge—a very detailed PDF.
All these can be found by going to:
https://www.faa.gov/regulations_policies/handbooks_manuals/

5. And *SKYbrary*.
https://skybrary.aero
This is a European site with safety-related information and a dictionary explaining in detail almost everything.

6. For the very latest and a historical view of "Incidents and News in Aviation" there is *The Aviation Herald*.
https://avherald.com

I. Military Related Information (Mainly US)
AIR & SPACE FORCES Magazine
Weapons and platforms
(https://www.airandspaceforces.com/weapons-platforms/)

II. *THE WARZONE*
Navies
(https://www.twz.com/category/navies)

III. *DEFENSE NEWS*
Air warfare (https://www.defensenews.com/air/)

7. Don't forget *Wikipedia*!
   Adding Wiki to a search term will help you find it.

8. *Quora* for the wider public, where a question is posed with an initial answer and people commenting, very often with personal experience or specialist knowledge. You can even search for items in a category, such as aviation.
   https://quora.com.

AWACS (Airborne Warning and Control System)

Airborne early warning and control (AEW&C) is a more general term for the general type of aircraft used in this role. For example, there is the Boeing E-3 Sentry, the aircraft developed under the United States Air Force's "Airborne Warning and Control System" program.

An AWACS aircraft very often used to be an airliner with a large mushroom-like housing on top for an antenna.

It would seem easy prey. However, in wartime it would be surrounded by perhaps hundreds of protective aircraft or drones.

Another thing in its favor would be that the signal from the hostile radar station trying to detect it has to travel twice the distance (i.e., there and back) to "see" it, while the AWACS has only to detect the one going one way from that radar station. That said, if the enemy in future were to have auxiliary ground stations further out or synced drones with radar, AWACS aircraft could become more vulnerable.

Axes

Aircraft can rotate about three axes:

1. **Yaw axis**—the perpendicular axis about which the aircraft turns to the left and to the right. The rudder has the greatest effect on the yaw though a roll normally induced by the ailerons has some yawing effect, and their combined effect can result in a so-called **Dutch Roll**.

2. **Transverse or pitch axis**—axis passing transversely parallel to a line from wingtip to wingtip. A positive pitch raises the nose and lowers the tail. A negative pitch is the reverse. The elevators and the trim have the greatest impact on the pitch though applying high power to low-slung engines can push the nose up too, and sometimes prevent recovery from a stall.

3. **Longitudinal axis**—passing from the nose to the tail with the movement about it referred to as rolling or banking. The ailerons control the roll though the rudder can have some effect and is sometimes required to recover from a spin.

Yawing to the right, pitching up, and rolling to the right are referred to as **positive**. Yawing to the left, pitching down, and rolling to the left are referred to as **negative**.

~ B ~

"B" Designates Bomber, Not Boeing

With many famous bombers such as the B-17, B-29, and B-52 made by Boeing, people can mistakenly assume the B stands for Boeing, when, in fact, it is for "bomber."

See **Designations [US Military]**.

Babies and Tots, Where to Sit with Them on an Airliner?

There are many complaints about the noise that babies are liable to make, with some passengers sympathetic, imagining how stressed the parents must feel. Others are not.

It has been suggested that parents with babies should sit towards the rear of the aircraft where it is noisier, and the toilets are conveniently near for toddlers. A changing table is likely to be nearer too. The noise from the engines not only masks the sound of the crying—or at least lessens the annoyance—but can have a calming effect on the child and make it easier for them to sleep.

Sometimes the crying can be due to air pressure differences in the ears which sometimes can be relieved by giving them a sweet to make them swallow. In slightly older children squeezing their nose and getting them to blow can relieve it (on landing). However, this should not be tried when too young as it can result in a burst eardrum.

BAE Systems Plc [Uk]—Pronounced "B" "A" "E"

The largest defense contractor in Europe and third largest by sales after Lockheed Martin and Boeing.

In selling its inherited stake in Airbus in October 2006, it ceased all civilian aircraft production. It concentrates on work related to the US, together with business with Saudi Arabia, India, and Australia. Enormous deals, particularly with Saudi Arabia, have featured prominently with questions asked as to how they were gained.

BAE is the successor to many famous UK companies, including Marconi, de Havilland, A. V. Roe, Hawker Siddeley, Vickers Armstrong, British Aircraft Corporation, Supermarine, and Vickers Shipbuilding.

Major projects include cooperation on the Lockheed Martin F-35 Lightning, the Eurofighter Typhoon, the Astute-class submarine, and the two Queen Elizabeth-class aircraft carriers built to provide jobs for Scottish workers. In collaboration with other companies such as Rolls-Royce and Leonardo S.p.A., it is working on a sixth-generation fighter to be called the *Tempest* with great emphasis put on the use of AI. However, it may be a decade or more before it enters service with the Royal Air Force.

"Bank Angle!" "Bank Angle!" "Bank Angle!"

This is a warning the pilots do not want to hear, especially when landing. It means the aircraft is banking dangerously. Should they bank further they could sideslip and enter a precipitous dive as happened to Korean Air Cargo Flight 8509 that took off from London's Stansted Airport in the early evening of December 22, 1999*.

Barrel Roll

A roll where the aircraft flies as if following the contours of a barrel. It looks risky, but a very skilled pilot can do it all the time maintaining 1 G.

It is not something to be done with an airliner, though in 1955 Boeing's chief test pilot, Tex Johnson, famously performed two unprogrammed barrel rolls with the **Boeing 367-80** prototype at a demonstration show over a lake at Seatle, before thousands of people. Hauled up before Boeing's angry CEO the next day, he was asked what he thought he was doing, and replied, "Selling aircraft," and indeed thereby had greatly promoted sales of the Boeing 707.

An "aileron roll" with the aircraft going straight ahead but rotating about its longitudinal axis might be done in the simulator but not in an airliner, with negative G causing all sorts of problems including misdirection of lubricating oil and fluids in the toilets.

Base Leg

Aircraft often must fly in a rectangular pattern before landing at an airport, and the base leg is the one before the last 90-degree turn to place the aircraft, on its final leg, in alignment with the runway for landing. See **Traffic Pattern**.

Bathtub Curve

The failure rate for products tends to be relatively high immediately on entering service. Then for a lengthy period called *useful life* it remains extremely low, before rising, but not as steeply as the initial fall, with the latter classed as *wear-out* failures.

The curve showing faults against passage of time (years in the case of airliners) happens to be like that of the traditional bathtub shape viewed from the side. That is, almost straight down at the foot of the bath, with the bottom almost horizontal, and the head of the bath at one's back curving gently upwards.

BATHTUB CURVE

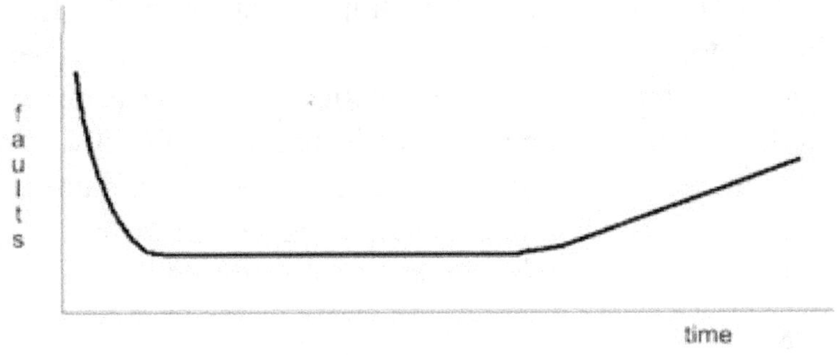

Purchasers of electrical white goods and computers are often inveigled into taking out insurance for just when it is not needed and virtually pure profit for the seller. The insurance is for three years, the first year failures being covered anyway by statutory guarantees with failures in the second and third years most unlikely.

Batteries

Rather than lithium-ion batteries with their danger of catching fire and which are getting near their limit as regards upgrading, so-called solid state batteries are on the horizon. Their development is being spurred by the demands of the EV (Electric Vehicle) industry.

Besides the advantage of ultra-rapid charging, they *may* be 70% smaller and lighter and of obvious value not only for EVs but also for aerospace. However, large-scale industrial production will most likely take until the end of the decade (2028–2030) to be realized.

Beartrap, Enabling a Helicopter to Land on a Ship in a Rough Sea

Officially called the Helicopter Hauldown and Rapid Securing Device (HHRSD), the beartrap enables helicopters to land on ships and aircraft carriers even in bad weather.

Hovering over the trap installed on the deck of the ship, the helicopter lowers a light line with a probe at the end to which a heavy line from the mouth of the trap is secured, and pulled up to the helicopter.

Once the line is secured to the helicopter, the beartrap winches down the helicopter using a sophisticated hydraulic system to regulate the tension in the line, with the pilot rotating the rotors at near full power to avoid the helicopter being struck by the deck should a large wave cause the deck to rise abruptly.

Behavior Detection Officers

A resource-demanding procedure used most of all, and very successfully, by El Al using highly trained officers (often in pairs so one can concentrate on noting reactions to their colleague's questions). They will question *anyone* at the airport boarding an El Al aircraft.

Apart from mundane questions about the person's travel history, purpose of their trip, and who they intend to meet and where they intend to stay (with, if possible, documentary evidence) the colleague might then come up with some quite unrelated anodyne question such as, "Did you replace your phone in the tray?" to throw the person off-balance and test natural spontaneous reactions.

El Al's safety record suggests the technique is successful, though knowledge of the likelihood of being subjected to intensive screening may have dissuaded a fair number of would-be suicide bombers from trying.

Other countries, including the US, have behavior detection officers looking out for suspicious people. Video cameras are also linked to software looking for certain traits, though in certain countries they risk being accused of racial profiling. In fact, crudely picking out people based on race as likely terrorists does not really work and can lead to time and resources being squandered.

However, as mentioned elsewhere, having pictures likely to offend those who might be hijackers placed on the route to the aircraft with cameras to detect reactions might be less costly and work occasionally, though perpetrators could be taught not to react. Nothing beats talented interviewers working in pairs, and behind the scenes intelligence gathering and exchange of information between all agencies.

See **Screening of Passengers, Luggage, and Freight**.

"Bingo Field"

The alternate (air)field the aircraft can go to with still *just enough* fuel to land. Pilots may be circling waiting to land on an aircraft carrier or airfield when one might say "Bingo Field," meaning their fuel is too low and that they should immediately proceed to their only option.

Bird Strike

When many airliners had four or three engines, the chances of all being completely crippled by a bird strike was less than now.

Great efforts are therefore being made to lessen their frequency. To deter birds from flocking to airports the contents of their stomachs are examined to see what they are eating, and attempts are made to eliminate that food source from the airport and its environs. This proves effective.

Contrary to what many might have thought, the large Canada geese that entered the engines of Sully's "Miracle on the Hudson" Airbus A320 were not local ones attracted to golf courses and the like but migratory ones from the far North of Canada, and this is the reason they were flying unusually high.

Not only are the engines tested to ensure they can continue to function after the ingestion of small birds but also be able to be shut down without damage to the aircraft systems after being crippled by the ingestion of large ones.

"Black Boxes" [Cockpit Voice and Flight Data Recorders]

Handy misnomer used by the media after a crash to describe the **CVR** (cockpit voice recorder) and **FDR** (flight data recorder), which unless damaged should have information about what happened on the flight prior to the disaster.

Colored *orange* rather than black to be easily visible, they are designed to withstand crashes, fires, and immersion down to 20,000 feet in water. They also incorporate radio beacons to broadcast (by pinging) their location for 30 days, even underwater. There have been crashes in deep oceans where the batteries have expired before finding the recorders, and it has been suggested they should last much longer. However, with this applying to new aircraft it would take some time for this to have an impact. A more promising approach is to have some data and the location transmitted by VHF radio or satellite before the crash. See **ACARS**.

In addition, many modern aircraft have quick access recorders (QAR) that record not only information like that recorded by the FDR but also such parameters as brake temperatures, brake torque, and tire pressures. As their name implies, they provide information for immediate use by maintenance personnel and the company. The QARs are not positioned or designed to withstand major crashes.

"Bleeds off" Takeoff [Boeing] / "Packs Off" Takeoff [Airbus]

Sometimes where the aircraft is taking off "at its limit," where all the thrust the engines can provide is needed, the bleed air they provide for say air conditioning is switched off.

There can be many reasons for this:

1. Short runway.
2. The temperature outside means engines are less efficient.
3. High elevation of the airport means engines are less efficient.
4. Noise abatement regulations or high terrain on the climb out necessitate an extra steep climb on takeoff.
5. Windshear might require extra thrust.

The relatively short duration of any lack of air conditioning means any slight discomfort for the passengers is short lived, and even unnecessary for the **APU (Auxiliary Power Unit)** can often be used to provide air for the air-conditioning packs as on the ground.

Blended Wing

The *idea* of a "blended wing" whereby the wing is blended seamlessly into the fuselage to produce a V-shaped aircraft has been around for a long time but difficult to realize because of difficulties of control since it has no natural stability. Now with computers able to provide stability they have successfully been employed, for example on the B-2 bomber.

Blended wing. NASA

One advantage of the blended wing is that it makes it easier to achieve laminar flow over the whole body, rather than just the wings.

Boarding would be quicker as there would be, say, five aisles.

Block Time

See **Flight Time**.

Boarding, the Most Efficient Way

In some cases where first class passengers are seated at the front of the aircraft, they are (or were) asked to board first, not because of favoritism but to prevent the aircraft tipping backwards when economy passengers are boarded starting with those at the back.

Various computer analyses have worked out the most efficient order in which passengers should board. One way is, say, to have passengers in window seats board first, those in middle seats second, finally those in aisle seats. One problem is that if passengers have excessive carry-on baggage those boarding last find little space left in the bins.

It has been argued that some airlines wanting extra revenue do not want to follow this efficient system because they want people to pay for priority boarding.

The problem with any logical boarding order is that there are always some passengers who do not obey—even one passenger blocking the aisle on a narrow-body can hold everyone up.

Identifying Those Boarding Out of Turn

In 2024, American Airlines began testing a simple technology at selected smaller airports designed to prevent passengers boarding the aircraft before their group is called.

When the passenger's boarding pass is scanned an "error beep" sounds when it is not that person's turn, with the gate agent telling them to step aside—apparently to the great satisfaction of other passengers.

BOEING

William Boeing made his money in the timber industry and found the expertise he had gained in designing wooden structures was invaluable for making airplanes. On July 15, 1916, he set up Pacific Aero Products Company, then on May 9, 1917, changed the name to Boeing Airplane Company. Located in the Seattle area, where wood was plentiful, the company employed ex-boatbuilders whose skills were well suited to the task of making planes.

The company's first big break was making 52 seaplanes for the US Navy, but when World War I ended there was a surfeit of aircraft in the US, England, and France, and many companies making them went bust. Boeing Airplane kept operating by making furniture as well, and over the years produced a series of aircraft, including small airliners, eventually creating an airline called United Airlines.

In 1934, the Air Mail Act forbidding manufacturers of airplanes owning airlines forced Boeing to split up the companies it controlled into the Boeing Airplane Company, United Airlines, and United Aircraft Corporation, which ultimately became United Technologies. William Boeing then sold his shares and left the companies.

The Boeing Aircraft Company had begun a long-lasting relationship with Juan Trippe (1899–1981), who had graduated from Yale in 1921 and first worked on Wall Street building up connections in high places. Trippe is regarded as one of the great driving forces behind civil aviation and the one who made the airline business what it is today.

Those connections enabled him to find funds to purchase small airlines and subsequently to create Pan Am, opening international routes with a significant degree of legislative protection.

In 1938 Boeing produced the Boeing 314 Clipper long-range flying boat, the largest civilian aircraft of its day, with a capacity of 90 passengers on day flights and of 40 passengers on night flights. Pan Am called its aircraft "Clippers" and began with a regular 314 Clipper service to the UK, following it up with others. Because they were seaplanes, he dressed his pilots in a kind of naval uniform, which became the norm for airline pilots all round the world!

Origin of the B-17 Bomber

Earlier, in 1935, Boeing had been spending funds working on a four-engine bomber. Though much more expensive than those of their competitors, Boeing entered it into a competition organized by the US Army Air Corps. Unfortunately, checklists were not used in those days, and before taking off the pilots forgot to disengage the gust locks (used to lock control surfaces of parked aircraft so that strong winds did not move them and damage the control lines). Rendered uncontrollable, the bomber crashed, and Boeing was disqualified. Nevertheless, by using a legal loophole, the US Army Air Corps placed a small order and, in fact, ended up with a much better aircraft, as in the meantime Boeing had made improvements leading to what was to become the B-17 bomber, which saw wide service in World War II and was called the "Flying Fortress" because of all the guns sticking out of it.

In 1938 Boeing finalized its 307 Stratoliner using the same wings, tail, and engines as on the B-17. It was the first airliner with a pressurized cabin, and since it was able to cruise at 20,000 feet (6,100 meters) it could fly above most turbulent weather. Only ten were produced, and though used by Pan Am, the war meant civilians were no longer traveling overseas in luxury as before.

During the war Boeing not only produced the B-17 but also the B-29 long-range bomber in a project that was said to have cost more than the Manhattan Project for the atomic bomb, which it also dropped on Japan.

Postwar

As mentioned under **Germany's Contribution to Aviation**, after the surrender of Germany in 1945, George S. Schairer, head of Boeing's technical staff, came to Germany as part of the US Air Force technical team and visited their secret aeronautics laboratory situated near Braunschweig. On seeing what the Germans had done, he realized their wind tunnel tests confirmed that a swept wing is best for high-speed jet aircraft. He immediately cabled his people back at Boeing telling them to halt work on the B-47 bomber and its wing in view of the information he had garnered.

Incidentally, Boeing's adoption of the swept wing in its jet aircraft designs such as the 707 gave it an advantage over Douglas's DC-8.

While we individually list the airliners Boeing produced below, here are the general details of how they evolved.

In 1949, Boeing produced a luxury twin-deck four-engine airliner called the 377 Stratocruiser. Though technically advanced, its engines and props were unreliable and only 56 were produced. Competing airliners such as the Lockheed Constellation were cheaper to operate.

This made Boeing realize jet engines were the future and they began work on a "Demonstrator" model called the **367-80 Dash 80**.

Since the 1960s Boeing has produced a whole series of commercial and military aircraft. See below.

In 1996 Boeing bought Rockwell's aerospace businesses, including what was once North American Aviation and Rocketdyne. Rockwell had once been a major player, but with the termination of the B-1 bomber and end of work on the Space Shuttle program, it had suffered a decline.

In August 1997 Boeing merged with McDonnell Douglas, partly because of the latter's military portfolio. Some say McDonnell Douglas absorbed Boeing with Boeing money. At the beginning the merger posed problems as Boeing was having a hard time facing down the challenge Europe's Airbus presented and then there came 9/11 with many orders for aircraft cancelled or put on hold. In September 2001 Boeing moved its corporate headquarters from Seattle to Chicago, separating the top management from the designers and engineers.

A few months before 9/11, Alan Mulally, who had led the development of the thereto very profitable Boeing 777, had been promoted to be president of Boeing's Commercial Airplanes Group. He took them through the crisis and the painful restructuring that followed.

He began by identifying and articulating a "clear and compelling vision" for the company, then developed a strategy to deliver on that vision. That strategy was simple and straightforward, and Mulally made sure everyone at Boeing, from the C-suite to the factory floor, understood it and saw their place in it. Then he made sure the company executed that strategy by creating a relentless implementation system called the Business Plan Review (BPR) process that tracked each business unit and function's progress against that plan.

Mulally was twice passed over for the post of CEO of the Boeing Company after the resignations of its CEOs, and in 2006 he left Boeing to become CEO of Ford Motor Company where he was very successful. This was possibly because he was a technical and Boeing man through and through and this did not fit in with the people from McDonnell Douglas.

After a period where Boeing seemed to be resting on its laurels and averse to risky new projects, it launched the 787 Dreamliner that entered service in 2011 with the idea of getting some of the suppliers such as those in Japan to share the risk. At first it was a costly nightmare with supply and integration problems. Then there were problems with the powerful lithium batteries catching fire, not to mention the fasteners. Boeing had to compensate airlines such as Qantas for delays.

Yet after all the travails, the 787 has proved a great success, with airlines finding the easier-to-fill fuel-efficient composite wing and composite fuselage jet not only profitable but one that enabled them to open long-haul routes that had not previously been viable.

Much has been made of the failure of Boeing to create a new aircraft to replace the 737 based on the 1960s design and have a final throw of the dice with the 737 MAX.

As Scott Hamilton of *Leeham News* has pointed out, Boeing's decision at the time to delay replacing the 737 series was not irrational for:

1. The 787 hadn't been delivered, and it was not only three years late but also billions and billions over budget.

2. The revamped 747, the 747-8 was late and billions over budget with not many airlines seriously interested.

3. The USAF air refueling tanker contract was proving problematic. Not only was it late, but billions of dollars over budget due to penalties and unexpected work. Ironically, it had been first won by Airbus based on the more suitable A330 platform, with a US company doing the finishing work. However, Boeing on lobbying won it back after Congress had objected to the original award, and the US Air Force had tweaked the terms so only Boeing could win with a bid only 10 percent less than Airbus for something less promising.

4. Boeing's engineering resources were stretched to the point where engineers were diverted to the 787 from other programs and engineering work was even being outsourced to India and Russia. The work from India on the 747-8 was poor and had to be re-done by SPEEA, further straining resources.

Also, Boeing had to bear in mind that when coming up with a new design, one needs to choose the sweet spot to benefit from current great advances in technology and not risk it rapidly becoming outdated.

With over 5,000 orders in 2019 and cheap to produce, the 737 MAX seemed to be the ultimate moneymaker. Boeing appeared to be on a roll until two new 737 MAX aircraft crashed due to an apparently "infantile" software design error that led not only to Boeing losing its reputation but also its power over the FAA, which had been under its thumb. With the FAA determined to regain its own reputation, it would take much longer and be more difficult for Boeing to get approvals—perhaps in the long term the costliest aspect of the 737 MAX debacle.

In our entry for the 737 MAX (and its crashes) we mention how Boeing spends millions lobbying Congress and federal agencies each year and other companies and lobbying groups in the aviation–defense industry.

Boeing is now moving its headquarters from Chicago to Arlington, Virginia, which will conveniently put its executives close to Pentagon leaders. Note that rival defense contractors including General Dynamics, Lockheed Martin, and Northrop Grumman are already based in the DC area.

Due to its clout with the FAA—including the revolving door between the FAA and Boeing—Boeing had got sloppy in the knowledge it could get almost anything approved without delay. Now the boot is on the other foot with the FAA trying to restore its reputation by taking even longer than it otherwise might to approve anything and even looking for trouble on the manufacturing side. That is proving to be the greatest downside to the MAX fiasco, with the extra cost likely to last for years and Boeing never able to regain the power over the FAA it once enjoyed.

Technically speaking Boeing has been benefiting from close collaboration with NASA, for example on work integrating raked winglets on the 787 and work on trials to explore the possible low drag advantages of a high-wing aircraft with a strut eliminating heavy wing roots and itself providing some lift.

However, in early 2024 an incident where the unneeded door made to look like a wall with a normal window blew out on an Alaskan Airlines 737 MAX raised doubts about Boeing's quality control. A major US airline said it was losing confidence in the company despite being a Boeing airline for many years and might consider switching to another supplier. Unfortunately, Airbus has a full order book, and it would take time to get aircraft from them.

~ Boeing 247 1933/75

The first serious airliner with many innovations able to fly coast to coast in the US, eight hours faster than its predecessors and enabling passengers to do it in a day without night stops. As it only carried ten passengers and a flight attendant, it was not commercially viable long term. Initially popular, sales were limited to Boeing-related companies, causing other airlines to turn to Douglas, ending up with the exceptionally successful DC-3.

~ **Boeing 2707/0, SST, Supersonic Transport**

On June 5 ,1963, President Kennedy, riled that Pan Am's Juan Trippe was placing an order for the Anglo-French Concorde supersonic airliner, with Trippe telling him it was because the US had nothing similar or better to offer, proclaimed the United States would develop an even better SST (supersonic transport).

In claiming it would fly at Mach 3 rather than just above Mach 2 like the Concorde, Kennedy had doomed the project to failure from the start for flying at that speed was not possible with the materials available at the time due to the high skin temperatures that would involve.

A competition was held with three entrants, Lockheed, North American, and Boeing. Though it had no experience with supersonic aircraft of any kind, Boeing won the competition with a very ambitious design with a swing wing.

With a $100,000 grant from Congress, money was not the problem for Boeing but rather the diversion of resources from other projects such as the 747.

Boeing was to encounter a whole series of intractable problems. First, the weight, which was increasing as things were added to meet airline requirements—interestingly, airlines were probably demanding a high degree of luxury in contrast to the Anglo-French Concorde that had a sparse cabin with narrow seats resembling those in today's low-cost carriers (BA made up for it by serving champagne and caviar). Worse, the mechanism for operating the swing wing was so heavy that it was abandoned, and another design was tried.

While flying around Mach 2 like Concorde was possible with special aluminum alloys, at Mach 3 it was a completely different ballgame due to the surface friction heat generated that could not be dissipated. Boeing spent a considerable amount of its own money playing around with titanium, which at the time people had little experience of working with.

Enthusiasm for the expensive project was cooling with people questioning the effect of the sonic boom. An anti-SST movement was growing, and Congress refused to provide further funds, leading to the project being abandoned.

~ Boeing 367-80 (Dash-80) 1954/1 (Demonstrator)

Prototype four-engine jet aircraft that first flew on July 15, 1954 and was the forerunner to the KC-135 Stratotanker and the Boeing 707. Ahead of its time, it was designed as a demonstration model to prove what a jet could do both as a refueling tanker and as a passenger airliner.

Impudently, at an air show its test pilot, Alvin "Tex" Johnston, much to the irritation of Boeing's CEO, twice performed a barrel roll in front of thousands of spectators with 367-80. It was not as risky as it appeared for he had taken care to maintain 1 G, so no gravity-dependent fuel or lubrication systems failed.

When the angry Boeing CEO asked, "What were you doing?" Tex replied, "Selling aircraft!" which was indeed true.

Unfortunately for Boeing, the Dash-80 proved to competitors such as Douglas what could be done, with Douglas designing its aircraft with a wider cabin able to seat six as opposed to five abreast.

~ Boeing 377 (Stratocruiser) 1949/56

Boeing's first postwar airliner. It was a true twin-deck airliner with a figure-of-eight cross-section. The upper deck had normal seats and the galley, while the lower deck had a rest area with a bar and bunks, couchettes, and toilet facilities. Though publicity-wise Pan Am credited itself for its development it was, as is often the case with Boeing, an aircraft that had evolved from a military aircraft, and notably the C-97 Stratofreighter, which owed much to the incredibly expensive to develop B-29 Superfortress.

It was technically advanced, luxurious, and pressurized but was not only more expensive than Lockheed's Constellation but prone to trouble with its four powerful engines and props. When superseded by jets, it languished on airport aprons while the Lockheed Constellations were still operating on lesser routes. A few were modified to create the **Pregnant Guppy** and **Super Guppy** to transport outsize items for the space program. See **Super Guppy**.

~ Boeing 707 (707) 1958/1,010
~ Boeing KC-135 Stratotanker 1957/800 Approx.

Based on design work for the 367-80 (Dash-80) prototype jet-engine midair-refueling platform, the Boeing 707 was not quite the brave betting-the-company venture it is often purported to have been, since there was an obvious market for the refueling tanker. Jet-engine bombers developed in the late 1950s had to lose height and slow to almost stalling speed for refueling by traditional piston-engine tankers. Having developed the refueling boom used by those piston-engine tankers, Boeing duly received orders for what it called the KC-135 Stratotanker. Surprisingly, it was the tanker rather than the airliner that was ultimately to prove a great moneymaker because the arrival on the commercial scene of competitors, such as Douglas with the DC-8, which had a wider cabin and greater payload, necessitated numerous expensive modifications to the 707 variant.

Unfortunately for Douglas, the blunt wing of the DC-8, though giving good performance at takeoff and low level, was not as good as the more swept back one on the 707—possibly designed with the help of engineers from the defeated Germany—at high altitude and could not give the performance Douglas had promised airlines. Various tweaks improved it, but it was to bedevil the program for years and proved costly to Douglas despite many versions of the DC-8 being sold.

The 707 entered commercial service in 1958 with Pan Am across the North Atlantic, just a few weeks after the much smaller and more-expensive-to-operate British Comet 4 with its strengthened fuselage and modifications to eliminate places where stress could concentrate—see **Comet**.

Pan Am's exclusive 707 jet flights had the then-unheard-of load factors of 90 percent.

~ Boeing 717 (MD-95/Boeing 717) 1999/156

Originally a sales flop, this 100-seat twinjet was the MD-95 before Boeing took over McDonnell Douglas and renamed it the 717. A derivative of the DC-9 family, it has now found a new lease of life in the secondhand market, with Delta and Hawaiian Airlines pouncing on any that become available. "They look so modern," said someone at Delta.

Interestingly, when Hawaii was threatened with a violent hurricane the airline flew all its aircraft back to the US mainland except for the 717, which did not have the range. When the aircraft have to go for major servicing in the US mainland the airline has time to remove some seats and install some fuel tanks in their place. The only other option would be to fly the other way round the world.

~ Boeing 727 (727 Trijet) 1964/1,832

When introduced into commercial service in December 1964, the Boeing 727 was the first trijet, and no one imagined it would become one of the most successful aircraft up to that time, with production continuing up until August 1984. As a short to medium-range airliner, it was widely used on US domestic routes via secondary airports. Overseas airlines also had many similar routes where it could be used to advantage.

For a passenger with a window seat above the wing, the sight of the deployment of its triple-slotted flaps and slats was astounding—the wing would seem to spread like that of an eagle, allowing you to see right through the middle. As Boeing said, these sophisticated, triple-slotted trailing-edge flaps and novel leading-edge slats gave the 727 low-speed landing and takeoff performance, unprecedented for a commercial jet, allowing it to service smaller airports than those the 707 required. The width of the fuselage was nevertheless the same as the 707's.

Pilots greatly appreciated the operational flexibility these triple-slotted flaps and slats provided but at first failed to realize they presented a danger. Extremely high sink rates could develop without them noticing. Thirty-seven percent thrust was required just to maintain level flight with full flap.

Recovery from a high sink rate, and especially at low airspeeds and low engine revolutions, took time, and until pilots remembered this, there were occasions when there was not enough time, with disastrous results. In those days pilots did not have the sophisticated automatic sink-rate warnings they have now—one reason why flying is so much safer today.

Other novel features that made it possible to deploy the aircraft at smaller airports with limited facilities included:

1. An auxiliary power unit (APU), providing electrical power without having to run the engines.

2. The ability to back up without a tractor.

3. An underbelly gravity-operated airstair to enable passengers to embark and disembark without provision of steps by the airport. (A hijacker famously deployed this air-stair mid-flight to parachute from the aircraft with $200,000 in ransom money. See **D. B. Cooper***.)

There was an incident in 1979 where a 727 climbed to 39,000 feet to escape from a 100 mph headwind that had been delaying progress and depleting its fuel. At that exceptionally high altitude, the aircraft suddenly yawed, flipped over, and entered a precipitous dive. Deployment of the spoilers/air brakes had no effect, and it was only by lowering the landing gear (undercarriage) at an airspeed which normally would be unthinkable—resulting in the housing being ripped off—that the pilot, a Captain Gibson, was able to recover and land safely*.

Though many thought Gibson a hero, the authorities accused him, the copilot, and the flight engineer of having intentionally deployed the slats in order to be able to fly better at that great height—it being alleged that pilots were in the habit of doing this surreptitiously. Gibson always maintained that one of the slats had deployed on its own with no input from the crew.

An excellent account of this muddied affair can be found in Stanley Stewart's excellent book *Emergency (Crisis on the Flight Deck*. According to Stewart, deploying the slats did not improve performance at great heights, and the whole idea of pilots doing this on a large scale must therefore have been largely untrue.

Whatever the truth, it was an early demonstration of how dangerous it can be if something untoward happens when cruising in thin air at very great heights, in other words in "coffin corner." It is also an early example of Boeing incriminating the pilots.

~ Boeing 737 (737) 1968/10,500+

On the way to being the most successful series of aircraft ever. It has passed through three generations and at one time was looking forward to an even more profitable fourth with the MAX version.

The 737, a narrow-body twinjet especially favored by low-cost carriers such as the US's Southwest Airlines and Ireland's Ryanair, was first conceived in the mid-1960s. By progressively improving on and adding to a basic design, a process called "grandfathering," Boeing had been able to proceed without waiting for "time-wasting" approvals and please airlines by telling them expensive retraining of pilots was unnecessary.

There were four main generations of the 737:

1. Original.
2. Classic.
3. Next Generation (NG).
4. MAX.

Though it is difficult to believe now, sales of the 737 were initially so sluggish that in 1970 Boeing even considered canceling the program and selling the design to the Japanese.

At first, it had relatively small-diameter Pratt & Whitney JT8D-1 engines under the wings, allowing it to sit close to the ground. This facilitated servicing of the engines and resulted in faster turnaround times, since baggage, food, refreshments, and so on did not have to be heaved high up in the air. However, this proved a disadvantage when later generations of the aircraft required larger and larger diameter engines to be efficient. Initially, the problem was solved by flattening the bottom of the nacelles and placing engine accessories to the side of the engines, which explains the squished oval-shaped nacelles seen on many 737s.

With so many 737s flying, what crashes there were had meant the safety record continued to look good, with Boeing always saying afterwards it was making a safe aircraft *even safer*. When a couple of 737s plunged into the ground in 1991 and 1994* Boeing doggedly blamed the pilots, but in its longest investigation ever, the NTSB was finally able to show they were caused by a freak malfunction of the power control unit (PCU) driving the rudder.

~ Boeing 737 MAX, and Crashes

It is a misconception that the ill-conceived **MCAS (Maneuvering Characteristics Augmentation System)** that led to two crashes and the grounding of the aircraft for over a year was an anti-stall system necessary because the much larger engines could not fit under the exceptionally low wing and had to be moved forwards and upwards, thus changing the flying characteristics. In fact, the aircraft was perfectly safe without it.

The seemingly childish mistakes in the final design of MCAS leading to the two crashes and its grounding are difficult to comprehend until one realizes it was conceived in two stages. Originally, MCAS was something that would only come into play at high speed in the rarest of circumstances with high g-forces that would *have* to be present for it to kick in. And since it would operate at high airspeeds, the change in angle of the trim only needed to be slight. Later, when it was found the larger engines placed further forward would make it easy for a pilot to pull up and, however unlikely, possibly stall the aircraft as there would be little resistance, MCAS was used to deal with that and give the same "feel" so special training would not be required. However, for MCAS to work at low speeds very much greater downward trim was needed. Furthermore, at low speeds simple g-forces were not strong enough to be factored in.

Unfortunately, Boeing did two things wrong. First, it made MCAS dependent on only a single angle of attack sensor (without the g-force protection). This apparently was because if Boeing logically made it dependent on two or more sensors it would be classed as a safety-critical feature requiring a lengthy approval process.

Second, Boeing gave MCAS overpowering authority applied in cumulative increments. On the first application it would just give a nudge by adding a few degrees of downward trim. If the single angle of attack sensor still showed a dangerous angle, MCAS repeated the process, adding the nudge to the downward trim again and again until total downward trim attained some 12 degrees! This was way beyond the ability of the elevators to counter. Had the FAA known the total, it would have recognized the potential danger and been obliged to investigate regardless of Boeing's clout with its management.

All it needed was for that sensor to signal too high an angle of attack for MCAS to push the nose of the aircraft down, time and time again until it crashed. No aeronautical engineer in their right mind would create such a thing.

According to Boeing, pilots faced with the nose being pushed down by MCAS—which originally had not been mentioned in the manuals—should be able to disable it by following the instructions for "runaway trim" in the manual and save themselves and their passengers. However, with hardly any time and all sorts of alarms sounding this was a difficult proposition and impossible if the airspeed was too high as in the second MAX crash.

This may be an example of how the inordinate time taken to get certification for a simple improvement, whether it be a cargo hold door latching mechanism or MCAS, means it is not even considered.

It is not widely known that Southwest Airlines was incidentally largely the instigator of the MAX debacle, in that as the key purchaser of the aircraft it was insisting no extra simulator training should be needed for the MAX, and indeed made Boeing agree to staggering financial penalties per aircraft should it not comply.

~ Boeing 747 (747 Jumbo) 1970/1,574

While for Boeing this represented the second "betting-the-company" airliner project, it represented a much greater risk than the 707, even though again based on the design for the military. The military aircraft was to be a large freighter (canceled in the early stages) able to be loaded from the front as well as from the rear, hence the perching of the cockpit (flight deck) of the 747 so high above the ground.

The result was an airliner whose size far surpassed that of any other. The diameter of the engines required to power the monster was so great that they tended to flex and have other teething troubles. In the early days some joked that the four-engine 747 was really a five-engine one, for some had an extra pylon to carry a spare engine.

Even the inaugural Pan Am commercial flight from New York to London—on January 1, 1970—encountered a problem, with a watching journalist writing, "The jumbo revved up its engines at the end of the runway for takeoff, only to limp back to the terminal like a wounded beast."

Yet despite its teething problems, the 747 went on to become one of the most successful aircraft ever, and to be called The Queen of the Skies, opening long-distance air travel to the wider public. With no direct competition, it was a steady source of income for Boeing for years, and some assert that the profits on the 747 enabled the company to offer keener prices for its smaller aircraft where the company did face competition.

Juan Trippe, the head of Pan Am, played a crucial role in making the 747 possible—as he had also done for the 707. This was not only in urging Boeing to pursue the project in the first place but also later in persuading the then president, Lyndon Johnson, that it made sense even in an economic downturn to keep alive a project that would open up long-distance air travel to the masses.

The hump, for which the 747 is renowned, is there because the cockpit is set exceptionally high up so the aircraft can be used as a freighter able to be loaded from the front. Over the years, its length has been gradually extended, with the upper deck proving popular with business class passengers. Boeing found that drag did not significantly increase so long as the hump was not extended beyond the wings. That is why it never became an almost full-fledged twin deck airliner.

The 747-400 series entering service in 1989 had several improvements including a so-called glass cockpit (with simpler instruments) allowing the aircraft to be flown without a flight engineer. It included the stretched upper deck introduced in the 747-300. There was even a little-known 747-400D where *D* stood for domestic. It only had limited range but more seats with a couple of extra windows on each side where galleys on long-haul 747s would be. Only 19 were sold uniquely in Japan to JAL and ANA.

~ Boeing 747-8 2012/155

When Airbus came up with its superjumbo A380, which threatened the position of the 747, Boeing at first said times had changed and market research showed passengers now preferred to fly point to point rather than via hubs, and that by implication a giant aircraft no longer looked viable.

Perhaps to avoid Airbus having the clear run in terms of price it had enjoyed with the 747, and because it thought that by "grandfathering" it could produce an aircraft to compete relatively cheaply, Boeing decided to offer a larger 747, involving a stretch of 5.6 meters (18.3 feet) over the 747-400 versions.

Things did not go smoothly, with several problems and production delays. The UK's *Flight International* editorialized that Boeing's problems with the 747-8 paralleled those with the first 747s. For in both cases Boeing had had to focus limited engineering resources on prestige projects. In the case of the 747, it was the subsequently canceled Supersonic Transport intended to trounce the Anglo-French supersonic Concorde. In the case of the 747-8, it was the 787 Dreamliner.

Another reason for the starving of resources in the case of the 747-8 may have been the lack of enthusiasm initially shown by the airlines, with Lufthansa being the only early customer for the passenger variant. The freighter variant has fared somewhat better, but production had to be cut back in 2016 because of a weakness in the air cargo market.

The 747-8 uses the General Electric Genx-2B next-generation turbofan engine, designed for medium-capacity long-haul aircraft, a version of which also powers the Boeing 787 Dreamliner.

Despite all the talk about how safe twin-engine aircraft are for the public in general, the US government wanted four engines for the next Air Force One to safely ferry the president of the United States. The US Air Force bought two 747-8s destined for the bankrupt Russian airline Transaero stored in the Mojave Desert for conversion to Air Force Ones. Boeing was awarded the contract to convert them for presidential use, something that is proving costly for Boeing, with multiple problems. See **Air Force One**.

~ Boeing 757 1983/1,050

Reflecting the enhanced reliability and power of modern jet engines and the fact that two are cheaper to buy and service than three, the Boeing 757 medium-range twinjet was designed to replace the successful 727 trijet. Loved by pilots but not so much by airlines being "over-powered" to be able to take off from short runways like the 727.

Avionics, controls, and handling were to be like those of the larger Boeing 767, developed around the same time, so pilots could easily switch from one to the other without type approval.

The British government, with the backing of the UK's Rolls-Royce, had tried to have the wings made in Britain, and though it failed, Rolls-Royce supplied the engines used initially and remained a serious player. While the 757 was a successful aircraft, its sales were largely restricted to the Americas.

Like its elder brother, the 767, two 757s featured in 9/11 as:

1. The aircraft deliberately crashed into the Pentagon, with the loss of 125 lives on the ground and 64 (including the 5 hijackers) on board.

2. The aircraft crashed in open country, with the loss of 44 on board (including 4 hijackers), after the flight was delayed, and the passengers, having learned by phone the hijackers' true intentions, confronted them.

~ **Boeing 767 (767) 1982/1,038+**

For no fault of its own the Boeing 767 wide-body twinjet also has the distinction of being involved in famous and infamous incidents.

This is largely because this commonplace medium- to long-haul aircraft has been in such wide use on relatively long-haul flights:

1. Two 767s laden with fuel for transcontinental flights purposely crashed into the North and South Towers of the World Trade Center in New York on 9/11*.

2. An Ethiopian Airlines 767 ran out of fuel after being hijacked, forcing the pilot to ditch it in the Indian Ocean in full view of bathers on a beach in near Comoros. There were survivors, and those that did not survive included some who inflated their life vests prematurely and were unable to get out when the fuselage filled with water*.

3. EgyptAir Flight 990, a 767, crashed into the Atlantic Ocean shortly after departing from New York's JFK for Cairo in 1999. After analyzing the cockpit voice recordings and other evidence, NTSB investigators concluded the first officer deliberately crashed the aircraft. The Egyptian side disputed this, saying it was an example of a case where cockpit video-recorder evidence would have been invaluable*.

4. In 1983 an Air Canada 767 ran out of fuel after confusion over metric conversion and managed to glide a considerable distance to land safely at a tiny airfield known to the copilot from his military service. Referred to as the *Gimli Glider* incident*.

5. In addition to a couple of incidents due to bad weather, there was one where the left engine thrust reverser on a Lauda Air 767 deployed in flight over Thailand, causing the aircraft to crash, with the loss of life of all 223 on board*.

With the production run for the 767 coming to an end, Boeing had high hopes of extending it by using it as the base for a bid for the next in-flight refueling platform, only to find itself beaten by a bid by Northrop Grumman based on the Airbus A330. Boeing used political and other pressure from lobbyists to have the decision revoked. The bidding rules were revised to make Boeing's position impregnable, with the result that Airbus declined to bid, and Boeing won a one-horse race, and the US Air Force ended up not only with a less than optimal tanker platform but one that has caused the US Air Force considerable trouble due to problems associated with its production.

~ Boeing 777 (777) 1995/1,689+ (Also Called Triple Seven)

Seating more than 300 passengers, the 777 is a remarkably successful long-range twinjet. It was Boeing's first fly-by-wire airliner and was for a long time notable for having the largest-diameter turbofan engines of any airliner. Publicity shots could have a friendly flight attendant standing inside one!

Originally, Boeing envisaged a 767 stretch, but input from several major airlines in the design process led the company to go for a new airliner with a wider-diameter cabin. The 777 was the first airliner the FAA certified for ETOPS operations right from the outset and owes much of its commercial success to the fact those regulations allow twin-engine airliners to fly oceanic routes as much as three hours' flying time from a diversion airfield.

With two engines being cheaper to operate and maintain, the new ETOPS regulations nullified the A340's long-range oceanic advantage.

~ Boeing 777X (777X) 2026?

Airlines always want better and more efficient aircraft, though the situation is complicated by the fact that their appetite for expensive upgraded variants can wane when the price of oil halves. Nevertheless, Boeing is offering a new version called the 777X, with more efficient General Electric engines, a wider wingspan, and other improvements for delivery probably in 2026. The project has been plagued by delays with approvals taking twice as long as for the original Boeing 777. Finally, the first test flight took place in January 2020.

One novel feature is its folding wingtips enabling it to be used at airports able to take the original 777.

It has received large orders from middle eastern airlines such as Emirates and even Ethiopian Airlines because their hub, Addis Ababa International Airport, is not only in a hot place but at an elevation of 7,625 feet (2,334 meters) where the extra thrust from the massive GE engine is a great advantage. It seems that the aircraft has been tailored for those airlines, with other airlines finding the extra power not needed and the aircraft too heavy due to that. Unlike the Boeing 787 and Airbus A-350 the fuselage is made of aluminum alloy not plastic composite which would have made it perhaps too expensive.

~ Boeing 787 (787 Dreamliner) 2011/918+

A radically new aircraft using composites for the fuselage itself rather than just for minor parts, such as the tail fins. When the 787 was first hyped, it seemed the "Dreamliner," as Boeing decided to call the 787, would immediately sweep all before it. However, as the British found with developing the world's first jetliner, the Comet, being first is not always best. The 787 program cost Boeing billions of dollars more than expected and left Airbus, coming second, somewhat wiser. However, unlike the Comet, the problems were solved, albeit at considerable cost.

Outsourcing much of the production (together with some of the financial risk), and with those outside companies delivering finished sections and parts for final assembly at Boeing's plants, led to many problems and delays. When it finally went into production, there were further delays because the powerful lithium batteries started catching fire—a novel feature of the aircraft was the dependency on electrical power for so many functions, including bringing air into the cabin from the outside.

Despite a costly three-year delay, with Boeing having to pay compensation to customers, the 787 began to prove itself, making it possible to open up long-haul links between lesser cities without enough traffic to justify the use of, say, a 777. It has finally been a great success though recovering the sums invested and spent compensating clients for the delays remains problematic.

After much hesitation, Airbus finally came up with a direct competitor, the A350, benefiting from Boeing's experience.

Quality Problems

Investigations in connection with the 737 MAX crashes revealed that Boeing staff reporting the use of parts that were not quality-approved on the early 787s, and other problems were not taken seriously by the company or the FAA. According to one Boeing engineer, problems arising from inferior parts or workmanship do not show up immediately but are liable to do so after some ten years, which will be soon. Let's hope no crashes arise therefrom.

However, until the Air India 787 crash in June 2025, it had flown for years and millions of miles without a crash, and even that crash where there seemed to have been a loss of power immediately after takeoff did not indicate anything structurally wrong.

Could Fly Faster

Interestingly, the Boeing 787 **in practice** cruises at Mach 0.85 like other contemporary airliners, although it was designed to fly at Mach 0.89 like the Boeing 747 thanks to its advanced features and its composite fuselage. This because doing so saves a significant amount of money, and does not make much difference with it having to fit in with air traffic control sequencing, which, of course, did not apply to the supersonic Concorde flying alone at 60,000 feet.

~ Boeing 797 New Midsize Airplane (NMA) Possible Project

This possible project to create a new design for an airliner to fill the gap between the 737 and the 787 and replace the much-loved (by pilots) 757 has been talked about for several years, with Boeing in 2022 saying it had been put off until the 2030s as it wanted to wait for the next generation of engines. Creating a totally new aircraft is costly, and manufacturers fear their aircraft might be already out of date when deliveries start.

Originally, Boeing envisaged a twin-aisle cabin with an elliptical shape using composites but following the success of the Airbus A321XLR seem to expect it to be single-aisle.

~ X-66 NASA-Boeing Collaboration on Trussed Wing Airliner

Project to investigate the feasibility of having a high supercritical wing with a truss underneath not only holding it and lessening the need for heavy wing boxes but also providing some lift.

TRUSSED WING PROJECT

[NASA]

Boeing/Airbus Duopoly Under Threat?

With McDonnell Douglas having merged with Boeing and Lockheed having withdrawn from the production of airliners following the losses incurred with the L-1011 TriStar, which though better and safer than the DC-10 failed to achieve a marketing foothold (due largely to the late delivery of the special engines from Rolls-Royce), we have ended up with a Boeing/Airbus duopoly as regards sizable airliners.

However, there may soon be a "new kid on the block," with COMAC of China able to sell its developing aircraft into a vast domestic market even larger than that of the United States, which should be deep enough to sustain it and make it a significant player.

Overseas sales are likely to be limited, at least in the short term, for political reasons. Getting certification and setting up servicing networks will be difficult to start with, though they could start with areas outside the US and Europe.

Just as in the case of electric cars where the experience and know-how gained in the manufacture of internal combustion engine (ICE) cars no longer gives the German manufacturers the advantage they once had, the possible switch from tube and wing to blended wing aircraft may mean some upstarts could start competing with Boeing and Airbus. Upstarts are already appearing in urban mobility eVTOL craft sectors, with even prospects for short-haul regional jet as batteries improve.

Boneyard

US term for various sites, mostly in dry, isolated places, used to keep unwanted aircraft. Usually, these are older aircraft, but sometimes they are relatively new ones or even brand-new ones straight from the manufacturer at times of a slump, such as after 9/11. The most well-known aviation boneyard is at Mojave in the desert about 95 miles north of Los Angeles, where over 1,000 aircraft are parked. They are stripped of parts and eventually taken away to be melted down.

Borderline Pilots

Once an airline has engaged a pilot and invested considerable funds training them and helping them develop, it is loath to "let them go" unless they are clearly a liability.

Yet there have been several cases, including a crash at Buffalo, New York (Colgan Air Flight 3407, February 12, 2009*) where pilots responsible for crashes have been found to have a checkered history as regards check rides, and with hindsight should have been let go.

Instead of being fired, such pilots are generally sent for remedial training, but one wonders whether any amount of remedial training can make up for a lack of innate aptitude or ability to analyze while under stress. At one time many commercial pilots in the US came from the military, where the weak ones were quickly weeded out.

In the case of the Buffalo crash the airline said it was doubtful that it would have employed the captain had it known how many checks he had failed at other airlines. The NTSB recommended that there should be a limit to the number of failed checks a pilot is allowed before their dismissal. However, the FAA considered that to be too arbitrary and impractical. Nevertheless, the FAA did introduce a system, however imperfect, to officially record pilots' failing checks that could be consulted by airlines thinking of hiring them.

Evaluation is not easy. The military did psychological tests to see how soldiers would fare in dangerous situations, only to find that those that had been classed as "the fearful" ones were more courageous—sometimes heroic—in actual combat.

Most pilots never have to face a crisis—and AFF447 would never have crashed had the pitot tubes not iced up—so are not tested in extreme circumstances. One retired BA captain told the author, "When children came up to the cockpit and asked me what my greatest worry was, I would jokingly tell them it was being halfway across the Atlantic and wondering whether I had turned off the electric fire before leaving home," which had some truth in it. At that time long-haul aircraft had four engines so having one fail was no big deal.

See **Pilot Aptitude Test**.

Brace Position Can Be a Lifesaver

The "Brace Position" can be a lifesaver though it is misunderstood in two ways.

1. First, people may believe the myth that it is intended to kill you outright so that compensation would be much less than paying for years of medical treatment coupled with compensation for disability.

2. Second, passengers whose mother tongue is not English don't understand what "Brace" means.

Nowadays instead of shouting out "Brace! Brace!" flight attendants may call out, "Head down, hands over your head," which more people can understand.

"Safest Position" might be a better term as it would help correct both misunderstandings.

The whole point of the "Safest Position" is to limit injuries when there is a sharp deceleration of the aircraft on impact.

The most important thing is perhaps to avoid tipping forward and smashing one's face against the seatback in front, and possibly breaking one's neck, or being knocked unconscious and then being unable to evacuate.

If in a Business Class seat, the obstacle in front may be further away with the harm caused to someone thrown forward much greater. In that case the seat may be fitted with a shoulder harness. In First Class, the obstacle will likely be so far away that the occupant would only, with the seat belt fastened, have to lean fully over to prevent them flicking dangerously.

The advantage of bending right down is that it lessens the likelihood of being struck on the head by the many displaced objects, including luggage from the bins, flying around.

Passengers may believe they should have their hands behind their heads with fingers interlinked, when it is apparently better to have one hand over the other so one still has one good hand left in the event of something falling on them!

Flight attendants sitting in backwards facing seats by the exits are usually told to protect their hands by placing them almost completely under their thighs.

Some people ridicule the instruction to have one's seatback in the fully upright position, not realizing having it reclined can prevent someone behind escaping.

Besides limiting injuries, correctly adopting the Safest Position can mean you are more likely to be in better condition to save yourself in the event of a fire.

See **Safest Seat**.

Brakes—Can Become Very Hot on Rejected Takeoff

Brakes can become extremely hot even during a normal landing. In summer in hot climates pilots doing quick turnarounds have to ensure the brakes have cooled properly before takeoff for there is always the possibility they might have to perform optimally in the event of a **Rejected** (aborted) takeoff at high speed close to V_1 with limited runway remaining and tremendous momentum due to the vast amount of fuel on board for a long flight.

Surprisingly, the latest carbon brakes wear less if applied continuously rather than intermittently. Because of their great expense, their use used to be restricted to long-haul aircraft where their lighter weight was an important consideration. Now carbon brakes are cheaper they are being used quite widely.

Brakes cool more quickly when not applied, and that is why the aircraft may be held in place with chocks placed in front and behind the wheels when parked at the gate.

As an option, Airbus does offer brake fans to cool overheated brakes, but as carbon brakes allow a rejected takeoff so long as they are below 300°C, there is hardly a need for fans to cool them to allow quicker turnaround times. Besides the extra cost, they add weight and require maintenance.

Bravo (B) Airspace

See **Airspace Classification**.

Airspace around major airports that is somewhat like an inverted wedding cake—in that it is of smaller diameter at the bottom and wider at the top where aircraft would be on their approach. One must have permission from air traffic control to enter it, which airliners under ATC control, of course, have.

Pilots of GA aircraft straying into Bravo airspace without permission will be on the receiving end of an irate controller's comments since what they are doing can be very dangerous. They are likely to be reported to the FAA.

Brevity Codes [Military Use]

These are particularly applicable in military combat situations, where time is of the essence and confusion must be avoided at all costs. One notable code used by NATO is "FOX" followed by a number such as "FOX TWO" designating the launch of an infrared-guided missile. This lets the pilot's wingman know what is happening so they can take necessary action.

While "FOX" happens to be a good word in that it is not easily confused and is short, it really stands for "F" for "fire," and F is Foxtrot in the NATO **Phonetic Alphabet**.

Examples of brevity codes:

1. "FOX ONE" initiation of a semi-active radar-guided missile launch.
2. "FOX TWO" launch of an infrared-guided missile.
3. "FOX THREE" launch of an active radar-guided missile.
4. "PICKLE" ordinary bombs away; "PAVWAY" guided bombs away.
5. "RIFLE" launch of an air-to-ground guided missile.
6. "BRUISER" launch of anti-ship missile.
7. "GREYHOUND" launch of cruise missile such as Tomahawk.
8. "MAGNUM" launch of High-Speed Anti-Radiation Missile (HARM). See "Wild Weasels."

There are other terms such as "Bogey" referring to an unidentified aerial contact that could potentially be a hostile aircraft, requiring investigation, and possibly action, but only if sure. There is also "Bandit" for an enemy aircraft not doing anything hostile that should not be attacked, and "Hostile" for an enemy aircraft that has already engaged or is engaging with friendly units and needs to be dealt with.

Bugs

Bugs are not only insects or listening devices but also sliding markers on the scales of mechanical instruments to show key points, just as one might have on one's thermostat at home. Now, with the glass cockpit displays, the critical settings for height, airspeed, and so on, are shown automatically by bugs (markers) on the scales on the primary flight display, without the need for awkward physical manipulation.

Bunker Buster

Modern bunker buster bombs can penetrate concrete of almost unbelievable thickness. This means aircraft in bunkers on airbases of known location become vulnerable, and this is a great argument in favor of **ACE (Agile Combat Employment)**.

The largest bunker buster bomb in the US arsenal, the GBU-57, was used to take out Iranian nuclear weapons facilities deep under a mountain and concrete in June, 2025. In fact, they used half their stockpile, with them carried from the US in B-2 stealth bombers, with the round trip taking 37 hours, it being fortunate that the aircraft has a toilet and facilities for preparing food.

Bunt, Throwing Hijackers off Their Feet

Maneuver whereby the pilots push the nose down so suddenly that weightlessness (negative G) results. A bunt, followed by severe rolling, can be used to try to topple hijackers.

Also used in training astronauts to enable them to experience the sensation of weightlessness.

Business Jet, Private Jet, Bizjet, Corporate Jet

These range from light jets, mid-size jets, super mid-size jets, large jets, long-range jets, and airliner-based VIP jets, not to mention turboprops.

Some Key Makers

1. Textron Aviation (US)
   Beechcraft, Cessna, and Hawker brands.

"Over more than 95 years we have provided customers with more than 250,000 aircraft in over 170 countries—totaling more than half of all general aviation aircraft worldwide."

2. Gulfstream, subsidiary of General Dynamics (US)
3. Bombardier (Canada)
4. Dassault (France)
5. Embraer (Brazil)
6. Cirrus Aircraft (US company but Chinese owned)
7. Honda Aircraft Company (US)

With electrically powered light aircraft on the horizon there are several start-ups with seemingly promising designs.

The Market for Business Jets

The market can be roughly divided into Operators and Private Owners.

1. **Operators** (Three Quarters)
   Charter companies, fractional ownership programs, and air taxi services.

2. **Private Owners** (One Quarter)
   High-net-worth individuals and corporations.

See https://market.us/report/business-jet-market/

Corporations and high-net-worth individuals though able to present themselves as private owners, may use operators to avoid the hassle of ownership, which includes having their own pilots always on tap and paid for, and because they do not use the aircraft enough to make it worthwhile. Another possible advantage is that they cannot be easily tracked as they would not always be using the same aircraft (see **Flight Tracking**).

For high-net-worth individuals, celebrities, or bands frequently on tour, private ownership offering exclusivity, complete control over the aircraft, and the ability to tailor the travel experience to individual preferences may be paramount.

Latest Superlative Business Jets

1. **Gulfstream G700**
   According to the company it is "the most spacious aircraft in business aviation."
   Received FAA type certification on March 29, 2024, with a certified balanced field length takeoff distance of 1,827 m and a landing distance of 969 m.
   7,500-nm range, 0.925 maximum Mach speed.
   Can seat up to 19 passengers and sleep up to 13.

2. **Gulfstream G800** ($72,000,000)
   While the G700 offers more space, the G800 gives more range, namely 8,000 nm (14,816 km) at Mach 0.85 and 7,000 nm (12,964 km) at Mach 0.90.

3. **Dassault Falcon 10X ($75,000,000)**
   Though for many years Dassault has done very well with its trijet Falcons with their great power enabling them to take off on short runways, their Falcon 10X is a twinjet with powerful Rolls-Royce Pearl engines with a range of 7,500 nm (13,890 km) at Mach 0.85. Maximum Mach Operating Speed 0.95. Spacious cabin with large windows. Advanced avionics and systems.
   Thirty-eight windows will line the fuselage making for the brightest cabin in business aviation.

4. **Bombardier Global 8000 ($78,000,000)**
   Though the cabin is slightly longer than that of the G800, the latter's is an inch taller and two inches wider. Also, it has a slightly higher top speed and can use slightly shorter runways. It has a very low cabin altitude of just 2,900 feet when at 41,000 feet. On the day of its inaugural flight in May 2025, Bombardier claimed it was the corporate jet with the longest range and the highest speed of Mach 0.94.

A step up from the above, we find airliner-based VIP jets such as the Airbus ACJ220 based on the **A220**, which was, in fact, a Bombardier product before being taken over by Airbus.

Other Aircraft, Including Tiny General Aviation Craft

For a list of the many aircraft with photos and (mostly) prices see: https://www.aopa.org/go-fly/aircraft-and-ownership/aircraft-guide

~ US Is the Best Environment for Business Jets

Steven Varsano, founder of **The Jet Business**, has pointed out that the US has by far the most business jets because there are 5,000 airfields enabling one to go very near where one wants to go without formalities such as passport control. Also, distances are such that they can save one much time.

~ C ~

C2 (Control and Combat) or (Command and Control)

A simplified acronym for Control and Combat.

Neater than C&C, it can also stand for Command and Control as in **JADOC2 (Joint All-domain Command and Control)**.

See **Acronym**.

C (Computer)

Readers should not allow themselves to be overwhelmed by the repeated appearance of the word "computer" and its abbreviation C, in describing controls, equipment, and software used in an aircraft. Whereas in the old days many functions used to be purely direct mechanical links, now almost everything involves IT, like the gas boiler for heating one's home. Many of the systems are subsystems. For instance, there is even an SEC (spoiler/elevator computer) and SFEC (slat/flap control computer), and most of what they do is to regulate how the components work together.

Though linked, the computers are usually separate entities to facilitate maintenance (by replacing a module) and because specialist hardware can take the load off central processors, rather like the graphics card in a PC taking much of the graphics burden off the shoulders of the main processor. In addition, failure of a component is not likely to lead to a generalized catastrophic failure.

C-Check for Aircraft as Part of ABCD Checks

The C-Check is a major maintenance check done every 18 months to two years. Done all at once, it takes one to two weeks, but is sometimes split into parts so the aircraft can be kept in use at busy times.

Cabin Air Is Cleaner than You Think

Contrary to what many believe, the quality of the air in most airliner cabins is much better than that found in many public spaces such as trains, buses, and even hospitals. However, see the entry for **Safest Seat** regarding infections from fellow passengers.

One little-known fact is that the air in passenger cabins generally comes down from the ceiling and exits via vents at the floor and does not generally travel horizontally through the cabin carrying viruses, etc.

Cabin Temperature Seems Lower Because Air So Dry

Sometimes passengers find the cabin cold even though a thermometer might not show it to be so. This is because the low humidity (dry air) makes it seem so. This phenomenon is why people can withstand high temperatures in dry air in deserts.

People are affected differently because those who do not sweat so much may be less affected in airliner cabins and not feel cold.

"Cabins of the Future" (Sensory Seats Checking Your Well-Being)

Ideas are abounding regarding ways the passenger cabin could be made more efficient, hygienic, and intelligent. One idea is to have seats that can monitor the passenger sitting there—weight, body temperature, breathing, heartbeat, and movement. Should the passenger not move around for several hours, they could be checked on and warned about the danger of deep vein thrombosis and advised to move. Cabin crew could better monitor the situation including temperature without constantly walking up and down. Another idea is to give tray tables, arm rests, and toilet door handles a coating that kills germs.

CAM (Cockpit Area Microphone)

Abbreviation used in transcriptions of cockpit voice recordings to indicate words said in cockpit. As there are several, these are sometimes referred to as CAM 1, CAM 2, and so on in accident reports.

Canards

Besides meaning a planted or false story, the French word canard means "duck."

In aviation, it means having the horizontal stabilizers and elevators sticking out of the fuselage in front of the wings near the nose instead of having them at the tail. Seen on some business jets and on the Wright brothers' machines at the beginning of the twentieth century.

Cannibalization

With some military aircraft such as the B-52 being used for decades, getting spare parts such as a valve can mean "cannibalizing" another B-52.

This stripping another aircraft of parts is a common practice but is not always possible, which is concerning as in the US, it is often difficult to get essential spare parts as too few are needed for old aircraft where there is not a civilian market as there is little chance of a significant profit.

In the case of the B-52, where there is talk of its life being extended to 2060, the spare parts problem should ease as parts of it, notably the engines, are to be changed to modern ones from Rolls-Royce.

Carbon-Carbon Composites (CCC)

Depending on how it is formed or treated, carbon can be in various forms with some able to have great strength and heat resistance.

CCC have the following features, among others:

1. Advantages:
   Light weight
   High strength at high temperatures, up to 2,000 °C, in nonoxidizing atmosphere
   Low coefficient of thermal expansion
   High thermal conductivity (>Cu & Ag)
   High thermal shock resistance

2. High fabrication cost
   Porosity
   Poor oxidation resistance
   Poor inter-laminar properties
3. Some applications
   Missile nose-tips
   Leading edges (Space Shuttle, SR-72)
   Aircraft brakes, etc.

Carbon fiber reinforcements are embedded in a carbonaceous matrix subjected to complex treatments to arrive at the final CCC. These CCC composites are a new material that is like ceramic but exhibits brittle-to-pseudoplastic behavior.

Carbon Fiber Composite Polymer (CFCP)

This is used on the Boeing 767 Dreamliner, where working with it posed several problems, for while it is exceptionally strong as regards tensile stress it is weak with respect to compressive forces. Also, it is not tough and any impact to a wing or fuselage made of it could damage it with internal delaminations that are difficult to see and detect, unlike in the case of metal where defects can be found electrically.

It took some time to learn how best to work with this exceptional strength to weight ratio material. It is said the Airbus A350 has 52 percent CFCP, whereas the older Boeing 787 Dreamliner has 50 percent.

Carbon Offsetting

Although aviation shares little responsibility for global warming, image is very important and airlines such as United Airlines, British Airways, and easyJet resorted to carbon offsets. This consists of partaking in schemes such as the planting of trees, which absorb carbon dioxide, or preventing deforestation, and so on, and deducting the carbon saved from that consumed on flights. Organizations such as Greenpeace claim this is not valid and sometimes untrue.

EasyJet has decided to withdraw from carbon offsetting and go in the long term for carbon neutral flying, say, with hydrogen.

Carbon Titanium Fan Blades

Rolls-Royce is using these on their new UltraFan engine. A thin titanium sheath is bonded to the leading edge to resist erosion, and to protect against foreign objects and bird strikes. These will be very much lighter, and with the composite casing mean a weight saving of 700 kg equivalent to seven passengers with luggage.

Cargo Hold Doors

Airliners carry cargo for supplementary income and the doors for this are on the righthand side of the aircraft. Victuals are also loaded from the righthand side straight into the cabin.

The type of cargo allowed on passenger-carrying aircraft is restricted to items that do not present a fire or explosion risk and can well be expensive fruits or berries, and this is why one can unexpectedly find them flown in from faraway places in the supermarket.

Dangerous items, and particularly lithium batteries, can only be carried in cargo planes.

Caravelle (Sud Aviation Caravelle) 1959/282

Having almost four years elapsing between the first flight and entry into service meant the Caravelle was well-tested and reliable. It was the first medium-range airliner with the distinction of having its two engines at the back, giving it a clean wing. Many of those working on the Caravelle went on to work on a much more ambitious project, the Concorde supersonic airliner. It was a great first step for its time.

Cargo Hub and Spoke [US]

Hub and spoke operations are still largely viable for cargo.

Major cargo hubs in the US are:

Louisville for UPS

Memphis for FedEx
Cincinnati for DHL

Cascade of Events

A situation where something minor leads to some disruption, that is then followed by something else minor, and so on, finally leading to something serious such as a crash.

See **Unintended Consequences of Guerilla Action.**

CAT (Clear Air Turbulence)

Turbulence in clear air cannot be detected as yet by the aircraft's weather radar. It can be very violent, with food carts, unbelted passengers, and flight attendants hitting the cabin ceiling.

Category of Runway (Visibility for IFR Approach)

To make clear what is possible when landing under instrument flight rules (IFR) in bad weather, runways are classified by the ICAO according to their nature and ILS equipment, and by the FAA as in the table below.

Categories of ILS approaches
Data from Aeronautical Information Manual, AIM, FAA.gov

Category	Minimum decision height	Minimum runway visibility range	Remarks
I	200 feet	2,400 feet	
103I	200 feet	1,800 feet	With touchdown zone and runway centerline lighting.

II	100 feet	1,200 feet	Half the minima of a standard Cat I approach.
IIIa	100 feet	700 feet	
IIIb	50 feet	150 to 700 feet	
IIIc	No DH	No RVR limitation	Completely blind landing—very rare.

The decision height is the height at which the pilot must decide whether to land or not, and the minimum runway visibility range (RVR) is the distance that can be seen down the runway—usually measured automatically by equipment alongside the runway.

Cayley, George (1773–1857)

English baronet who studied birds and carried out experiments using an airfoil attached to the end of a rotating spar to measure lift as it passed through the air, and thereby found flight was possible with forward motion through the air without flapping.

According to Phil Scott, who wrote a book called *The Pioneers of Flight*, Cayley made his coachman the first adult to fly in a heavier-than-air craft by putting him in a glider and rolling it downhill on a windy day in 1853—50 years before the Wright brothers. Years before, he had worked out how one could make a machine that could fly but had put it on the back burner because of the impossibility of finding an engine with the necessary power-to-weight ratio at that time. Was he the true groundbreaker?

CDL (Configuration Deviation List)

Like the MEL (minimum equipment list), this lays down what deviations from the norm are acceptable for the aircraft to be allowed to depart.

CDU (Control/Display Unit)

Intimidating-looking cockpit instrument resembling an expensive, old-fashioned calculator used for inputting and checking data in the so-called glass cockpit.

Ceiling

Ceiling has two meanings:

1. The height of the cloud base when the sky is overcast, or clouds are virtually continuous.

2. The maximum height at which an aircraft can operate.

Handling aircraft at extreme heights can be very tricky, both as regards flying characteristics of the aircraft and performance of the engines. Fortunately, modern computers can keep aircraft cruising for hours within tight limits, without getting tired.

The practical ceiling for airliners is lower than what is physically possible because the aircraft cannot be so high that it cannot in the event of decompression get down to a level such as 10,000 or 12,000 feet where passengers can breathe without assistance in 4 minutes.

See **Maximum Height an Airliner Can Fly Mach Number**, and **Coffin Corner**.

Celestial Navigation

This is seeing something of a comeback, especially in military circles as stars cannot be interfered with, unlike satellites launched from Earth. Also, GPS signals can be spoofed, or otherwise interfered with, not to mention their being taken out by an **EMP (Electromagnetic Pulse)**.

In the 1930s celestial navigation was the key to navigation over oceanic routes with no beacons, with it being less precise in daylight with no visible stars, as Emelia Earhart* and her navigator, Noonan, found to their cost.

Center (Air Route Traffic Control Center)

In the US, air traffic control centers manage high-altitude air traffic, and notably through traffic, over a wide region. In ATC exchanges the controllers merely refer to themselves as "center," with the name of the area covered—for example, New York Center or Boston Center. See **Airways**.

CESSNA AVIATION

Cessna at one stage encountered difficulties that could not be overcome by having some manufacturing done in China. After many layoffs, Cessna became part of Textron Aviation.

Cessna 172 1955/45,000+

Many small aircraft are used by flight schools of which there are many, especially in the US. However, these are propeller driven, with the Cessna 172 being the most widely used.

Interestingly, the Cessna 172 is the most produced aircraft ever with some 45,000 made since 1955.

As an aircraft for training novices it has many pluses, making it the king for flying schools in the US and abroad. These included a well-proven record with only 0.56 crashes per 100,000 flying hours, tricycle landing gear facilitating landings, and a high-wing configuration giving good all-round visibility and inherent stability.

It was also reliable, reasonably priced, and easy to maintain.

Secondhand ones are available at a reasonable price for personal use. See **Business Jets and Pipeline Flying**.

Cessna 182 1956/23,000+

Also produced in several variants.

CFIT (Controlled Flight into Terrain)

CFIT simply means the aircraft flying normally under control flies into terrain or water.

There are warning systems in the cockpit triggered by the radio altimeter that make a whooping sound and blare out, *Terrain! Terrain! Pull up! Pull up!* if there is a danger of the aircraft hitting the ground below.

Now that airliner computers have GPS telling them exactly where they are and data regarding the height of the terrain ahead in their memories, they can also warn the pilots if they are about to fly into a mountain or cliff in front of them. Another example of advances in technology making flying so much safer.

See **GPWS**.

Chaff

A radar countermeasure whereby aluminum foil, paper strips coated with aluminum foil, metalized glass fibers, and the like are emitted from the target, which can be an aircraft or ship. When emitted from an aircraft, they slow, allowing modern radar using doppler analysis to filter them out. To counter that the target may illuminate the chaff with a signal having a doppler frequency indicating high-speed motion.

Even **ICBM**s can release chaff during the latter part of their trajectory.

Chapter 11 of The US Bankruptcy Code

Chapter 11 is a form of bankruptcy in the US that allows companies to continue operating with them protected from claims from creditors and not subject to agreements like labor union contracts.

The aim is to give the company time to reorganize, and many emerge stronger from Chapter 11.

With major US airlines often operating under Chapter 11 and hence unshackled from commitments undertaken in historically more profitable times, non US carriers, unable to benefit from such legislation, claim they are disadvantaged.

Emirates, Etihad, and Qatar Airways, accused by American Airlines, Delta, and United of benefiting from government subsidies, claim this unique bankruptcy law is itself a form of subsidy.

Charlie Bravo (CB)—Meaning Cumulonimbus

"CB" is the meteorologist's code for cumulonimbus and Charlie Bravo is "CB" in the NATO **Phonetic Alphabet**.

Though in the US pilots might say "Unable because of weather," internationally they would or should say "Unable because of Charlie Bravo."

CB are towering storm clouds and very dangerous.

Checklist Paper/Digital

Checklists are important tools for ensuring everything is done correctly. There are ones for various stages of the flight and special ones for dealing with problems.

Checklists for dealing with problems have—apart from Boeings' one concerning the MCAS system on the 737 MAX—been created with input from a whole range of experts who can see dangers that a pilot would not usually think of.

For instance, where the instruments show there must be a fuel leak from an engine, the checklist for dealing with the problem will say they should shut down that engine and divert to the nearest suitable airport. The engine most likely will be running perfectly, but fuel could be leaking from a damaged fuel hose or connector. This might just streak out from the aircraft at cruise but might not only cause a fire when the configuration changes with the lowering of the flaps but cause a fire on the ground if reverse thrust were use. Even after the aircraft comes to a stop it could fall onto very hot brakes and ignite. All good reasons to shut down the engine and cut off the fuel supply.

Checklists can be on paper or digital.

See **Crew Alerting Systems**.

Check Captain

A management captain who carries out the regular pilot competency checks mandated by the authorities and the airline.

Checking-in and Security at the Airport

Going through Security has been a nightmare ever since the unmasking of a *Liquid Bomb Plot** in 2006 to bring down several airliners flying from England to the US with explosive material hidden in innocuous-looking soft drinks "bottles."

For example, liquids in hand luggage have been severely restricted and must be less than 100 ml and in a transparent bag. However, thanks to advances in technology all is changing.

The new expensive (and very heavy) carry-on baggage scanners being installed at major airports can see the inside of bags in three dimensions rather than just two and be truly able to detect anything dangerous, including liquids. The tiresome restrictions on the volume of liquids will no longer apply at airports that have installed the new equipment and trained their staff on the use thereof.

See **SSSS (Secondary Security Screening Selection)** sometimes printed on one's boarding pass in the US.

Chimes (Cabin Chimes)

Chimes are used to draw the attention of passengers—say to the "fasten seat belt" sign being switched on—and more importantly to liaise with cabin crew and even a pilot who might be located anywhere in the aircraft. Their nature can signify anything from the need for the purser to contact the cockpit to what to do in an evacuation.

The meaning of many can be deduced from the behavior of crew according to the stage of the flight such as flight attendants (strapped in their seats by the exits after takeoff) waiting for the ping on reaching 10,000 feet telling them they can get up and attend to their duties.

Chimes vary from airline to airline, with there being perhaps one for someone smoking in the toilet, a sick passenger, and so on where they do not want to trouble the passengers. There may be one telling the cabin crew to congregate (at the galleys?) for instructions in a possible emergency landing.

CHINA

Unlike the US, China does not have a long history of making planes starting from before World War II, and then during the war, and even after the war, benefitting from technology from Germany (see **Operation Paperclip**). However, apart from having much local talent, it does have a huge domestic market and military demand able to sustain a domestic aviation industry.

While it was thought that by hook or by crook Chinese airframers could begin making good aircraft it was believed their weakness would always be that they would always need "Western" engines, since engines depended on savvy engineers with years of experience. However, it seems that even in that domain China is catching up and beginning to produce good engines, at least for their military. [It should be noted that not only performance but reliability to qualify for ETOPS may still be a problem and one for Chinese hoping to sell aircraft for long haul.]

Interestingly, with Russian airspace no longer an option for Western airlines, Chinese airlines have the advantage of still flying through there to Europe much to the consternation of airlines such as BA, which has stopped certain routes such as between London and Shanghai as it cannot compete having to fly longer southern routes.

~ Civil Aviation

Just before Covid-19 disrupted air travel, some 500,000,000 people were flying to, through, or from China each year, and the number was growing rapidly. Soon China would have an aviation market greater than the US.

Some $35,000,000,000 worth of aircraft were sold to China in 1918. Roughly 47.2 percent of the aircraft were from Boeing and 45.3 percent from Airbus.

China was the largest user of the 737 MAX with 97 operating there compared with 72 in the US just before they were grounded after the two crashes. At the time Chinese airlines also had some 500 MAXs on order.

Airbus has a significant final assembly plant in China, and Boeing has just established a lesser one there for work on interior fittings and so on. However, these may not be enough to cement their position in China, for China is developing its own aviation industry. See **COMAC**.

With China having such a large internal aviation market it can keep going on its own for years without needing to threaten the Airbus/Boeing duopoly.

The "Big Three" Chinese Airlines are Air China (the flagship airline), China Eastern, and China Southern. While they are on a solid financial footing that is not true of many of the others.

~ Military Aviation

The US defense budget greatly exceeds that of China, but the significance of that should not be overplayed for all may not be as it seems.

First, the US military must cover commitments all over the world, whereas China focusses its military activities on specific areas. It also pays its military personnel much less.

Second, the cost of manufacturing, say, a ship, is very much less in China, with its yards unlike those in the US competitively providing ships for the whole world.

The discovery of the construction of numerous ICBM silos in China's desert regions was noted in 2021 and concerning. Some, or many of these, may be decoys with no missiles. There were even suggestions that corruption meant some missiles were filled with water and were not functional.

Compared with the US and Russia, China's nuclear arsenal once seemed to be limited to what would be necessary to preclude a first strike, and not for them to make a first strike themselves. Though Mao-Zedong is alleged to have said that China with such a large population does not worry about a nuclear attack as there would still be so many people left! However, the country with so much of importance on the coast is rather vulnerable geographically compared with the US and Russia.

It is difficult to be sure about what is happening in China as regards military aircraft with China even mentioning seventh-generation aircraft on the horizon!

The capability—constantly improving—of Chinese military aircraft should not be underestimated. The Radar Cross-Section (RCS) of the American F-35 and F-22 is possibly of the order of 0.0010, whereas that of the Chinese J-20 Mighty Dragon stealth fighter is poor by comparison at perhaps around 0.1. However, the Chinese aircraft might beat the F-22 as regards air-to-ground attack, and with the F-22 being of an older generation the J-20 has the advantage of **IRST (Infra-Red Search and Track)** which the F-22 lacks. Also, it can carry more missiles than the F-35B in its three internal bays because of the space taken for the VTOL engine on that version of the F-35.

The Chinese who had reverse-engineered a Russian fighter might have been caught out when one was downed in Ukraine and debris sent to the US and UK giving away the frequency it used.

~ China and US Requirements for Next-Generation Aircraft

With the most foreseeable conflict between China and the US being over China's underlying wish to take over Taiwan, the two sides have quite different requirements as regards next-generation aircraft.

The Chinese would be fighting next door and would not need long-range aircraft, whereas the US would. The US would have to have their aircraft carriers well away out of range of China's hypersonic missiles, and even islands such as Guam would be vulnerable. See **ACE (Agile Combat Employment): Dispersed "Austere" Airbases**.

The loss of an aircraft carrier with up to 1,000 sailors and airmen aboard would be devastating and have ramifications that neither side would want.

~ DF-17

Solid-fuel, road-mobile, medium-range ballistic missile destined to carry the DF-ZF hypersonic glide vehicle.

Range approximately 1,120–1,550 miles (1,800–2,500 km) at Mach 5–10.

The DF-ZF hypersonic glide vehicle is said to attain speeds between Mach 5 and Mach 10. Besides being used to deliver nuclear weapons, it could carry out precision strikes with conventional warheads, perhaps rendering US aircraft carriers vulnerable despite their layered defenses.

China Aerospace Science & Industry Corporation (CASIC)

China's main missile maker.

China Airlines (CI)

Airline based in Taiwan, not to be confused with **Air China (CA)**.

Suffered a series of disastrous accidents, attributed by some to the double-punch—like South Korea—of:

1. The deference to elders once, and often still, the norm in Asian societies.

2. Many pilots come from the military, where the querying of a senior man's orders is unthinkable.

Chine

A term taken from boatbuilding, where there is a longitudinal line on hull where the cross-section is maximal as shown on the flank of the **SR-71**.

CHIRP [UK] (Confidential Human Factors Incident Reporting) [UK]

A confidential reporting system in the UK by which professional pilots and ATC staff report incidents arising from human errors for analysis by the RAF Institute of Aviation Medicine at Farnborough. Initially, this was set up to deal with the problem of some overbearing but fallible ex-military captains. See **EUCARE** and **ASRS**.

CFDF (Cockpit Flight Data Recorder)

Flight and data recorder combined in one unit.

CFDS (Centralized Fault Display System)

Circular Error Probable (CEP), Accuracy of an ICBM

A mathematical term used in ballistics to portray the precision of weapons and in particular ICBMs such as the US Minuteman.

The latest version of the Minuteman is said to have a CEP of 120 feet, meaning that 50 percent should fall within a circle of radius 120 feet centered on the target. However, that center may be wrong, in which case the difference is called the bias.

CEP was alluded to in the movie *Clear and Present Danger* about a computer running wild and about to launch nuclear Armageddon.

ICBMs use inertial navigation, since GPS cannot be relied on in wartime. Submarine-based ICBMs may also use celestial navigation as the launch position to be entered in the inertial navigational system is not precisely known.

Cleanskin

This term was originally used by the security services to describe an operative—their own or their enemy's—with no compromising history, connections, or antecedents that might draw attention and engender suspicion.

It is now increasingly being applied to homegrown subversives fitting that description, who may stab or shoot people, plant bombs, hijack airliners, and so on. It is exceedingly difficult to preempt acts by such individuals since they are almost impossible to identify for monitoring in the first place. This is especially true if they are lone operators.

Clear Air Turbulence (CAT)

While relatively rare, clear air turbulence can be frightening and cause injury because the aircraft may suddenly drop in clear weather with passengers who have not taken the advice to keep their seat belts loosely attached hitting the ceiling—sometimes very hard—and falling back on others. Flight attendants moving around the cabin with carts that go up in the air are liable to be injured as they fall back.

The aircraft's weather radar cannot detect clear air turbulence because unlike in the case of thunderstorms there is no moisture. The only warning pilots get is from pilots experiencing it ahead, though avoiding the edge of jet streams where it tends to occur is a good policy. See **PIREP**.

Clearance

Before taking off the aircraft must obtain clearance from the air traffic controllers. There are three ways in which this may be done:

1. Voice.
   This is the traditional way whereby the pilots obtain it vocally from the air traffic controller. Still used at some airports.

2. **ACARS** (Aircraft Communications, Addressing, and Reporting System).
   Outside the US this is the most common method.

3. **CPDL** (Controller Pilot Data Link).
   Mainly used in the US.

Clearway

An area free of unforgiving obstacles before and beyond a runway. In the interest of safety, there are recommended distances, but with conurbations encroaching on airports, or vice versa, it is not always possible to comply.

With the decrease in other types of accidents, runway overruns, especially when landing in bad weather, feature increasingly in incidents and accidents. Where there is insufficient space, a specially devised arresting zone made up of particulate material that squashes under the wheels called **EMAS** can be installed to improve safety significantly but at some cost not only to install but also maintain since it can only be used once.

Cluster Bombs or Cluster Munition

A type of air-dropped or ground-launched weapon that releases or ejects smaller submunitions.

It can include:

1. Explosive bomblets to take out personnel and destroy vehicles grouped together.

2. Incendiaries combined with explosive devices to destroy the roof of buildings to allow the fire generated in.

3. Crater-making bomblets intended to destroy runways.

4. Conductive streamers to short-circuit electric power transmission lines.

5. Bomblets dispersing chemical or biological weapons.

6. Bomblets scattering mini land mines.

7. Devices to disperse leaflets.

Collateral damage from cluster bombs can be significant involving civilians not only during the attack but also afterwards. Unexploded bomblets can kill or maim people—especially children—long afterwards, hence efforts to ban them or renounce their use. Locating and removing them is difficult and costly.

Unfortunately, they are a very effective multipurpose weapon and particularly effective against massed tanks, armored vehicles, and grouped personnel.

Cockpit

In a letter to *Flight International*, reader Anthony Jones wrote that one of his pet hates was the use of the term "cockpit," especially by Americans, for what should be called the flight deck. For Jones, the cockpit was a pit or hole in fuselages used long ago to house the engine cocks or controls. However, others suggest the term has nautical origins.

Nevertheless, he is possibly fighting a losing battle, with "cockpit" used also in such terms as *glass cockpit*, where many of the traditional dials are replaced by computer screens and where the same one can be set to show different things. Also, one must bear in mind that fighters do not have flight decks anyway.

Astonishingly, someone at the FAA has been recommending that the word cockpit be dropped on the grounds that it undermines female staff!

Cockpit-Centric (as Opposed to Tower-Centric)

The future NextGen ATC system envisages a more cockpit-centric approach, with pilots less dependent on the tower for information and instructions.

Although NextGen is going to require considerable investment, the philosophy applies at the simplest level as well. For example, *Air Transport World* reported the award to Sensis of a contract for runway status lights at the busiest airports, which would automatically come on should it be dangerous to cross or enter a runway without the controller being in the loop. There would also be *takeoff hold* lights working on a similar principle.

Cockpit Confidential

A highly readable and a very successful book by Patrick Smith, an airline pilot, who for many years had a witty aviation blog on the US website salon.com. He is very outspoken, and his interesting character radiates through his excellent and provocative writing. He puts things in perspective.

Cockpit Instruments and Controls

In the modern *glass cockpit*, the once individual key instruments showing airspeed, heading, altitude, attitude (bank/pitch), and rate of descent are integrated on a single screen called the **PFD (Primary Flight Display).**

Alongside, there are other screens showing the terrain ahead, the weather, engine performance, and advisories. A feature of many of these screens is that the same screen can be switched to show different things.

Cockpit Video Recorder?

Pilots once opposed **Cockpit Voice Recorders (CVRs)** and they were only made mandatory in the UK following the crash of a BEA Trident airliner taking off from London's Heathrow Airport in 1972 where it was impossible to know who in the cockpit did what. They are now even opposed to cockpit video recorders.

A decade ago, the case for having cockpit video recorders was surprisingly much stronger than it is today for they could have shown what the cockpit instruments were showing and what controls were activated. Nowadays **Cockpit Data Recorders (CDRs)** are so sophisticated and able to log thousands of parameters that all that can now be known.

Inside the cockpit, a cockpit video recorder would be invaluable in the exceptionally rare case where a suicidal action by one is possible by one of the pilots is thought as in the Air India crash on takeoff in 2025.

The pilots' unions are very much opposed to cockpit video recorders, with the US pilots' union calling them "fool's gold," since in its opinion they would not improve safety. In addition, the union said that the assurances regarding privacy and data protection it received before the introduction of CVRs have proved worthless.

However, there might be a strong case for video cameras with a 360° view able to record collisions and even **UAVs (Unmanned Aerial Vehicles)** tailing or following alongside the airliner, be they civil or military.

Cockpit Voice Recorder (CVR)

Records:

1. Pilots' conversations between themselves.

2. Radio communications with air traffic control and so on.

3. Extraneous sounds captured by cockpit microphones (CMs) placed at various points on the flight deck.

The sound data is stored in a so-called black box (painted orange), usually situated in the tail, and designed to withstand crashes, fire, and immersion in deep water.

Extraneous sounds, and particularly the pitch of the engines recorded by the cockpit microphones, can show accident investigators the rate of rotation.

Belated Introduction of CVR in the UK

As surprising as it may seem today, UK pilots' unions once tenaciously resisted the introduction of cockpit voice recorders on **the grounds of invasion of privacy.**

Despite CVRs being already mandatory in the US they were not in the UK in 1972 when a British European Airways (BEA) Trident crashed shortly after takeoff from London's Heathrow*.

A postmortem showed the captain was probably dead before the mishaps, due to a heart attack induced most likely by a violent pre-flight argument with a pilot in favor of strike action at the airline to which he was totally opposed.

Someone in the cockpit had moved the droop (slat) control lever thereby retracting the droops before the aircraft had gained enough speed on its climb-out. This resulted in a change-of-configuration stall with the aircraft pancaking and all on board losing their lives.

Without a CVR, investigators were unable to determine whether it had been the captain, his rookie copilot, or even someone else, who moved that droop lever. As a result, the British government faced down the pilots' union and made CVRs mandatory as in the US.

For privacy reasons, CVR recordings are overwritten after two hours as a compromise between the air crash investigators and the pilots' unions as that is thought to yield enough information for analysis of most crashes. It used to be half an hour, and that still applies for some small aircraft and helicopters.

The problem with two hours is that in near-misses and narrowly avoided accidents the aircraft continues in service with the recording of the event being overwritten. A case in point was an incident that could have been even worse than the worst-ever multi-aircraft crash at Tenerife.

It was on July 7, 2017, at night at San Francisco International Airport where one of two parallel runways (the one on the left) was out of use with lights turned off. This meant it was not immediately obvious where the out of use runway was.

Air Canada Flight 759, an Airbus A320, almost landed on top of four fueled-up airliners on a taxiway parallel to the right-hand runway. The pilots of one of the waiting aircraft, shocked at the sight of the A320 descending towards them switched on their aircraft lights as a warning while air traffic control told the Air Canada flight to go around, which it had already started.

The Air Canada Airbus came within 14 feet (4.3 m) of one of the waiting aircraft and as low as 59 feet (18 m) above the ground.

Following the go-around the Air Canada flight landed safely on the runway it should have landed on in the first place. After refueling it went on its way and the NTSB was not informed until July 9, by which time the CVR had been overwritten several times.

One reason for the pilots mistaking the taxiway for the runway was that they were not using the instrument landing system (ILS) later made mandatory when one runway is unlit at night. The Air Canada pilot with perhaps a tired voice said he could see something on the runway only to be told by air traffic control that it was clear—why they did not go around there and then is incomprehensible.

While a key factor was thought to have been pilot fatigue, lack of a cockpit voice recording made this difficult to ascertain.

Technically, CVR recordings could be kept for a week or more.

Coffin Corner

This is the tricky high-altitude flight level where there is a tiny margin between flying too fast and too slow. Too fast leads to difficulties as parts of the aircraft exceed the speed of sound. Too slow leads to the risk of stalling in the high-altitude thin air. Fortunately, it is normally not much of a problem, for the computers (with the aircraft on autopilot) fly the airliner more precisely than humans, and do not get tired and forgetful.

Cognitive Incapacitation

Incapacitation of pilots can take various forms, partial incapacitation, which is the trickiest as it is not always evident to the other pilot, and full incapacitation such as a heart attack, which is rare but does happen even with the person dying. There is also cognitive incapacitation that can in some ways be likened to computer networks where a spike in the amount of data can overload them and they "collapse."

This can happen to a pilot, especially where something unexpected happens. They freeze up mentally, doing things unnecessarily and without informing the other pilot, with this leading to further problems.

Cold Dwell Fatigue/Cold Creep Fatigue

The fan-hub on an engine of an Air France A380 disintegrated over Greenland with pieces falling to the ground. With considerable effort much of it was recovered showing the fan hub had failed due to a little-known type of fatigue called "Cold Dwell Fatigue" or "Cold Creep Fatigue."

Colgan Air Disaster Reforms, a Safety Watershed

This crash* where the captain stalled his Colgan Air turboprop putting it into a fatal dive on February 12, 2009, as it approached Buffalo, NY, killing 50 people, was never fully explained. However, the investigation publicized aviation shortcomings, forcing Congress to act, resulting not only in changes to pilot training and the management of pilot fatigue but also ensuring the

FAA set up a database—that airlines are able to access before hiring a new pilot—logging cases where pilots such as the one involved in the Colgan disaster have failed checks and had to undergo extra training.

Collaborative Combat Aircraft (CCA), Loyal Wingman

A collaborative combat aircraft (CCA), that often used to be called a Loyal Wingman, is an uncrewed combat aerial vehicle (UCAV) that can operate in conjunction with crewed aircraft.

At the 2024 Farnborough Air Show CCAs were the subject of much discussion with participants wondering whether they would mainly be:

1. Very basic and very cheap (see **Attritable ("Cheap Enough to Sacrifice")**, links with crewed aircraft, even being in a swarm as decoys.
2. Quite sophisticated, not nearly as costly as a crewed aircraft, but besides being able to execute many functions including intelligence, the launching of weapons, and full integration with crewed aircraft, satellites, and ground stations.
3. Ultra-sophisticated and very expensive, though not costing more overall than a crewed aircraft taking into consideration the human cost—training and value of the crew's lives.

CCAs are seen to raise air combat potential at a less crippling cost and operate often in large numbers relative to the associated manned aircraft.

See **Loyal Wingman**.

COLLINS AEROSPCE

A subsidiary of RTX Corporation (formerly Raytheon Technologies). Raytheon is the world's fourth largest **Defense Contractor** and Collins Aerospace represents much of that.

Collins is not only a large player in defense work including electronic warfare systems but also a maker of all sorts of components, systems, and equipment ranging from the very sophisticated to airliner seating. Also provides services in many areas as well as training.

Supplies the ultra-sophisticated **Head-Up Display (HUD)** helmet for the pilot of the F-35 Lightning.

COMAC (Commercial Aircraft Corporation of China)

The Comac C919 may receive its European certification in three to six years, according to **EASA (European Union Aviation Safety Agency (EASA)** Executive Director, Florian Guillermet, in April 2025.

Like the US, China has a vast aviation market able to support the production of airliners, so even though it may take time for COMAC to get certification abroad that will not block its development into a threat to the Boeing-Airbus duopoly.

Comet—World's (ill-fated) First Jetliner

Britain's de Havilland Comet, the world's first jetliner (first revenue passengers May 1952), stole a march on the Americans and, with its elegant design, looked as though it would be a tremendous success, despite its limited number of passengers (initially 44, in great comfort) and rather limited range, which it made up for with its much greater cruising speed of 460 mph (740 km/h), compared with 325 mph (523 km/h) for piston-engine airliners, such as the Argonaut.

Some initial crashes and overruns on takeoff were dismissed as being due to overconfident pilots not appreciating the handling characteristics and rotating too early when taking off in hot conditions. It was later found that the problem was more complicated than it first appeared, and that there was a known design fault with regard to airflow over the leading edge of the wing and inability of the engines to gulp in air, both being evident at high angles of attack. Unfortunately, these were the least of its problems, since there followed a couple of mysterious disasters where on reaching cruising altitude, aircraft suffered catastrophic failures and broke up in midair.

Winston Churchill, the prime minister at the time, ordered that everything be done to find the cause. Engineers subjected an entire Comet airframe to pressurized water tests to simulate the cycles to which the Comet airframe was subjected in operation. Finally, these tests revealed a failure point due to fatigue and, in addition, that stress concentrations around the square windows were much greater than expected.

The Comet 1 never saw service again, while the Comet 2 (with small round windows as opposed to the previous square ones) was used without incident by Britain's Royal Air Force, who operated it for years but with lower pressurization.

The rather larger Comet 4 later entered service over the Atlantic but was superseded by the arrival only three weeks later of the Boeing 707, which was much larger and more efficient to operate. Had de Havilland not been first, and the British government not been so open, other manufacturers could well have suffered the same fate.

Command Responsibility

"Command responsibility" is the tenet that the captain of a ship or aircraft has the ultimate responsibility for ensuring their passengers' and vessel's safety, whatever the pressures put on them by management or circumstances. There are no excuses. It is responsibility to a higher God, or professionalism in the best sense of the word.

Composites

New materials, such as carbon fiber, are replacing parts traditionally made of metal, such as aluminum alloy, to give a better strength-to-weight ratio. Usually, materials are combined to try to exploit the qualities of them all. Aluminum alloy netting or foil may be included to avoid the "creep" associated with plastics and to help cope with lightning.

Composites in aircraft consist of a matrix (binder) reinforced with fibers. The matrix, when cured, gives the part its shape and the fibers provide much of its strength. The matrix is usually a type of epoxy thermosetting resin because of its strength and good high temperature characteristics. Sometimes fiberglass is used for the fiber, and though cheap and easy to work with, has the disadvantage of being heavy, so usually carbon fiber is chosen.

In the Boeing 787 Dreamliner, metal netting was incorporated in certain parts to prevent damage from lightning, which though unlikely to cause a crash could result in damage that would be costly and time-consuming to repair.

The greatest worry has been that of checking for cracks since finding them in plastics is much more difficult than for electroconductive metals. See **SHM (Structural Health Monitoring)**. The great bending strength of composites, though an advantage, is a disadvantage in that it can obscure a break underneath.

Certain aluminum alloys form a voltaic cell when adjacent to carbon fiber composites, and Boeing had trouble finding fasteners for its 787 Dreamliner without using expensive titanium. Airbus had trouble with the copper netting to deal with lightning at the rear of the fuselage of its composite A350 leading to a lengthy dispute with Qatar Airways. Finally, Airbus decided to use perforated copper sheeting, which had the advantage of being lighter.

The conflagration that broke out when an Airbus A350 collided with a small aircraft that had entered the runway when it should have waited at Haneda Airport in Japan at the beginning of 2024 burned for hours, destroying the Airbus. However, all on board the A350 survived, even though a number of exits were unusable, and evacuation was delayed because the pilots were initially not aware of the fire. This good outcome was perhaps because the composite fuselage was more resistant to penetration by fire than aluminum would have been.

Compressor Stall (Engine May Emit Loud Bangs)

Not to be confused with aerodynamic **Stall**.

A situation produced by the abnormal flow of air through a turbine or jet engine, caused by several factors, including ingestion of birds and side slipping. Sometimes it results in loud bangs and flames coming out of the back frightening some passengers.

While there is usually no damage to the engine, it can sometimes flame out, be irretrievably damaged, or even catch fire. Some stalls affecting the entire engine are referred to as compressor surges. Though somewhat frightening for passengers, they are usually not serious but can result in major damage to the engine.

The usual procedure is to reduce thrust on the engine in question until the trouble stops. Sometimes the engine can recover and be powered up again. If possible, the pilots will avoid shutting down the engine completely just in case it is needed—in our air crash book there is the account of the Kegworth crash* in England where the crew shut down the good engine by mistake and, flying too low to restart it, crashed on a motorway embankment just outside the airport.

Common Traffic Advisory Frequency (CTAF)

A frequency used by pilots at non-towered airfields (because they are so small or shut down at night, for example) in the US and Australia so they can converse with other aircraft taking off or landing there and avoid collisions.

Sometimes it shares the frequency with UNICOM for contacting airfield staff regarding services.

These frequencies are shown on charts.

Commonality

Commonality is the concept whereby the manufacturer uses not only the same parts but also whole sections of the airframe over a whole range of different aircraft.

Computer-Aided Design (CAD)

Aircraft manufacturers depend on computers for much of their design work.

Less well known is the role computers play in predicting the behavior of the materials making up the aircraft and engines, saving considerable expense carrying out physical trials and theoretically improving safety. They have vast databases showing how metals for example behave, something not available to the makers of the de Havilland Comet, the world's first jetliner, which crashed due to metal fatigue.

These programs work well, except when over-extrapolated, as in the case of the Space Shuttle Challenger disaster, where foam cladding breaking off from the fuel tank during launch fatally damaged the protective tiles making up the heat shield. When the shuttle reentered the Earth's atmosphere at high speed several days later, the hot gases generated by the friction penetrated the hull where the tiles of the shield had been dislodged due to impact from the foam. Disaster followed.

The Boeing computer program created and calibrated to predict the effect of tiny *particles* on the tiles had been over-extrapolated (not by Boeing) to mis-predict the effect of large chunks.

Concuspuity, Electronic Concuspuity (EC)

A new term referring to electronic techniques to make aerial vehicles visible to other craft in uncontrolled airspace and thus prevent collisions.

The specifications are still being worked on.

Configuration

Describes the setup of the aircraft as regards deployment of flaps, slats, undercarriage, and so on.

A "change-of-configuration stall" is a stall that occurs without a change in airspeed simply because the flap or slat settings are (inadvertently) changed.

See **Clean Configuration**.

Confirmation Bias

Situation where the pilot selectively picks out clues to confirm what they believe and dismiss any evidence contradicting it.

This has happened in many other situations, and notably in the Kegworth Air Disaster* in England where the pilots mistakenly shut down the good engine and found they could not restart it before crashing just short of the runway. In that case, the captain was about to review the situation but was interrupted by ATC.

Apart from being a lesson in confirmation bias, it led to ATC being taught not to distract pilots in emergency situations with nonessential calls and demands to change frequencies.

Decisions should always be reviewed. Confirmation bias has led to pilots landing at the wrong airport where two airports are situated close together.

Connectors (Airlines Based on Strategic Hubs)

On June 3, 2010, *The Economist* called Emirates, Qatar Airways, and Etihad "Super-Duper-Connectors from the Gulf," saying they were exploiting their hubs and establishing themselves as major players, even threatening the predominance of erstwhile flag carriers, such as British Airways and Air France–KLM.

Now, five years on, the US majors are trying to persuade the US administration to limit the operations of those carriers allowed under the Open Skies agreement, alleging unfair competition, which those connectors dispute.

While airlines such as Emirates relying on their Middle East super hubs still seem to be doing well, other hubs in cooler places where aircraft can take off with their engines less stressed may prove better placed. Aircraft such as the Boeing 787 and Airbus A350 can reach them from far away with passengers and fan out on smaller aircraft from there. One such is Istanbul served by Turkish Airlines with flights to and from Australia's East coast. It is already very busy.

Contaminated Runway

The term used to describe a runway where something on the surface prevents the tires getting a grip. That something can often be ice, but a deep layer of water can be equally pernicious, since at high speeds the aircraft tires will aquaplane, making the brakes virtually useless, as happened in the case of the 100-mph Qantas overrun at Bangkok* in 1999, fortunately with no fatalities thanks to the soft water-laden ground helping slow the aircraft.

Contested Airspace

US military aviation, including special ops operations using rotary craft, is having to reckon with operating in contested airspace at greater distances.

Contrails/Vapor Trails

Vapor trails formed way behind the aircraft due to the added water produced from the burning of the fuel, pushing the humidity of the air over 100 percent, so that water droplets and ice crystals are formed.

These differ from the vapor trails seen coming off the flaps and wingtips when aircraft are landing or taking off in extremely humid conditions. These are simply the result of condensation due to the cooling brought about by the drop in pressure as air compressed under the flaps comes out from under them, and from pressure drops occurring in the wingtip vortices.

Now that aircraft including business jets fly higher there are concerns their contrails last longer and by reflecting heat back to the ground contribute unduly to global warming. Hydrogen-powered aircraft will produce water and hence contrails meaning they might not be so perfect environmentally as envisaged.

The European Commission is putting forward a Monitoring, Reporting, and Verification (MRV) for non-CO_2 emissions (contrails) with its scope limited to Europe much to the disgust of budget carriers such as easyJet, Ryanair, and Wizz Air. They maintain that studying long-haul flights over the North Atlantic would be much more meaningful.

Control Area/Zone

"Control area" is airspace where all aircraft are subject to ATC control. "Control zone" is similar except that it applies specifically to a designated zone encompassing an airfield and its environs.

Controlled Airspace

The ICAO defines classes of airspace from A to G, with A being the most restrictive as regards flight rules, and with F and G being essentially unrestricted airspace. It has been suggested the number be reduced to three, namely those roughly equivalent to the current C, E, and G. In fact, a number of countries, including the US, work with a reduced number. Finally, there is special use airspace (SUA), where the letter-based classifications may still apply but where only certain aircraft, say military, can operate.

CONVAIR

The company was formed in wartime in 1943 by the merger of Consolidated Aircraft and Vultee Aircraft. In 1953, it was purchased by General Dynamics and operated as its Convair Division. In 1994, most of the company was sold to McDonnell Douglas and Lockheed, and what remained shut down two years later.

It had a promising beginning producing the Convair B-36 Peacemaker, the Convair B-58 Hustler strategic bombers, and other military aircraft. It also produced Atlas rockets for Project Mercury.

Venturing on the civilian side, it developed airliners such as the Convair 880 and Convair 990, which though not profitable led to profitable subcontracting work producing fuselages. In this respect General Dynamics seemed to be calling the shots, and its Convair division was doing the work. This was true at least in the case of the cabin for the DC-10 where General Dynamics was not happy with the cabin floor to which the hydraulic lines and other services were attached. Decompression could arise not only from the notorious locking mechanism for the cargo hold door that led to the DC-10 disaster near Paris but also other causes including a terrorist bomb. However, it was a tricky situation, for General Dynamics was engaged in delicate negotiations with McDonald Douglas, and it was feared that making too much of the problem could make it seem responsible for any disasters even years ahead. In the end, General Dynamics did not officially convey its doubts to McDonald Douglas.

~ Convair 880 and 990, 1960/65 and 1961/37

The Convair 880 was a jet airliner designed by the Convair Division of General Dynamics to compete with the somewhat larger Boeing 707 and Douglas DC-8. General Dynamics terminated the project after building only 65, losing an enormous sum of money and demonstrating how risky the creation of airliners can be. The company also produced a stretched version, the Convair 990, which also failed to sell as hoped, with only 37 built, thus compounding the loss made on the 880.

The Convair 990 even to today is notable for being the fastest subsonic airliner ever. In fact, its sales point was to be its promised speed compared to the Boeing 707 and Douglas DC-8. Convair had great difficulty tweaking the design to achieve not quite that promised speed, for drag increased sharply in the transonic zone. Passengers were not willing to pay extra to compensate for the greater fuel consumption and sales never took off. The Convair Division withdrew from the construction of airliners and concentrated on subcontracting and work for NASA.

Coordinated Flight

Simply means flying the aircraft with appropriate banking so that it feels as if it is level when turning—somewhat like a car going around a corner with the road cambered perfectly for the given speed. This is achieved mainly by the correct coordination of rudder and aileron movements but also partly by applying some elevator to stop the aircraft pitching down and sinking due to the banking.

A balance indicator, operating somewhat like a spirit level, enables the pilots to achieve this. Pilots will generally use the autopilot to make changes of course perfectly, with the passengers not noticing anything unless looking out of the window.

Coordinates

The geographical position on the Earth's surface, or above that point on its surface, usually expressed in terms of latitude (degrees and minutes of arc north or south of the equator) and longitude (degrees and minutes of arc west or east of Greenwich, in the

UK). Note that one nautical mile equals one minute of arc. (Therefore, the circumference of the Earth is 360 × 60 = 21,000 nm.)

Though the entering of coordinates into navigation systems enables pilots to reach their desired destination or waypoint almost effortlessly, there is the danger, just as with motorists' satnavs, that erroneous coordinates may be entered, with disastrous results. Thus, airlines insist pilots cross-check when entering coordinates and use other means en route to confirm they are where they should be, unlike the two Korean airliners* shot down over Russia.

Coriolis Effect

It is a myth that the Coriolis effect makes water flowing out of the plughole in your bath rotate in a clockwise direction in the Northern hemisphere and anticlockwise in the Southern hemisphere. However, this is because on such a small scale the effect is too weak to overcome other factors such as the fact that the water would be swirling in one direction in the first place and the topography around the plughole. However, if one fills an elevated large-diameter kids rubber pool outdoors on a support so the bottom is perfectly flat and leaves the water for 24 hours to settle after filling, and releases the water via a tap in the pipe well away from the pool, the Coriolis effect can be demonstrated—clockwise in the Northern hemisphere!

The Coriolis effect is why the prevailing wind in the Northern hemisphere's middle latitudes is westerly—from West to East.

Tropical storms seem to rotate anticlockwise in the Northern hemisphere and clockwise in the Southern because they depend on air sucked up into the eye.

See **Prevailing Wind**.

Corruption—an Ever-Present Danger?

In the US, lobbying is acceptable, and airlines and defense contractors spend enormous sums on it. (See **Lobbyists**.)

One could even say the influence of senators and representatives in Washington has resulted in too many flights to

Washington's Ronald Reagan Washington National Airport, and perhaps the collision there between a military helicopter and an airliner in 2025.

Since airliners are now usually certified by reliable bodies, the choice of one over another is not normally a matter of life or death. However, the choice of something as simple as a component might be.

Boeing's undue influence over the FAA obtained by arranging jobs for officials with associated companies on retirement might have led to the two 737 MAX disasters, something that has notoriously backfired, with the FAA becoming extra rigorous to restore its tarnished reputation meaning Boeing is having to wait very much longer to get aircraft certified.

On another level, there was the recent scandal in Pakistan where it was found many Pakistan Airlines pilots had obtained their pilot's licenses fraudulently resulting in the airline until recently being banned in Europe.

While spare parts for airliners need to be certified and approved, some companies have been supplying illegitimate ones.

Finally, one must remember corruption is a two-way street in that there is the company offering the bribe—in some countries it is expected or demanded—and the airline personnel, government officials, or military officers receiving it.

Corrosion

With some airliners being 20 or 30 years old, corrosion—or rather, checking for corrosion—can be costly. There is also the phenomenon called *stress-assisted corrosion* whereby the flexing of the material produces microscopic cracks on the surface, which increases the rate of corrosion.

In the case of metals, electrical testing methods can be used to detect potential points of failure. However, these methods do not work with composites, unless conductive material is incorporated.

Cospas-Sarsat (Satellite-Based Search and Rescue)

International program with satellites picking up signals emitted by beacons indicating people in trouble on land, at sea, and in the air. See **Emergency Locator Transmitter, Distress Tracker**.

Cost of Fuel for Airliners

Overall, with more efficient aircraft and particularly engines, the cost of fuel represents less than 20 percent of the cost of a ticket, though it could increase to 30 percent for ultra-long-haul flights. However, when profit margins are tight it can make the difference between a route being viable or not.

Cost of Weapons, US Aircraft and Weapons Too Expensive

Hypersonic glide vehicles and many other missiles are incredibly expensive and for the cost of just one launched, hundreds of drones could be launched in a swarm with a good likelihood of one or several reaching the target and able to detonate their ordinance.

One tactic is to trick the enemy into using up their expensive and difficult to replace weapons and missiles before major combat. Possibly the far greater cost of producing weapons not to mention ships in the US as opposed to China is going to leave the US at a disadvantage. In China most of the companies making the aircraft and weapons and their components are state owned, whereas in the US, shareholders and senior executives must be paid on top of manufacturing. Also, salaries are lower.

Already, thousands and thousands of drones are being used by each side almost every day in the Ukraine. Some on the Ukraine side are even being made cheaply in garages using cheap components from China and Taiwan. The parts from there can so cheap because of the of the volumes also produced for civilian use.

While the US is blocking supplies from China to produce them domestically. The volumes without the export markets may not be enough.

CPDL (Controller–Pilot Data Link)

CPDL uses VHF and SATCOM and ACAS to transmit air traffic control clearances and messages from the ground to the aircraft, and vice versa. There are monitors in front of each pilot, and a printer. It may feature the ability to input a new route received through CPDL straight into the FMS rather than print it out and then retype it. See **Clearance**.

It is claimed that air traffic control in the US is antiquated and reliable CPD should be used much more for communications. Voice communications can get garbled or indistinct with the risk of people talking over each other with an urgent instruction not received at all.

CPL (Commercial Pilot License)

Basic license required to pilot an aircraft for hire. It is issued for the different categories of aircraft and, importantly, with various ratings specifying what the pilot is allowed to do.

Passengers have sometimes been alarmed to hear flight attendants say a pilot cannot land the aircraft at such and such a place when all they mean is that they are not certified for landing in such weather conditions.

Except for those trained in the military, pilots generally obtain a Private Pilot License (PPL) first. Note that in the US the term "certificate" rather than "license" is used. The highest level is the Air Transport Pilot License (ATPL) or in the United States the Airline Transport Pilot (ATP) certificate. The latter is now mandatory for all pilots in the US, not just captains.

Crabbing (Landing Crabwise)

When landing with a significant crosswind, the pilot points the nose a little into the wind so as not to be blown to one side in a move called "crabbing." In effect, the aircraft is tracking in a straight line down the runway but with the airframe and wheels at an angle to it. At the last moment, the pilot will straighten up so as not to damage the undercarriage when the wheels touch the runway. The pilot may also dip the wing into the wind to produce a sideslip, which is another way of compensating for the crosswind. The pilot would then level the wings at the last moment before touchdown. In most cases it is a combination of these two maneuvers.

Crew Alerting Systems

EICAS (Engine Indicating and Crew Alerting System) [Boeing]

ECAM (Electronic Centralized Aircraft Monitor) [Airbus]

These are electronic systems that alert pilots to abnormalities or problems, even recommending steps to be taken, and bringing up (displaying) the relevant checklist to deal with the problem.

The systems can prioritize problems and have many advantages leading the FAA to mandate that aircraft certified after 2022 be fitted with them. However, this has posed problems over the certification of the Boeing 737 Max 10, with the FAA finally making an exception for the 737 after airlines and unions had argued that with so many in use, having pilots switching daily from model to model including the 737 MAX 10, there was the danger of them getting confused and making mistakes.

Criminalization (Antithesis of No Blame)

Most aviation accidents can be attributed to a whole host of factors. Even a blatant error by a pilot may be attributed to failures in training and even their selection or retention after lapses in the first place. The accepted view in the industry was always that causes—and particularly potential causes—should be sought rather than blame attributed, since this was the only way that pilots and mechanics would come forward, and the truth be revealed.

Recently, there has been a tendency for judiciaries to seek criminal prosecutions, which many in the aviation community consider counterproductive. The French always launch a parallel judicial enquiry, which sometimes reveals surprising facts kept from the public. For instance, the fact that the captain of Air France Flight **AF447 ONLY A MINUTE TO SAVE THEM** that crashed into the South Atlantic in 2009* said that *he had only had an hour's sleep and that it had not been enough.* He had expected his admission to be erased as the cockpit voice recorder was on a two-hour loop to protect privacy. The BEA, the French aviation investigators, had concealed this, considering it to be a private matter!

CRM (Crew Resource Management)

The idea of using techniques from business management to govern the way pilots interact was formally established as a mantra for pilots by United Airlines following an incident at Portland in 1978* in which the crew of a DC-8 were so preoccupied by a possible but nonexistent problem with the undercarriage that they failed to notice they were running out of fuel.

This was despite the crash six years earlier in 1972 of Eastern Air Lines Flight 401* at night in Florida's Everglades where the crew were preoccupied by a failed indicator bulb meant to show whether the front undercarriage was down and locked. The captain had nudged the control column so hard on twisting to reach the bulb that the autopilot passed vertical control to him without him or the first officer noticing. With everyone preoccupied, they sank lower and lower until striking the swampy ground with its snakes and crocodiles.

Originally CRM stood for "**cockpit** resource management," but it was subsequently widened to mean "**crew** resource management" to imply it included all concerned—flight attendants, mechanics, and so on.

The idea is that each person has their designated task, for instance, at least someone should be flying the aircraft while the other pilot seeks the cause of the problem, and each person can have their say. It is applicable in many other domains, and especially medicine where juniors now dare to speak up when concerned about what the very senior head surgeon is doing.

Recently, psychiatrists and psychologists have had input, saying CRM should not just be a simple matter of good procedure with emphasis on everyone being able to speak up but also have a mental health component resulting in everyone being rendered in a better mental condition and therefore more likely to act appropriately. They claim that pilots, partly due to the nature of their work, suffer more from depression than the population in general and suitably tailored CRM could be a great help. Interestingly, military pilots suffer less from depression.

See **Mental Health of Pilots**.

Croc in Carryon Luggage

A small aircraft inexplicably crashed in Africa in 2010 killing all but one of its 21 occupants. On recovering, the sole survivor came up with a reason for the crash that no one could have ever imagined.

A crocodile hidden in someone's hand luggage had got free in the cabin causing the terrified passengers to flee to the front. This shifted the center of gravity so far forward that the aircraft became uncontrollable and crashed (source: @andyyjiang).

Crossing a Runway

Anyone, even a controller, can make a mistake.

When taxiing and having been given permission to cross a runway, savvy pilots usually look both left and right to make quite sure no aircraft is about to take off or land.

Cuban Missile Crisis

When the Russians attempted to place long-range missiles in Cuba just off the US coast most US military chiefs recommended bombing and invading the island. What they did not know was that there were already short-range nuclear tipped missiles set up there and that they would no doubt be used in the event of an invasion. This could well have escalated into nuclear exchanges with disastrous consequences. Many officials were surprised the "young" president stood up to the "older" senior men.

Cylindrical Fuselages

Though it is not best for giving maximum width to fit passengers in comfortably, fuselages are mostly perfectly cylindrical as that means there is no point where stress is concentrated as the cabin expands and contracts when the aircraft ascends and descends.

~ D ~

Dado Panels in Cabin Floor and Skirtings

These are grills situated mainly on the cabin floors and skirtings to allow air to pass through should there be a great differential between the air in the cabin and the space below. They usually have weak plastic shutters that only open when there is a significant pressure difference, meaning passengers are not subject to draughts through them.

They were introduced after the crash in 1974 of a DC-10* that had taken off from Paris where the cargo hold door opened due to poor design of the locking mechanism. The difference in air pressure between the passenger cabin and the hold below caused the floor to buckle downwards, severing the hydraulic and control lines attached underneath. This rendered the aircraft uncontrollable, resulting in it crashing into the forest below minus six passengers already blown out through a gap in the floor.

Dark Star

See **SR-72**.

DARPA (Defense Advanced Research Projects Agency)

This US agency to promote and encourage technologies with possible military applications was created in 1967 by President Eisenhower, reacting to the shock of the Soviet Union launching the first satellite, Sputnik 1. Another reason was that Eisenhower was incensed by duplication on projects due to each branch of the military wanting a share.

Wernher von Braun was shortlisted to head the agency but deemed unacceptable as he insisted on bringing in fellow Germans (also brought to the US under **Operation Paperclip)** tainted even more than he by their Nazi past.

Though much of DARPA's work is kept secret for many years it has been responsible for the promotion and development of advanced technologies including the Internet, RISC computing,

GPS, stealth technology, UAVs, and micro-electro-mechanical systems (MEMS), which are now used in everything from air bags to ink-jet printers to video games like the Wii. Its achievements are too many to list even roughly, though a number are mentioned elsewhere in this book.

Though applications were originally for the US military, the agency's advances have played a central or partial role in launching sizable industries.

Data Mining

Data mining is the sifting by computer of large quantities of data using algorithms to pick out trends, abnormalities, or signs of specific outcomes. It is used in all sorts of domains, including intelligence gathering such as TIA (Terrorist Information Awareness), business, and industry.

In aviation, it has shown its worth as a means of predicting failures in the absence of obvious signs of anything being amiss.

The greatest problem with data mining is that little details, insignificant on their own but revelatory in the context of hundreds of aircraft or events, may not be thought worth recording in the first place. This may be particularly true where human factors are involved. Unlike engineering and maintenance data, human reactions and actions are not usually recorded and therefore cannot be automatically downloaded.

D. B. Cooper Mystery Where Hijacker Parachutes from Aircraft

On November 24, 1971, at Portland Airport in the United States, a 40-ish man with receding hair boarded a one-third-full Boeing 727 under the name Dan Cooper for a half-hour flight to Seattle. Once in the air he passed a note to a flight attendant sitting in a nearby jump seat. She ignored it as an attempt at a pass and dropped it unopened into her purse. He then leaned over and told her she had better read it as he had a bomb, as said in the note, and she should come and sit next to him.

Through her he conveyed his instructions to the pilots, saying he would release the passengers at Seattle if the airline paid $200,000 in ransom money in unmarked notes and provided four parachutes (two primary and two reserve). Throughout, he was calm and gracious, even ordering a second bourbon and telling the flight attendant to keep the change.

Though the banknotes he was given were unmarked, they were all issued in San Francisco and thus identifiable. After taking off with a skeleton crew en route for Reno (Nevada), where the aircraft was supposedly to refuel before flying on to Mexico City, they were told by Cooper to fly at approximately 100 knots with the gear down with 15 degrees of flap, and no higher than 10,000 feet.

With the crew including the one flight attendant told to stay in the cockpit, Cooper soon parachuted out using the rear airstairs—an almost unique feature of that airliner. Two fighters trailing the 727 could not see where he had jumped, since it was nighttime and there was a storm, making the subsequent search for him or his body very difficult, even though considerable resources were devoted to it.

Despite the offer of a considerable reward, none of the notes were ever found in circulation. In 1978 a placard containing instructions for lowering the aft stairs of a 727 was found by a deer hunter within the purported path of the 727. Then in February 1980, an eight-year-old boy digging a fire pit on a riverbank found three packets of the cash, amounting to $5,800, with the rubber bands still on them. Nothing else was ever found, and though many suspects were identified, DNA and other evidence on the clip-on tie left on the aircraft made the FBI discount them.

The man and the mystery are referred to as D. B. Cooper because that was the name of the first suspect cited, and the press latched on to it.

In some ways Cooper appeared very astute. For instance, by asking for parachutes for two people he gave the impression he might jump with a member of the crew and thus dissuaded the FBI from sabotaging the chutes. Also, by choosing the day before Thanksgiving, with a four-day weekend ahead, he could get back home and return to work with no one noticing his absence. In other ways he seemed less astute—for example, by choosing an obviously dummy parachute as his reserve.

In the area where he is said to have jumped, D. B. Cooper became something of a folk hero, with D. B. Cooper trinkets on sale, and this is possibly why the piqued authorities put so much effort into trying to find out who he was and what happened to him.

Following that incident, checks started being made of passengers' luggage, spy holes were put in cockpit doors so pilots could see what was happening in the cabin, and a system was implemented to stop the airstairs on the Boeing 727 being deployed in midair.

DCDU (Datalink Control and Display Unit)

Digital lnk between pilots and air traffic control, allowing them to receive important updates and instructions, and not transmit information, such as situation and flight plans but also get approvals. Particularly useful on oceanic routes where normal VHF communications not possible.

It has a display screen and controls, and may be linked to a printer.

Defense Contractors

"Western" defense contractors in order of volume of arms sales:

1. **Lockheed Martin**
2. **Boeing**
3. **BAE Systems** [UK]
4. **Raytheon**
5. **Northrop Grumman**

Deadheading

Refers to aircrew and cabin crew traveling on a flight to reposition themselves to take up duties elsewhere. Not to be confused with off-duty personnel enjoying the free or discounted air travel perquisites of working for an airline.

Although the analogy is not strictly correct, the term "deadheading" is said to be derived from the practice in the theater world of filling up empty seats with nonpayers (called deadheads) to give a better ambience. The term is also applied to airport tractors without trailers and even empty railway trains being repositioned.

Deadheads normally travel in the passenger cabin as passengers since the jump seat in the cockpit is quite uncomfortable and would only be chosen were no seat available.

In the movie *Catch Me If You Can* the character pretending to be a Pan Am pilot did not know what "deadhead" meant.

Dead-Stick Landing

Landing with no power available from the engines. The term evokes the leaden feeling of the controls (joystick in the old days).

Death on Board

People can die anywhere when traveling, and do so in trains, buses, and cruise ships. Aircraft are no exception. While captains consult specialist medical advisors on the ground and any doctors or nurses on board regarding the need to divert when someone is taken ill, when someone has died there is no point, and indeed it may be better for all concerned to continue to their destination where there is a greater likelihood of there being family and friends, not to mention fewer problems over documentation such as visas.

Singapore Airlines did have a special facility for a body on its original world's longest A340 flight from Singapore to New York but usually the airlines simply try to handle the matter discretely—often moving the deceased to Business Class if there are lie-flat seats and covering the body as if asleep.

One passenger traveling First Class on British Airways did not really mind a dead body being placed near him but found the nonstop wailing of the female Indian relatives who had come to pay their respects excruciating. When he later asked the airline for at least some compensation for the thousands of pounds he had spent to enjoy peace and quiet, BA claimed he lacked respect for the dead!

If the aircraft is from an area, say, with Ebola, a death on board can result in scary medical personnel dressed in bio-hazard suits coming on board on landing. This is usually the authorities just being careful, with people allowed to proceed after giving their contact details and itinerary so they can be contacted if there is a problem. Recently, people without certain immunizations such as for measles—the most contagious but not the most dangerous virus—have caused concern when coming from areas where there have been outbreaks.

Deceleron

A control surface combining air brake and aileron functions, with the panels split into sections.

Decision Height

When aircraft are landing, they come to a point—or rather, height—at which the pilots must decide whether to proceed or engage in a missed-approach procedure. The latter might be because they cannot see the runway or find an aircraft or vehicle is still on it. Each runway will have its standard missed-approach route to avoid the risk of the aircraft subsequently wandering into the paths of aircraft taking off or hitting obstacles.

Decompression (Loss of Cabin Pressure)

Depending on the rate at which the air pressure inside the cabin falls, decompression can be gradual, rapid, or explosive, again depending on the height of the aircraft and the nature of the failure that allowed air to escape. As an aircraft can never be hermetically sealed—the seals around the doors are never perfect—cabin pressure can progressively drop if the pilots by mistake do not set the pressurization to automatic. This happened once with the pilots not realizing their mistake because the warning was the same as for something else. They lost consciousness, and the aircraft flew on to its destination on autopilot, running out of fuel and crashing.

As narrated in our *Air Crashes and Miracle Landings*, lessons have been learnt from some notable explosive decompression incidents, such as:

1. The **de Havilland Comet***, the world's first jetliner, where the fuselage failed, and three aircraft broke up mid-air.

2. The **Turkish Airlines DC-10***, where after taking off from Paris in 1974, a cargo hold door blew out producing such a great difference in pressure that the cabin floor buckled downwards, severing the hydraulic control lines attached underneath it and making it impossible to control the aircraft.
   A similar thing had already happened to a DC-10 in the US, but in that case the cabin floor had been specially reinforced to support a piano for a special event, and the floor did not buckle enough to sever all three hydraulic lines, and the pilots were left with just enough hydraulic control to land.

3. The **Japan Airlines 747** where, in the worst-ever single-aircraft disaster, in 1985, the rear bulkhead, which had been improperly repaired by the maker's very own engineers following a tail strike on landing a few years earlier, failed, blowing off most of the tail. The uncontrollable aircraft flew around drunkenly for 30 minutes, with the pilots trying to maneuver it by adjusting engine power while passengers wrote their wills on their boarding passes*.

4. The **Aloha Airlines 737 in 1988*** where a weak spot in the roof of the passenger cabin split open with the panel around it, and specially designed to let the air out safely, blowing out. Unfortunately, a female flight attendant standing just below was sucked upwards into the aperture and momentarily blocked it like a

plug in a bath, with the shockwave in the air blowing off the roof. The photo below shows the aircraft having landed with the passengers in the forward section sitting in the open as if on an open-top sightseeing bus.
With no leveraging tensile force from the cabin roof, the 737 had been flexing with the cockpit dipping as if it might break off. Only the rails along the cabin floor for attaching the seats held it together!

ALOHA AIRLINES INCIDENT

Nowadays, some simple measures to equalize pressure have greatly reduced the potential consequential risks associated with explosive decompression. These include incorporating vents and pressure-sensitive relief panels between compartments, and in the floor between cabins and the cargo holds below.

Decoys

An aircraft can launch decoys or even tow them so as to trick notably a missile into attacking them instead.

De-icing (Anti-icing)

Even a slight layer of snow or ice on a wing can change the way it behaves and drastically reduces the amount of lift it provides. As a result, the aircraft may have to be de-iced by spraying with a hot de-icing fluid before takeoff, and it may even return to be de-iced again should the wait in the queue for takeoff in cold, snowy conditions be prolonged. Though their essential ingredient is the same (ethylene glycol or, more usually, the less toxic propylene glycol), de-icing fluids often contain thickening agents so that they form a gel on the treated surfaces for longer protection, in which case they may be called anti-icing fluids.

Once the aircraft is airborne and fast moving, vulnerable and critical areas can be warmed enough by various means to prevent icing. Some scientists believe studying how insects keep their backs clean may provide ideas for ways ice could be more easily shed from the wings and control surfaces of aircraft.

Delamination of Stealth Coating above Mach 2

The top speed of some stealth aircraft was limited to just above Mach 2 because heating due to friction with the air risked causing delamination of the stealth coating.

Demonstrator vs. Prototype

With reference to the 6th Generation Fighter and similar projects, reference is frequently made to the supposed creation of a *demonstrator* using computer design techniques, speeding up the process.

A demonstrator is not a prototype, but an aircraft or, say, a drone to evaluate a new concept or technique. Boeing created the **367-80 (Dash-80) 1954/1** from which the Boeing 707 evolved. It also demonstrated to other manufacturers, and particularly Douglas, the viability of such a jet airliner.

"Densification"

The term for airlines cramming more seats into the cabin to make more money. This policy can include having smaller toilets and reduced galleys. Even legacy carriers are doing this.

Denver International Airport, Longest Civilian Runway in US

Opened in 1995, its $4.8 billion construction cost was almost three times the original estimate. It has the longest publicly used runway (16R/34L) in the US with a length of 16,000 feet (3.03 miles; 4.88 km). However, its elevation of 5,434 feet (1,656 m) coupled with often high temperatures is the reason for this exceptional length as aircraft require extra distance to take off when it is "high and hot." Some other runways are long, too.

It is the third or fourth busiest airport in the US and could see its six runways eventually be doubled to 12.

Depleted Uranium (DU)

With a density of more than one and a half times that of lead, depleted uranium is sometimes used where extremely high density material is required.

Apart from its well-publicized military use in munitions by virtue of its weight, and in armor-piercing rounds by virtue of keeping sharp when disintegrating and catching fire once inside the tank, its civilian uses range from radiation shielding in medical equipment to providing ballast for the keels of sailboats.

In airliners depleted uranium has been used as a counterweight to attenuate vibrations in certain parts of aircraft. Early Boeing 747s in the 747-100 series could have as much as one and a half kilos (three pounds) of depleted uranium, much of which was incorporated in the outboard engine nacelles to prevent a nasty flutter at high speed.

Designations of US Military Aircraft

With many very significant bombers in World War II made by Boeing, such as the B-17 and B-29, the "B" in their designation seemed to mean Boeing rather than bomber. Occasionally, bombers such as the F-117 Night Hawk have been given an "F" because it was feared pilots would think it demeaning to fly a "B" (bomber).

Here are the main designations based in part on http://www.globalaircraft.org/definitions.htm:

A (Attack)—e.g., A-10 Warthog

B (Bomber)—e.g., B-17 Flying Fortress (WWII)

C (Cargo transport)—e.g., C-5 Galaxy

E (Special electronics)—E-3 Sentry

F (Fighter)—e.g., F-15 Eagle

O (Observation)—e.g., OA-10 Thunderbolt

P (Maritime patrol)—e.g., P-3 Orion

However, in World War II, US fighters had the prefix P for "Pursuit" rather than F—e.g., P-51 Mustang

R (Reconnaissance)

S (Antisubmarine warfare)—e.g., S-3 Viking

SR (Strategic Reconnaissance)—e.g., SR-71 Blackbird

T (Trainer)—e.g., T-38 Talon

U (Utility)—e.g., U-2 Dragon Lady

X (Experimental/research)—e.g., X-15

D (Drone control)—e.g., DC-130

V (VIP/staff transport)—e.g., VC-137C, a presidential C-137C transport

H (Search and rescue)—e.g., HH-60 Jayhawk

K (Tanker)—The K comes from "Kerosene"—e.g., KC-135 Stratotanker

L (Cold weather)—e.g., LC-130H, a ski-equipped C-130H

M (Multimission)—e.g., MH-53E Sea Dragon

Q (Drone)—e.g., QF-106A, a remote-control-equipped F-106A

W (Weather observation)—e.g., WC-130J, a weather reconnaissance C-130J

A letter suffix may specify a particular variant of an aircraft, such as the F-16A, F-16B, F-16C, and F-16D Falcon.

Detent

When pilots are manipulating cockpit controls, such as the throttles, their task is facilitated by having them snap into certain positions without the operator having to align them painstakingly.

These are called "detents." Throttle detents include *idle detent* and *takeoff detent*.

"Detent" should not be confused with the political term "détente."

Deviation, Errors Made by Pilots

The rather innocuous-sounding word "deviation" can cover a multitude of sins ranging from a landing approach seriously below or above the glidepath in the air, to failing to hold short of an assigned clearance limit on the ground.

Pilot deviations, including ones that must be reported to the FAA, can occur in several different ways:

1. Airborne deviations
   can result when pilots stray from an assigned, heading, altitude, or instrument procedure, or if they penetrate controlled or restricted airspace without ATC clearance.

2. Ground deviations
   can happen while taxiing, taking off, or landing without clearance, deviating from an assigned taxi route, or failing to hold short of an assigned clearance limit.

"DF" Chinese Missiles

Chinese missiles are often called "Dong Feng" meaning "East Wind," hence the terminology DF-xxx.

"Dial a Yield," Especially for Tactical Nuclear Weapons

A wide range of yields can be selected (set) according to need. For instance, a low yield causing little destruction would be used to clear ground ahead of troops.

DIED WAITING FOR HELP

Lives can be lost in a crash not only because passengers collect their luggage and obstruct people behind them, but also because responders take too long to reach the crash site.

Sometimes it is because the wreckage cannot be located or even if found cannot be accessed due to it being at an isolated spot, say in tree-covered mountains where helicopters cannot land at night.

Search aircraft may mistakenly report there could be no hope of survivors with authorities hesitant to put the lives of the rescuers at risk in dangerous terrain at night when it is deemed pointless.

Here are three cases where survivors died due to delay but were also notable for other reasons:

~ Girl Survived 2-Mile Free Fall in her Seat and 10-day Hike

Lansa Flight 508* (Peru 1971)
Death toll: 91, Ultimate survivors: 1

The Lockheed Electra, broke apart in a terrible storm over the Peruvian jungle with a teenage girl called Juliane Koepcke surviving after having been thrown out still strapped in in her seat and falling from a great height of perhaps some 2 miles (3 km).

Being at the end of her row of seats, her fall had been slowed by it spinning like a mulberry leaf (which takes time to reach the ground). Furthermore, her landing was softened by coming down on high treetops intertwined with foliage leading down to the ground.

When she regained consciousness, she found she had a broken collar bone, a nasty gash on her arm, and without her glasses, which was a particular concern as she was very shortsighted.

She looked in vain for other survivors and then sought water, for her father had told her the way to find people when lost in the jungle was to follow water downstream. First it was a stream, then a large river down which she half swam, and half floated despite there being caimans (a type of crocodile).

After ten days, and semi-conscious, she came upon a hut on the bank and sought refuge there. When three woodsmen later returned, they thought she was the goddess said to haunt the area, for such was her emaciated form and her completely bloodshot eyes caused by the decompression. On hearing her explanation about the crash that they knew about, they tended to her and took her by boat to civilization where she could get more care.

She was finally flown by a light aircraft to somewhere with doctors, and on recovery was able to tell officials where the aircraft had come down.

By the time search teams reached the crash site, some 12 days after the crash, the bodies of at least six likely initial survivors including perhaps that of her mother were found. It was difficult to be quite sure what state they had been initially for they had been partially dismembered by jungle predators and ravaged by bugs, with them also having begun decomposing in the heat of the damp jungle.

[A Czechoslovakian flight attendant, Vesna Vulovic, is often cited as holding the record for falling 33,000 from an aircraft on which a bomb had exploded. Though her case was truly remarkable and it is said she survived due to naturally having an exceptionally slow heartbeat, falling on trees and then soft snow keeping her cold. However, she had been trapped inside in NZ broken off part of the tail section and not outside in her seat like Juliane Koepcke.]

~ Survivors Ate The Dead becoming Cannibals

Chartered Uruguayan Air Force Flight 571 (Andes 1972)*
Death toll: 29, Ultimate survivors: 16

The Uruguayan Air Force aircraft with 45 passengers and crew had flown off course into a snowcapped mountain in the Andes at a height of 11,710 feet (3,570 meters) on October 13, 1972.

Most of the occupants were members of a sports team and physically very fit.

Twelve people were killed on impact, soon followed by several more on account of their injuries and the freezing temperatures.

As the wings had broken off, search aircraft overflying the wreckage could not pick out the white fuselage against the snow, and after eight days the search was called off, something that those still alive learnt over their battery-powered radio.

They became ever more desperate as they nibbled at anything such as bits of leather resembling food, as people continued to expire. Finally, after much deliberation and soul-searching, someone tried the frozen part of someone long dead, with others following and later saying they would not mind being eaten if they died.

Knowing they could not stay there indefinitely, and with the weather improving, two of the fittest survivors set off—with some "human food"—for help. They climbed to the rim of the volcanic crater and hiked 38 miles (61 km) for ten days to get to civilization and set in motion the rescue of the remaining 14, some 72 days after the crash.

See *Society of the Snow* that appeared on Netflix in 2024 which shows what happened and how the moral dilemmas were handled by different individuals.

This crash is unlikely to set a precedent as these were truly exceptional circumstances—the freezing conditions, the extended period of time with people progressively dying off, the physical fitness and camaraderie of the sports team, and so on. Though those that were eaten may have been dead for days, they were completely frozen not decomposed, and not lifelike.

~ Worst-Ever Single-Aircraft Crash

Japan Airlines Flight JL123 (1985)*
Death toll: 520, Ultimate survivors: 4

The aft pressure bulkhead of the Boeing 747 jumbo specially configured to carry many passengers failed shortly after takeoff from Tokyo. The rush of air blew off part of the tail and severed all three hydraulic lines, rendering the aircraft uncontrollable other than by adjusting the thrust of the engines.

By cleverly doing so the pilots managed to keep the massive aircraft aloft, with it flying around drunkenly and passengers writing their wills and messages to loved ones on their boarding passes.

After half an hour, it crashed in an isolated mountainous area not very far from Mount Fuji, bouncing off a ridge and coming to rest further down on another. It was late evening, and search aircraft reported there could be no hope of survivors.

With the crash site inaccessible even for helicopters because of the tree-covered mountains, responders only reached the wreckage the following morning to be surprised to find four people alive some way away from the main debris in the broken-off tail section.

They were a mother and her 8-year-old daughter, a 12-year-old schoolgirl with only a strained wrist, and an off-duty flight attendant. Unable to move with a broken pelvis, the flight attendant had been fitfully aware of the sound of young children, their cries and words fading as their injuries, shock, and cold of the night took their toll.

She especially remembered how one boy said that being a man he would be brave! He and others would surely have survived if help had come sooner.

Perhaps only the US forces in Japan with their expertise of plucking soldiers from such places could have done something. The author tried asking them whether they could have helped but found it so politically sensitive that it could only be dealt with at a higher level in Washington. [The situation is somewhat different nowadays in view of the Japanese appreciation of the help the US forces in Japan gave in connection with the Great East Japan Earthquake that occurred at 14:46 JST on March 11, 2011.]

Anyway, with high-ranking officials absent during the Obon holiday it would be difficult for junior staff to seek help from the US, particularly when it would seem there could be no survivors.

****

The lesson from these three disasters and others is that one should not too easily abandon hope of survivors. Of course, there are some crashes, notably where the aircraft has plunged into the ground leaving a deep crater, where there obviously cannot be any.

Diego Garcia—US Indian Ocean Military Base

Perhaps the sole benefit for the US from the Vietnam War. Allegedly, the British government ceded use of the atoll in exchange for not participating in it. It proved especially useful for operations in the Middle East and now has bombers able to reach the South China Sea.

"Difference Training" When New Version of Aircraft Introduced

Though pilots may be able to fly on all generations of an aircraft such as the Boeing 737 there are subtle differences, and they should undergo training to learn about the differences. This is called difference training.

There was the Kegworth Disaster* where the captain shut down the good engine instead of the failing one because he believed the smell of smoke came from that one. It would have been a clever thing to do except that the new version of the 737 he was flying brought in the bleed air from the other side.

Just as they were coming in to land the operating engine failed completely with them flying too slow and too low to restart the good one, resulting in them crashing on an expressway just outside the airport.

With thorough difference training that accident might not have happened, though there were other factors responsible, such as the decision to shut down that engine not being fully reviewed due to interruption by air traffic control.

In the case of the Boeing 737 MAX, Boeing did not even in the manual mention the **MCAS (Maneuvering Characteristics Augmentation System)** responsible for the two crashes. Lion Air, the operator of the first Boeing 727 MAX to crash had requested "difference training" arrangements only to find Boeing stonewalling them lest it set a precedent, leading other airlines to want it.

Digital Radio Frequency Memory (DRFM) Systems [Electronic Warfare]

In electronic warfare (EW) the move to digitalization has led to great advances.

For instance, all sorts of information about the enemy's radar obtained on earlier missions and processed using AI can be held in a database for reference and incoming tracking radar signals can be digitized, processed and altered, and then transmitted back after conversion back to analog like the original signal. Being a coherent representation of the original signal, the adversary's radar will not be able to distinguish it from other legitimate signals and will recognize it as a real target. DRFM can be used to create false range targets both behind (reactive jamming) and ahead of (predictive jamming) the asset that it's protecting.

Dihedral

For stability, most civilian aircraft wings are dihedral—that is, their wings slope upwards as they extend from the fuselage, like a prized-open letter V.

When the aircraft rolls to one side, the lower wing is almost horizontal while the other wing points significantly upwards. The resulting difference in effective angles of attack means the lower, more horizontal wing produces more lift, and the opposite one less so the aircraft automatically rights itself. The aircraft thus has natural stability.

See **Anhedral**, the opposite.

Dimming Cabin Lights on Takeoff and Landing

This is done so people's eyes become accustomed to the dark in case there might be need for an emergency evacuation. It also makes it easier for flight attendants to see fire outside and not open the exit there.

"Direct Flight" or "Nonstop" Flight?

Until recently the term "direct flight" simply meant the flight would be going all the way to the destination cited, either non-stop or with intermediate stops, all in the same aircraft. Worse, in airline speak it could even mean the flight number remains the same throughout, with the need to change planes.

However, with Qantas referring to its planned nonstop Sunrise flights from Australia's East Coast to London and New York starting in 2026 with the A350-1000, as "direct flights" it seems one is fighting a losing a battle regarding the proper use of "direct."

Many other airlines, including Delta, refer to upcoming non-stop flights over great distances that will be possible with the Airbus A350-1000 as "direct flights.

Though "direct flight" now is generally accepted to mean non-stop, if you want to fly nonstop, make sure it is truly so by double-checking that the number of stops is zero and the overall flight time is commensurate.

Directed Energy Weapons (DEW), Ray Guns

Comic-book ray guns may finally be about to become a reality and might be used for:

1. Cheaply triggering explosions from a distance.
2. Destroying incoming artillery rounds, drones, and missiles at short range.
3. Destroying enemy missiles at long range during the boost phase when they are moving slowly and emitting easily detected heat.
4. Purportedly attacking enemy vehicles and materiel at short range on the ground—it being politically sensitive to acknowledge it would be enemy troops (who might even boil) that would be the main target.
5. Bringing down aircraft.

The more demanding applications, such as the destruction of missiles at long range, would use chemical lasers, which can be much more powerful than electrical ones.

The fact that they strike with the speed of light and the cost of each "round" is minimal compared with that for weapons using explosives would make them seem a potential wonder weapon.

However, their performance is greatly degraded by fog, clouds, falling snow, and rain, not to mention by distance. This means there must always be a backup.

See **Electromagnetic Field Generator**.

Disinsection (Destroying Insects)

Routine inflight/prearrival disinsection of aircraft, with flight attendants walking up and down the aisles spraying insecticide over people's heads, is virtually unheard of these days, apart from cases where there has been a severe outbreak of, say, malaria. However, some countries, and notably Australia, may insist on *residual-disinsection* (of the aircraft!) prior to the boarding of passengers using a residual-insecticide aerosol, plus additional in-flight treatment with a quick-acting spray shortly before landing.

Sometimes there is regular application of residual insecticide to all internal surfaces of the aircraft, except those in food preparation areas.

Tests have shown these to be harmless though in some quarters concerns have been raised that co-exposure to the tricresol phosphates found both in engine oils and many hydraulic fluids, which can contaminate cabin air, may make people unusually susceptible to the pyrethroids found in insecticides.

Dispatch

"Dispatch" is a much more important function at airlines than people realize and involves much planning and paperwork. At major airlines many people work on dispatching aircraft.

Dispatcher, Aircraft Dispatcher [US]

In the US, the dispatcher must be certified, and just like the captain has the legal right to say the aircraft cannot take off. A US dispatcher will monitor several aircraft over their entire flight.

Displaced Threshold

For various reasons—the presence of high buildings near the beginning of the runway or due to the impact of recurrent landings there over the years having weakened it—aircraft may be required to land farther down than they normally would. This threshold farther down is called a displaced threshold.

Aircraft taking off can still start their run right from the beginning of the runway and use it for rolling out after landing.

Distance to Theater of Operations

With hypersonic missiles making both **Aircraft Carriers** and islands such as Guam increasingly vulnerable, US defense chiefs are concerned at the greater range over the Pacific required for future aircraft and drones.

Distances

In drawing up flight plans, distances are measured as those at a height of 32,000 feet, considering the curvature of the Earth.

Diversion

Airliners may have to divert for technical reasons or for an extremely ill passenger.

It can prove very costly, especially on long-haul flights flying over numerous countries. Even when simply "offloading" a sick passenger the aircraft may not be able to simply fly on, for permission from the various countries might be necessary for a new flight plan, and the crew's maximum flying hours might not allow it anyway. This would mean putting passengers up in hotels. Also, there is the problem of finding flights for passengers meant to be on the return flight.

Also, if the diversion airport is one the airline does not use, it will have no staff to handle passengers and no technical people to do even the minimal servicing, and no people contracted to do these tasks.

British Airways has been known to divert a flight back to its homebase at London on finding a technical problem when almost at its destination. Not easy for passengers to understand when they have flown for ten or so hours with only one hour to go! However, replacing an engine in London is much cheaper than flying one out with technicians able to change it.

DME (Distance Measuring Equipment)

In conjunction with appropriate equipment in the aircraft, DME beacons can give pilots a readout of their distance from them and even give groundspeed and ETA.

The beacon is a transponder and replies to an interrogation by the aircraft, with the delay indicating the distance. Note that the distance indicated is the slant distance (hypotenuse). Thus, if the aircraft is right above the beacon, the distance indicated will be the height above it.

DNIF (Duties Not Involving/Including Flying)

Acronym used by the US Department of Defense for the situation where pilots are grounded perhaps for medical reasons in the very broadest sense.

"Do Not Pair" List

For one reason, or another, pilots may not get on with each other and an airline may allow them to list up to three they prefer not to be paired with. More would imply they have a problem themselves!

DOC (Direct Operating Costs)

Immediate operating costs, not considering other factors complicating the picture. The purchase of fuel is a simple direct operating cost that is easily calculated. On the other hand, in the case of maintenance there is the direct cost of performing the task and the more diffuse costs and benefits in terms of aircraft availability, reliability, and even safety, dependent on it being performed in a timely manner or even being managed so well (by sophisticated monitoring) that less maintenance is needed.

Dogfight

This term for aerial combat, exemplified by simple encounters in the Battle of Britain in World War II, derives from the fact that dogs are said to try to get on each other's tails when fighting.

The whole approach to aerial combat is evolving and changing. There are now fighters, not only US ones, with scanned-array-radar able to follow multiple targets 100 or more miles away, liaise with other fighters or unmanned drones via computers, with identifying done automatically by those computers.

The elimination of the target can be done remotely, so there is no need for a dogfight in the traditional sense. However, there are still skirmishes and probes where there would be a dogfight situation, and cases where the adversary would be a second ranking power militarily. Though it would seem impossible to evade a locked-on missile, aircraft do have various countermeasures.

As seen in Ukraine, there are even dogfights between drones, albeit relatively small ones.

Doolittle Raid (World War II)

A daring raid on Tokyo in early World War II in which bombers took off from a US aircraft carrier to bomb Tokyo and fly on to China. It was hailed as a great success and boosted US morale, with Doolittle hailed as a hero. The downside was that tens of thousands or more Chinese were killed in retaliation on suspicion of having collaborated.

To keep the weight of the Doolittle Raid bombers down, wooden broomsticks were substituted for the rear machine guns!

Doomsday Plane/National Airborne Operations Center (NAOC)

See **Air Force One**.

Doors and Overwing Exits

For an airliner to be certified it must be proven that a full complement of passengers and crew can evacuate within 90 seconds with half the exits unavailable. For this to be possible outwards-opening emergency exits are added over the wings, which allow some more people to evacuate. However, the primary evacuation route is via the plug doors where there are flight attendants to open them, supervise, and sometimes push those hesitating onto the slide.

Doors (and Slides)

Airliner doors are of the plug type and cannot be opened at altitude because the difference in air pressure makes that physically impossible, despite the fears of passengers seeing some crazed person trying to do so.

So that an emergency evacuation can proceed as quickly as possible, the emergency chute deploys and inflates automatically in such an eventuality. Because the slides deploy automatically in the event of an emergency, great care must be taken that that does not happen when the doors are to be opened, say, on arriving at the stand for passengers to disembark. Normally the senior flight attendant gives the order "Doors to manual and cross-check." The flight attendants at the doors then set theirs to manual to disarm the slide and cross-check that the attendant at the opposite door has done it too. When the aircraft is about to depart, with the doors closed, the reverse happens with the senior flight attendant ordering "Doors to automatic and cross-check."

Deployment of a slide by mistake is not only very costly but can be dangerous.

Doppler Effect

The phenomenon whereby something, such as a sound wave, emitted from something coming toward one has a higher frequency than when coming from something going away. It used to be explained to children in terms of the varying pitch of a railway train's whistle when passing at high speed through a station. The redshift of stars traveling away from Earth would be another example.

Applied in conjunction with radar, the Doppler effect can instantly show the air traffic controllers the speed of an object, such as an aircraft. Ground proximity warning systems (GPWS) in aircraft depend on the phenomenon to determine dangerous sink rates and warn the pilots.

Weather radar can use the Doppler effect to determine movement rates of droplets and warn of storms ahead but unfortunately cannot detect clear air turbulence as there are no water droplets. (See **Lidar**.)

Douglas Aircraft Company

The Douglas Aircraft Company was an American aerospace manufacturer founded in 1921 by Donald Wills Douglas Sr.

Douglas peaked in World War II with a 160,000-person workforce consisting of 40 percent women. The aircraft included more than 10,000 C-47s, the military version of the civilian DC-3 which entered service in 1936 and was called the Dakota in the UK. Another Douglas aircraft of great note was the SBD Dauntless dive bomber that sank three Japanese aircraft carriers in the Battle of Midway.

After the war Douglas was involved in various projects but no longer had the political clout it previously had to ensure a share of military programs. In 1967 it accepted an offer to merge with McDonnell Aircraft to become McDonnell Douglas. The merger seemed a good fit because McDonnell had military orders and Douglas was having problems with its civilian work accentuated by difficulties getting supplies of parts for its DC-8 due to the Vietnam War. (See **DC-8**.)

McDonnell Douglas in turn merged with Boeing in 1997.

~ Douglas DC-8 (DC-8) 1959/556

Douglas, the US airlines' favorite supplier of airliners, had expected there would be an interim period between piston-engine airliners and jet-engine airliners filled by cheaper-to-run but unromantic turboprops. Once the 707 entered service, Douglas whipped its people into action to get its new DC-8 pure jet certified by the FAA.

It did well to make the cabin wider than that of the original 707 based on the Dash-80, enabling six-abreast seating rather than five-abreast. However, the blunt wing chosen with less backwards sweep than the Boeing meant that it could not meet the performance and efficiency promised to the airlines. Various costly tweaks did improve things though it could not match the 707 at high altitude.

Attesting to the qualities of the DC-8 such as somewhat more space for freight is that many more DC-8s than 707s remained in service when new airliners were developed. Many were converted to freighters with some as freighter–passenger combis. Because of the number of different versions including stretches, it was not profitable with Douglas finally having to merge with McDonnell with its military business to create McDonnell Douglas.

~ Douglas DC-9 (McDonnell Douglas DC-9) 1965/976

The DC-9 was a short- to medium-range twin-engine airliner like the British BA-11, which despite having a two-year lead only achieved sales half those of the DC-9. Benefiting from cooperation with France's Sud Aviation, which had developed the remarkably successful rear-engine Caravelle, the DC-9 had very clean lines, its fuselage having much in common with the DC-8s. It evolved into the MD-80, MD-90, and then the Boeing 717. The grand total attained almost 2,500 units. Thus, it can be said to be one of the most successful airliners ever even though in the early years it had a rather checkered history.

Drag

Resistance to the forward motion of the aircraft through the air, the causes of which are multiple. It is a key factor that aircraft designers must consider because it greatly affects the amount of fuel consumed.

Height and airspeed also have a great effect on drag, something pilots must consider.

Drag can be broken up into three basic types:

1. **Parasitic Drag**
   Roughly defined as drag arising from things other than drag induced by lift. There are three types of parasitic drag:

a) **Form drag**
Generated by the shape of the aircraft. Sometimes one wonders why some aircraft have cramped cabins with so many seats abreast. The reason is that merely increasing the diameter of the fuselage by a few inches can greatly increase the drag and the amount of fuel consumed.

b) **Friction drag**
Resulting from friction of the air against the skin of the aircraft. The smoother the skin the better. For example, rivets must be countersunk perfectly. Furthermore, composite wings can be made smoother than metal ones.

c) **Interference drag**
Arising from interference where two different parts, such as the wings and fuselage, meet.

2. **Induced Drag**
Resulting from lift—the force required to support the aircraft in the air.

3. **Wave Drag**
Arising from the shock waves on the skin of the aircraft found at transonic and supersonic speeds.

These different types of drag vary greatly depending on airspeed, angle of attack, altitude, and how near one is to the speed of sound, which incidentally decreases the higher one gets (see **Mach**).

The is the **Flare on Landing,** where at a height of about 35 feet the pilots fairly sharply raise the nose to brake the rate of descent and add drag and considerably add drag so it slows and inevitably sinks gently onto the runway.

Conversely, if the pilots raise the aircraft nose too much on taking off, the drag can become so great it may never gain enough airspeed to fly lift off properly. This is especially true if the flaps are not configured for takeoff or if an engine fails.

Jet aircraft fly more efficiently in thin air at high altitude where friction drag is less. Circling low down near an airport can use up a lot of fuel and that is why in working out the fuel reserves required there must be enough to circle for 30 minutes at 1,500 feet, among other things.

"DragonFly" [Airbus]

Airbus likes to apply things that happen in the animal kingdom to new concepts, an example being **Fello'fly, Flying Like Migrating Birds to Save Fuel**. In the case of "DragonFly," it is the all-seeing dragonfly's eye "that can follow everything that is happening that is the key."

According to Airbus:

DragonFly could be a game-changer when it comes to de-risking emergency operations. Its focus is on three key areas, each one drawing on a combination of data captured during flight and a vast corpus of flight information to promote automated yet intelligent decision making.

DragonFly offers a solution to help ensure safe flight and landing. If the crew are unable to control the aircraft, the onboard function detects the issue and automatically selects the most suitable airport to redirect the aircraft towards. But of course, flight paths and external factors are complex and changing.

A dragonfly scans its surroundings and adapts its journey accordingly. Our DragonFly demonstrator does much the same thing, taking into account external factors such as flight zones, terrain and weather conditions as it chooses where to land. But unlike a regular dragonfly, our DragonFly also benefits from a constant channel of communication between the aircraft and both Air Traffic Control and the Operations Control Centre of the airline to ensure a safe and coordinated approach.

In theory, it could save the situation where the pilots become incapacitated.

Drift

Phenomenon whereby the wind blows the aircraft laterally from the direction in which it is heading (pointing). The resultant true course over the ground is its track.

Drip Stick/Float Stick for Measuring Amount of Fuel

In the old days, the fuel level in an aircraft wing tank could be measured by having a tube set in the bottom of the wing that when stowed and locked reached right to the top and always above the fuel. When the tube was pulled down from below the wing, fuel would enter it when the open top reached its surface and drip out from the bottom indicating the level. Charts would enable the amount of fuel to be calculated from markings on the stick.

Modern aircraft have what is called a float stick with a magnet surrounding the tube floating on top of the fuel. When the stick (with a magnet at the top rather than a hole) is pulled down it half-locks in reaching the floating magnet. The amount the stick protrudes indicates the amount of fuel, in the same way as the drip stick but without the messiness of fuel dripping down.

Drones, Also called UAVs (Unmanned Aerial Vehicles)

[See also US ~ Drones.]

An unmanned aerial vehicle can be almost anything ranging from a balloon to a winged asset like an aircraft. In fact, some of the newest fighters and bombers are being designed so they can also be deployed both with and without pilots. Drones can also operate under the sea. As regards airborne UAVs there are in general the following:

~ The Different Types

1. Lighter-than-air UAVs
   Balloons, blimps (airships without a rigid structure where the pressure of the gas within maintains their shape), and airships (with a rigid structure). They can be tethered with the tether sometimes used to download video and data. Tethered ones can be used

to monitor what is happening at a particular spot such as at Chernobyl.

2. Flapping wing devices (ornithopters)
   These fly like birds and have been used by the military for tiny spying devices, some even being little larger than bees. Their great drawback is limited range due to the great energy demand for flapping. As technology improves, they may start appealing to the public.

3. Rotorcraft
   Usually four or more rotors lifting them up vertically. Having more than four rotors provides redundancy should one fail. Some powerful ones have eight counter-rotating coaxial rotors in the four-rotor configuration with the one underneath rotating in the opposite direction.
   They are relatively slow-moving and can even be shot down with a rifle or captured by another drone with a net. However, in Ukraine, many of those downed are retrieved and repaired for re-use.
   These have multiple uses, both civilian and military, being great for aerial photography, spying, delivering packages, explosives, illegal drugs to dealers, and even prisons, nerve gases, and bacteriological weapons material.
   Though such drones have many good uses, systems to stop them engaging in harmful activity such as shutting down airports are becoming serious business.

4. Fixed wing craft
   These craft can come in all sizes and types ranging from a model plane through a medium size to an actual aircraft that normally would have a pilot. For military use they can have long endurance and carry munitions or cameras. There are even ones with solar cells and rechargeable batteries that can fly very high and stay aloft for a month or more.

~ First-Person View (FPV)

Drones can be flown in line of visual sight or with the remote pilot seeing what the drone sees by what is called **First-Person View (FPV)**, where the drone has a camera transmitting the view to the pilot on the ground who can be near or sometimes a continent away.

Another feature found in both civilian and military drones is "Return To Launch (RTL)" whereby the drone will return to the position from which it was launched if so ordered or automatically if contact is broken.

~ Civilian Drones

While drones weighing under 250 grams are generally classed as toys and not regulated, heavier ones are under increasingly strict controls.

The US's FAA, as the regulatory body, makes a great distinction between drones for personal use and those used commercially with the latter including the taking of photos for sale.

There are too many civilian uses of drones to mention with many obvious such as aerial photography, the tracking of criminals, and the search for missing persons such as children, and the delivery of packages including medicines to people living in remote and inaccessible areas. They can even be a life-saver in built-up areas by delivering in, say, seven minutes the opioid antidote naloxone to someone who has overdosed and is about to die.

~ Makers of Civilian Drones

By far the largest maker of drones is the Chinese company DJI. It specializes not only in small drones but also in cameras and steadying gimbals to support them. It has recently taken a majority shareholding in the famous professional camera company Hasselblad.

Its drones are used all over the world including by police forces in the US. While they have over 50 percent of sales other players might be around 5 percent. In the US they are being constrained by official fears that Chinese products might be "spying." Also, the fact that DJI supplies drones used by the Chinese authorities to monitor Uighurs in the north-western region of Xinjiang is held against the company leading some to ban its products, even though under the Chinese political system the company could hardly refuse.

To protect its reputation, DJI has tried to prevent its drones being used by terrorists and going to places where they shouldn't such as the White House lawn. This is mainly done by incorporating special software. Its drones are being used by both sides in Ukraine, and, in fact, any civilian drone can be converted to military use.

~ Military Drones

Drones ranging in size (and almost in cost) from that of full-size aircraft to almost that of bees are playing an ever-increasing role in warfare. There are also undersea drones, large and small, with the Russians having developed a nuclear-powered torpedo that can roam for months without surfacing and have a conventional or thermonuclear warhead able to destroy a ship, a task force of ships, or even do considerable damage to a coastal city or port by detonating nearby.

One problem with the drones controlled in real time from somewhere, say, miles away on another continent is the time-lag with the incoming and outgoing signal having to travel up and down from satellites. One way of overcoming this is to preprogram the drone and combine it with artificial intelligence using data from its sensors, inertial navigation systems (used where GPS fails), and so on. AI can even be incorporated in small kamikaze drones allowing them to search out their own targets.

The future of aerial warfare (**Integrated Sixth Generation Combat**) envisages a manned aircraft working in conjunction with a cohort of drones integrated electronically with one or more working as a **Loyal Wingman**, sometimes even being required to take the hit in the place of the manned aircraft. See **Attritable**.

Besides their use in actual combat, drones excel as surveillance tools, for unlike satellites—whose passage the "target" can predict, and which are expensive to reposition—they can be deployed flexibly and cheaply.

Satellites can detect a great deal but have two disadvantages in that repositioning them is complex and extremely expensive, and in routine operation the "target" can predict when they will pass overhead and hide what it is doing.

Now that drones can stay aloft for as long as 48 hours after which a relief can be sent to take over surveillance, this makes them more suited to tactical intelligence-gathering situations.

See **Wide-Angle Motion Imagery (Gorgon Stare)** where a drone can simultaneously track the movement of thousands of people in say in a town.

Also see **Replicator**.

~ Mothership Drones

When one talks about mothership drones, enormous platforms such as alien spaceships come to mind when, in fact, even quite small delicate drones can be motherships. A mothership drone using a wing has a number of advantages, including taking drones to the battlefield with them individually not needing the fuel or a wing for the range, and with quadrotors to hover on site with the drones not requiring a powerful easily disrupted signal, and controlling them with the mother drone relaying their individual FPV (that is what the drone sees) back to the controllers.

An example of small mothership drones is the Ukrainian domestically developed Dovbush T-10 fixed wing drone acting as a mothership for as many as six first-person view (FPV) kamikaze quadcopters. With the T-10 relaying the signals from the child drones including their FPV signal each can carry a bomb.

See https://www.twz.com/air/watch-a-ukrainian-uncrewed-aerial-mothership-launch-kamikaze-drones.

An example of large mothership drones is the 10-ton Chinese drone mothership nicknamed Jiu Tian (Nine Heavens). It is jet powered and seems to incorporate normal facilities such as electro-optical and infrared cameras and radar.

See https://www.twz.com/air/new-heavyweight-chinese-jet-drone-looks-to-be-a-swarm-mothership

See **US AIR FORCE ~ DRONES.**

~ Fiber Optic Drones

The war in Ukraine has been a great proving ground for drones, with the first ones controlled from the ground using radio waves that were subject to jamming even though they used frequency switching to confused the jammers. Now combatants are resorting to fiber optic drones linked to the controller through a fiber optic cable. There is little defense against these apart from having nets over roads and facilities. The problem is the limited range, of six miles or so. One way this is being overcome is to have a large mothership drone controlling drones linked to it via fiber optic cables.

~ Smuggling Drones into a Country for Launching Anywhere

In June 2025 Ukrainian agents destroyed many expensive Russian bombers parked at air bases, some of them thousands of miles away from the frontier using drones smuggled in inside portable cabins hauled by semitrucks and launched them near airbases. The drivers of the vehicles did not have to know what they were carrying.

Apparently, the event had been devised 18 months earlier, and will possibly be difficult to repeat in view of the precautions the Russians are likely to take.

Valuable Western aircraft are at risk and need to be dispersed or kept under expensive fortified cover. Google shows many of them as sitting ducks at airbases and airfields. However, the Russians would not have been expecting drone attacks and ready to shoot them down whereas the Americans and other countries are now likely to be better prepared.

~ Drone Makers for Military Apart From US

Though the US and other countries produce top-of-the-range military drones, the Chinese are surely the most prolific exporter of ones ostensibly for non-military use. One of those, perhaps a *Mugin-5 Pro 5000mm Carbon Fiber UAV Platform* ($9,499.00) or copy of it, was possibly used by the Ukrainians to target a faraway Russian oil refinery in 2022.

~ The Mugin-5 Pro has:

Maximum speed: 150 km/h
Cruise speed: 120 km/h
Stalling speed: 55 km/h
Maximum takeoff weight: 85 kg
Payload: 25 kg
Takeoff distance: 60 m
Landing distance: 85 m
Maximum flying time: >7 hours

We just cite this as an example of what is available on the open market—a flying time of 7 hours and a payload of 25 kg offers many possibilities at relatively little cost. It is a dangerous world.

Talking of less sophisticated drones, we might mention the Chinese are said to have ones made to look so much like birds that real birds fly alongside them!

The Turks are marketing drones specifically with military applications, with, for instance, the Bayraktar TB2.

~ Bayraktar TB2

Maximum flying time: 27 hours
Maximum height of 7,600 m (25,000 ft)

Besides using it for intelligence gathering, it can launch laser-guided missiles.

Costing about $6,000,000, it is about a sixth of the price of the latest version of the US's MQ-9 Reaper. Though not so high-tech or able to carry such a heavy payload, its relatively low price is a great consideration. The US is thinking of loosening its conditions for exporting drones to share some of this business, but prices and their strict conditions as regards deployment will likely deter purchasers.

Iran, which managed to bring down a US drone and copy some parts of it, is supplying Russia with drones for deployment over Ukraine.

The Ukrainians are developing great expertise in drones and even missiles and in 2025 produced a short-range one, the Sapsan, better than the ones the Americans were allowing them to have.

Drone Operator

Entity or person responsible for a drone but not necessarily licensed to operate it [US].

Drone (UAV) Pilot

With drones ranging from tiny ones for private use in the civilian domain to ones in the military domain ranging from line-of-sight bee-sized ones to aircraft-like ones such as the MQ-1 Predator and MQ-9 Reaper, being a drone pilot can mean many things.

At the top end, the US Air Force was having a problem retaining drone pilots in large part because of the continuing stress they were under. First, the hours were long, even 900 hours a year, while fighter pilots flew an average of 250 hours in that time. While not having the prestige of fighter pilots—stigmatized as the "Chair Force"—their work is complicated with them having to switch the nature of their tasks, say, from reconnaissance to combat to rescue, and even country in one day. Some suffer from PTSD having seen close-up collateral damage in attempting to "take someone out," which fighter pilots high up do not see. Nevertheless, the role of the fighter pilot may be diminishing and that of the chairborne drone pilot increasing.

British Royal Air Force drone pilots sometimes operate with a lawyer alongside to confirm the legitimacy of the target.

Drones as "Airborne Mine-Shields"

It has been suggested that swarms of drones with proximity fused explosive charges could loiter above sensitive areas as a shield rather like a minefield. They could not realistically protect a city but might be effective over limited areas.

Drug Testing after Incident

Even though a pilot may have acted "heroically" saving everyone on board, they will usually be subjected to a simple drug test and taken off duty. They may even be offered counselling.

Though there may be no suspicions, should questions be raised later people may ask, "Why wasn't the individual drug tested?"

Duct Tape?—Really Very Strong Aluminum Speed Tape

Passengers are sometimes shocked to see their airliner seeming to be held together by what appears to be household duct tape when, in fact, it is extremely strong, and very expensive, aluminum speed tape. It resists extremes of temperature and can hold things even more strongly than perhaps the missing bolt it is compensating for.

It is often used to hold, say, a cover or cowling safely in place pending proper repairs by fully qualified mechanics.

Dummy Attack/Phantom Attack

A simulated attack, sometimes computer generated and designed to baffle satellites in space, with the aim of getting enemies to use up their weapons before the real attack. The Chinese have claimed to have been exploring this concept.

Dutch Roll

An oscillatory roll-cum-yaw where the aircraft yaws and rolls to one side until the dihedral effect of the wings brings it back to centerline, only for the same course of events to be repeated in the other direction.

This unpleasant phenomenon usually results from insufficient directional stability. Interestingly, the cure may not only lie in providing greater directional stability but also in having computer software provide suitable control inputs to preclude the phenomenon developing. As mentioned under **Anhedral**, computers can even make intrinsically unstable military aircraft appear stable. One of the Airbuses found to have a mini-oscillatory problem (a kind of flutter) was cured very simply with the help of a German software programmer.

Dryden Flight Research Center (NASA)

NASA's Flight Research and Atmospheric Flight Operations Research Facility is situated at Edwards Air Force Base in the western Mojave Desert (California). It is now renamed the Armstrong Flight Research Center, but many references use the former name.

The facility makes a great contribution to aviation, including research in the area of cockpit displays and safer navigation. See **Intelligent Flight Control System [IFCS]**.

Durability and Reliability [Engines Off-Wing]

With airlines demanding more and more fuel efficiency, manufacturers are pushing their engines to their limits with higher and higher temperatures and pressures. As a result, problems, not immediately obvious, are appearing.

These can first appear as reliability issues but several years later as durability ones. In either case, the engine most likely will have to be taken off-wing with the aircraft unusable unless a replacement is immediately available.

Pratt & Whitney GTF (geared turbofan) engines but also some CFM LEAP engines had reliability problems. Interestingly, the geared turbofan mechanism itself has behaved extremely well with the problems lying elsewhere, and likely to be resolved entirely by 2025. However, just as these reliability problems were being solved, durability ones came into play, especially as some engines were introduced from 2015 onwards.

As regards durability, some countries, and notably India and the Gulf states, have been harder hit than others. This is probably due to the temperatures and more importantly particulate matter and notably sand sucked in. This was being solved by the rearrangement and redesign of the holes needed to provide a cooling layer of gas in the combustion chamber.

~ E ~

Earhart, Amelia 1897–1937*

The world-famous American aviatrix, born in Kansas, who disappeared on a round-the-world exploit. Sometimes described as "America's favorite missing person."

EASA (European Aviation Safety Agency)

http://www.easa.europa.eu

European Agency based in Cologne, Germany (established in 2003 and only now operational), responsible for safety matters, with actual regulatory power, as opposed to the JAA (Joint Aviation Authorities), which preceded it. Its first major act was the certification of the Airbus A380. (See **JAA**.)

easyJet

A successful British low-cost carrier (LCC), helped by having purchased aircraft at greatly discounted prices during an industry slump, recently introducing highly professional management, and lately tailoring its early-morning and late-evening departure times to suit business travelers.

While its rival, Ryanair, flies to airports situated sometimes a long way from major city centers, easyJet flies to many airports likely to suit businesspeople. A sophisticated software program, allegedly with as many as 14 different fare levels for each flight, enables it to maximize profit while taking care not to overdo the milking to the detriment of the brand.

According to the French magazine *Capital*, on a typical flight only 10 percent of passengers may be paying the advertised cheapest fare, and even though some passengers may be paying as much or more than they would have on airlines like Air France, the easyJet brand's reputation for being cheap is so strong that passengers fly with it regardless. The same would apply for

Ryanair, which is trying to present itself as a more caring airline than before.

ECAM (Electronic Centralized Aircraft Monitor) [Airbus]

Boeing's equivalent system is called **EICAS**.

The term Airbus use for a system that keeps an eye on all systems and presents pilots with information in order of urgency, with chimes as appropriate. First introduced on the A320, where computer control had been something of a revolution, and then in other Airbus craft.

It shows on the screen what it thinks is the most important first and what action should be taken with some others, if any, below. This works fine in normal circumstances, but is a problem when there are very many to show.

Interestingly, having consulted with Quantas Captain Richard de Crespigny following the QF32 incident (See **! >>> Famous Feats of Airmanship**) where there were one hundred faults, Airbus realized the need to have an indication of the number not shown. They applied this on the A350 and perhaps trying to make it possible to set the ECAM to show the condition of things needed say for landing.

ECM (Engine Condition Monitoring)

The term used by Airbus for a system that keeps an eye on the performance of the engines.

Echelon

An arrangement for the gathering and pooling of communications intercept intelligence by the US, the UK, Canada, Australia, and New Zealand, called the "Five Eyes." There had been a suggestion that Japan might join.

Tapping into connections used to be relatively easy when many communications were via simple cable, satellite, and microwave links. Now, with fiber-optic cables and VoIP (Voice over Internet Protocols), such as Skype, this is less true, and it is less easy to see who is talking to whom.

Other countries have similar systems, with business intelligence a profitable vein for all. Talking of business intelligence, the US once warned some of their top businesspeople that some seats in First Class on a certain European carrier might have microphones in them!

EFC Time (Expected Further Clearance) or Release Time

The time the pilots expect to be allowed to leave the holding pattern.

EGPWS (Enhanced Ground Proximity Warning System)

Nowadays, the latest ground proximity warning system called Enhanced Ground Proximity Warning Systems **(EGPWS)** can be linked to a **RAAS (Runway Awareness and Advisory System)**. In 2001 this enhanced version was mandatory in the US for airliners with more than six passengers. Now there are even more advanced systems that are described under **TAWS**. See **GPWS**.

EGT (Exhaust Gas Temperature/Turbine Gas Temp.)

An indicator of the overall performance of the engine (compressors and turbine) is based on the core flow temperature, which is measured at the turbine exit.

EFIS (Electronic Flight Instrument System)

Electronic (rather than electromechanical) display of flight data. Essential units are:

1. **Primary Flight Display (PFD)**
2. **Multi-function Display (MFD)**
3. **Engine Indicating and Crew Alerting System (EICAS)** [Boeing] and **Electronic Centralized Aircraft Monitor (ECAM)** [Airbus].

EICAS (Engine Indicating and Crew Alerting System)

Term used by Boeing for a display, usually situated above the throttles, showing complete information about engine performance, fuel management, and alerting pilots of system malfunctions, even including airframe failures. Should a malfunction occur, pilots receive an aural alert, with not only the screen for that part of the system automatically coming up but also a list of corrective actions.

Airbus has an equivalent system, called **ECAM**.

Ejection Seat

When Martin-Baker in the UK was working on ejection seats to save pilots on early jet fighters after World War II, it was mainly considering cases where there would be a fair amount of time and aircraft moved comparatively slowly. This meant pilots did not need to be subjected to such powerful forces.

Nowadays, they have to be very much more powerful and sophisticated to be able to cope with situations ranging from crises at liftoff—such as in the most expensive crash ever where blocked sensors caused a B-1 bomber on Guam to point upwards and stall at takeoff with the two-man crew ejecting safely—to ejections at high speed and high altitude.

Though the modern ejection seats have saved innumerable lives the stresses they place on the human body can be so great that the occupant can suffer spinal injuries, whiplash, organ damage, or even succumb. The g-forces can attain 14 g, with an ejection speed of four or more hundred miles per hour. They are very complex, and if any part does not function as intended, can be disastrous. They are called 00 ejection seats because they work at zero altitude and zero airspeed.

ELAC (Elevator Aileron Computer)

The computer on the Airbus A320 and other Airbus **aircraft** the controls the **ailerons** and **elevators** firstly in accordance with the input from the **sidestick** and secondly in accordance with the autopilot and the **FMC (Flight Management Computer)**.

Electrifying Aircraft

In June 2025, *Aviation Week* had an **interesting** piece on how Wright Electric in Malta, New York:

> It exemplifies the vision—if not the reality—of the electrification of aviation. On the stand is a fan module from a Honeywell LF507-1F turbofan, which is to be powered by Wright's WM2500 2-megawatt motor to test the performance of an electric jet engine. Wright is targeting the retrofit of regional jets and turboprops. An initial target is the BAe 146/Avro RJ family of 80-100-seaters, powered by four LF507s.

But in reality, electrification is starting much smaller, with two-seat trainers and nine-seat commuters, because of battery limitations and certification challenges."

We are almost there. With weight critical, small aircraft with no more than seven passengers so as to avoid needing two pilots are likely to be a start. However, we are waiting for advances in battery technology. (See **Batteries**.)

Electric motors are much simpler than jet engines and should be more reliable and need much simpler inspections and maintenance. Also, unlike jet engines they are equally, if not more, efficient at low altitudes, them more likely to be suitable for short hops.

There is even talk of retrofitting aircraft with electric engines.

The most probable solution for more normal length flights might be a combination of fuel cells for cruise and quickly rechargeable capacitors (or, when they are developed, Gen 3 solid-state batteries) for surge power required for takeoff and climb. Fuel cells have the advantage that their weight decreases as the flight progresses.

The trouble with producing hydrogen for fuel cells is that it is presently done from natural gas and that overall end-to-end efficiency is only about 25 percent considering the compressing. Furthermore, the extra weight of the strong tanks to hold the hydrogen is a negative. When some time in the future surplus electricity from carbon neutral sources can be used to create hydrogen, it may be a different story.

Electronic Flight Bag (EFB)

The electronic flight bag is not a bag in the traditional sense but an electronic device, usually a laptop replacing the increasingly voluminous paper documentation, such as the aircraft operating manual, aircrew operating manual, navigational charts, airport plans, and so on that pilots lug around in their carry-on bags. Before electronic flight bags, updating manuals and other material for a flight involved laboriously substituting pages with updated ones. Electronic flight bags can even be used for calculating fuel requirements, rotation speeds, and so on.

The latest versions can even supplement aircraft instruments by incorporating moving airport map displays with an indication of "own ship position," which could almost halve runway incursions and virtually eliminate such tragic errors as when a Singapore Airlines 747 took off from a disused runway at Taipei in 2000*.

The electronic flight bag can also be used for calculating takeoff parameters, such as the V-speeds and engine power settings.

However, in a near-disaster at Melbourne, Australia, on March 20, 2009, the first officer on an Emirates Airbus A340 bound for Dubai with 357 passengers and 18 crew, a 14-hour flight, mistakenly entered a takeoff weight **100 tons too low** into the EFB computer. Because of distractions (nonsterile cockpit) and other reasons the captain and first officer failed to pick up on the mistake. As a result, the engine power settings and takeoff V-speeds were too low. As they came towards the end of the runway, with the A340 refusing to lift off, the captain pushed the throttles to **TOGA**, and they lifted off beyond the end of the runway, suffering a severe tailstrike, hitting lights, and ILS equipment. After jettisoning fuel, they returned to Melbourne safely.

Such a banal error is unlikely to be repeated in that Emirates are tightening up their procedures including insisting on the use of *two* EFBs and developing software to pick up on lower than expected takeoff run performance. Other airlines are doing likewise, and Airbus has enabled the flight management system to identify such inappropriate settings and alert the pilots.

Electromagnetic Aircraft Launch System (EMALS)

The steam catapult system, though used successfully on aircraft carriers for many years, has several drawbacks, the foremost of which is that there is no feedback control. Without it, there often can occur large transients in towing force that can stress and reduce the life of the airframe. Furthermore, because the system is unpredictable extra force is always added, which overstresses the airframe unnecessarily.

The great volume of the steam catapult system not only takes up considerable high-value space on the carrier, its great weight of almost 500 metric tons situated mostly topside compromises the stability and righting moment of the ship. The need for even more powerful systems would necessitate even more bulk and weight.

The steam catapults are also highly maintenance intensive, inefficient (4–6 percent), and their availability is low. Another major disadvantage is the present operational energy limit of the steam catapult is approximately 95 MJ. The need for higher payload energies will push the steam catapult to be a bigger, bulkier, and more complex system.

The size and manpower requirements of steam catapults place limits on their capabilities. A newer approach is the electromagnetic catapult, such as **Electromagnetic Aircraft Launch System (EMALS)** developed by General Atomics.

Electromagnetic catapults place less stress on the aircraft and offer more control during the launch by allowing gradual and continual acceleration. Electromagnetic catapults are also expected to require significantly less maintenance.

Linear induction motors have been tried many years before, but only now with nuclear-powered aircraft carriers with abundant electricity have catapults powered by linear induction motors and electromagnets begun to be seriously considered. Even gas-turbine-powered carriers with an electromagnetic catapult could dispense with a separate steam boiler for generating catapult steam. The US Navy's new Gerald R. Ford-class aircraft carriers and the Chinese Navy's Type 003 aircraft carrier feature electromagnetic catapults.

Electromagnetic Field Generator with Compact Fusion Reactor

Electronic Counter-Counter Measures (ECCM)

Systems, designed to defeat the countermeasures, like chaffs, flares and jammers, employed by an aircraft.

Electronic Flight Instrument System (EFIS)

A cockpit instrument display system where the display technology used is electronic rather than the traditional electromechanical.

Electronic Warfare

Electronic warfare is paramount these days and has led to electronic countermeasures to combat electronic threats, and in turn to electronic counter-countermeasures (ECCM) to combat those countermeasures!

Electronic warfare concerns not only aircraft but troops on the ground protecting them from onslaughts of guided bombs, missiles, and drones.

It is a battle for superiority in the electromagnetic domain, involving "highly sensitive detection, analysis, classification, and digital jamming techniques, to counter the enemy's communications, radar, navigation, and other activities in the electromagnetic spectrum."

Elevation

The elevation of a point is its height above mean sea level (MSL) as opposed to height above the surrounding terrain (AGL). On charts for airmen, the elevation of high points (say, buildings near an airport) are given, with the height above the surrounding terrain (AGL) in parenthesis underneath.

Elevon (Elevator + Aileron)

Instead of having elevators (in the tail) to solely control pitch, and ailerons in the wings to solely control rolling (banking), some aircraft, such as the late supersonic Concorde, with its delta wings, have the two combined at the trailing edge of the wings. If both go up or down together in equal amounts, they act as elevators; if the amounts differ, they act like ailerons; and, of course, any combination of the two is possible. However, the input from the pilot remains traditional (as if elevators and ailerons were separate), with complex mechanisms, or more often now computers, producing the simulation.

EMAS (Engineered Material Arresting Systems)

A rarely installed but highly effective bed of specially engineered particulate material beyond the end of a runway into which the wheels of an overrunning aircraft sink deeply enough to crush them. This rapidly slows the aircraft but not so sharply that it risks flipping it over. Military fields in constricted areas may have arresting wires or nets at the end of runways that can be flicked up to arrest a dangerously overrunning aircraft. This is the case at Lajes in mid-Atlantic, but fortunately they were not needed when Captain Piché miraculously landed there after his almost 80-mile glide—it seemed that there was a cliff down to the sea at the end of the runway, but that was an optical illusion*.

EMBRAER

Embraer was founded 1969 by the Brazilian government to help meet the needs of the Brazilian Air Force, but once on its feet began producing small regional and commuter aircraft for export.

It was privatized in 1994, and became a public limited company in 2000. It has gone from strength to supplying business jets and many commuter jets to the US. However, in a recent case they found that improving performance by installing a more powerful engine made a new model just above the cutoff to comply with the US airlines Scope agreement with their unions. See **"Scope" Agreement Limiting Seats Regional Airlines Allowed.**

The arrangement for Boeing to take an 80 percent stake in Embraer's commercial division fell through with ill-feeling but was probably a blessing in disguise in view of Boeing's troubles.

Emergency Descent

In the event of an indication of an onboard **Fire** (often due to smell or detection by sensors of smoke) or cabin depressurization with the oxygen masks dropping down, the pilots may have to make an emergency descent. Some newspapers will say some passengers considered it a major drama when it was nothing major.

Emergency Frequency and VHF Guard 121.5

In the US, in addition to the radio frequency they are using to communicate both ways with the controller handling them, pilots are obliged to listen in on another one for use in emergencies as follows:

1. The Aircraft Emergency Frequency or VHF Guard 121.5 MHz.
   A military aircraft intercepting a suspicious airliner would use this frequency.

2. There is also a Military Air Distress (MAD) Frequency also called UHF Guard on 243 MHz.

Something similar applies in other countries.

Emergency Locator Transmitter, Distress Tracker (ELT-DT)

According to the ICAO **Global Aeronautical Distress and Safety Systems (GADSS)** recommendation, and a European Union mandate, all new aircraft delivered from January 2023 must be able to autonomously report their location anywhere in the world and determine the end of flight location to help rescue teams rapidly locate the aircraft and recover flight recorders.

Airbus has decided to use Orolia's automatic Ultima-Distress Tracker transmitters in all its new aircraft to meet the latest requirements of ICAO starting in January 2023. The Emergency

Locator Transmitter autonomously acquires the aircraft's location and sends a 406 MHz message in real time, including the accurate location, to the **Cospas-Sarsat** distress alert organization. Furthermore, it will be integrated with the aircraft's avionics and at a sign of distress (aircraft in a state that, if its behavior is left uncorrected, may result in an accident) will start transmitting while still airborne.

See **GADSS (Global Aeronautical Distress and Safety Systems)**

EMP (Electromagnetic Pulse) [Nuclear War]

A burst of electromagnet energy able to melt or burn objects can be something as natural as lightning, or purposely man-made to disrupt communications and damage unprotected electrical equipment and systems.

Generating a widely effective EMP could be relatively simply done by exploding a nuclear device at high altitude using a missile to achieve the necessary altitude. It would give off an EMP, which would trigger a much stronger one lower down. Most of the United States could be impacted.

Limited range—say, six miles—EMP-generating munitions exist, and it is thought that the Russians might use some they have in Ukraine.

It could disrupt a whole country's electric network with all sorts of consequences ranging from impacting water supplies, railways, telecommunications, banking, and much else. Hospitals have backup generators but those would only work for as long as fuel for them lasted, and even then, they might not be able to provide power to all the monitors to which patients in the wards might be attached.

In the United States, for example, they have very specialized transformers for electricity grid interconnections, which could take a year to replace. Meanwhile, much of society would be breaking down.

EMP-hardened solar panels are being developed that people particularly in the country could use with, say, 2 to 15 panels supported at the correct angle on the ground.

There are some myths such as that all cars and trucks would stop working, when only 10 percent might. Small devices with no lengthy wiring such as cell phones and laptops should be okay but not their charging devices.

Empennage

The empennage is the tail assembly of an aircraft.

The term derives from French, where it means the tail feathers of an arrow. Depending on the aircraft, the empennage includes the **rudder** with the **vertical stabilizer**, and the **elevators** with the **horizontal stabilizer**. The empennage often figures prominently in news photos of crashed aircraft, leading some airlines to try to have their logo painted out to avoid bad publicity.

See **Ruddervators/V-Tail**.

Endurance

The number of minutes the aircraft can theoretically stay airborne with the fuel on board.

Engine Failure

There are checklists for what to do when an engine fails to deal with the problem, but they may not cover how one should maneuver the aircraft.

If an engine fails or is seriously misbehaving on one side and one needs to make a significant turn, one should ensure the turn is away from the weak side with that wing high up. That is to say, if it is an engine on the left that has failed one should turn to the right.

There was a case where the aircraft crashed because it went into a spin through not doing this.

Engine Maker

A term used to distinguish the engine maker from the aircraft manufacturer, increasingly referred to as *airframer*. The engine is a major determining factor in how successful an airliner will be.

Engine makers often make little or no profit on the initial engines but hope to make their money from spare parts, maintenance, and subsequent engines.

In the West, the key engine makers are General Electric Aviation, Pratt & Whitney (a subsidiary of Raytheon Technologies), and Rolls-Royce. There are also joint ventures with other companies, notably General Electric with France's SNECMA called CFM International.

Pratt & Whitney also has a joint venture called International Aero Engines (IAE) with the MTU Aero Engines of Germany and Japanese Aero Engine Corporation. And General Electric has a joint venture called Engine Alliance selling a range of engines for aircraft such as the Airbus A380.

Engines

Aircraft engines are numbered from left to right facing forwards. On a trijet, the engine in the middle at the back is the Number 2 engine.

The arrival of the jet engine with no main reciprocating parts brought much greater reliability. Now we are almost getting towards the end of the turbofan engine with other types on the horizon with even less to go wrong in the absence of turbine blades.

Engines Next Generation

The major manufacturers are working on what might be called a new generation of engines. At one point, before the arrival of its new CEO, Boeing was saying it did not want to be caught out producing a new airliner for an out-of-date engine.

Engines Under the Wings or at the Back?

Airliners used to have their engines set in the wings. Then there seemed to be a fashion to have them at the tail, giving a cleaner wing much favored by Russian designers. Now engines are mostly placed under the wings.

This has several advantages, including:

1. Easier servicing.
2. Wider choice of engines, since only the pylons have to be changed.
3. Easier recovery from a stall, although the application of sudden power at low airspeeds can push the nose up dangerously if the engines are very low slung.
4. Engines can be placed far out on the wing, which itself shields some of the noise, thus reducing cabin noise.
5. As explained in the **Maximum Zero Fuel Weight** entry, the wings do the lifting, so there are structural advantages in having them lift the heavy engines at the same time.
6. Engine pylons are often designed to allow the engines to drop off in the event of a catastrophic failure or on encountering an obstacle, including water, on ditching.

This has several disadvantages, including:

1. Does not allow such a clean, aerodynamic design, thus incurring penalties regarding drag.
2. Low-slung engines pick up more extraneous matter.
3. Engines slung under the wings are liable to snag and cause the aircraft to cartwheel (or worse) when ditching in water. In theory, this should not happen if coming down with wings level, since the pylons are designed to break off when the engine pod is subjected to an extreme backwards force when striking an object.

Enhanced Airborne Flight Recorders (EAFRs)

Enhanced Airborne Flight Recorders (EAFRs) were mentioned in the context of the Air India Boeing 787 crash in June 2025. These can do more things with each recorder also recording what the other recorder is recording.

Envelope "Pushing the Flight Envelope"

The flight envelope is a high-sounding but useful expression usually meaning what occurs *within the boundaries (envelope) of what the aircraft or space vehicle can sensibly be expected to do.*

There can be some confusion regarding its use, since test pilots can also be said to be "pushing the flight envelope" to prove what the aircraft can do in the most extreme situations.

Engineers can tag it on to certain words to indicate specific aspects of the envelope, such as airspeed envelope or altitude envelope.

EpiPen

There was an incident where a doctor on board an aircraft more than an hour's distance from a diversion airport was faced with a passenger about to die from anaphylactic shock and found the medical kit had no EpiPen. He cleverly managed to save him using similar medication they had for someone whose heart had stopped which had an enormous needle, unlike the EpiPen.

It became big news on the Internet and the doctor was asked to help persuade the FAA and airlines that airliners should have EpiPens with the result that perhaps some 70 percent now do. The trouble is that they only have a limited life and must be replaced. Also, one might not be enough.

See **Allergies: Great Downside Risk for Airlines**.

EPR (Engine Pressure Ratio)

One measure of the thrust generated by turbofan engines is the ratio of the pressure at the outlet (after the last turbine) to that at the inlet (just before the fan). This is called the engine pressure ratio *EPR.*

However, with most of the thrust produced by modern high-bypass turbofan engines coming from the fan in the front, the speed *N1* at which the fan and to a lesser extent the speed *N2* at which the core is rotating are often seen as more significant indicators of thrust. However, in the event of a bird strike damaging the fan blades, the thrust might be much less than suggested by N1, and the EPR a more reliable measure.

ETOPS (Extended-Range Twin-Engine Operations Performance Standards)

Piston engines with their many moving parts were unreliable and after World War II twin-engine airliners had, according to the FAA's "60-minute rule," to stay within 60 minutes of a diversion airport, though there was some latitude with special approval.

In the 1950s the ICAO recommended a 90-minute diversion time for all aircraft.

With the advent of jet airliners with their much more reliable engines, things began to change radically. While at first four-engine airliners were the norm for flights over water far from diversion airfields, now twin-engine airliners meeting strict criteria can fly as far as 180 minutes or even 330 or 370 minutes from the nearest diversion airport.

Interestingly, with the turbines themselves so reliable it is often the ancillary equipment that is likely to fail, and sometimes has to be duplicated.

For both engines to fail it is likely to be due to a problem with the fuel—contamination, ice, or lack thereof due to it running out.

EUCARE (European Confidential Accident Reporting)

System based in Berlin for the confidential report of incidents. Like CHIRP for the UK and ASRS for the US.

EUROCONTROL

Despite its name, EUROCONTROL is not an EEC institution but an autonomous organization originally founded in the 1960s by Belgium, France, Germany, Luxembourg, the Netherlands, and the United Kingdom to control upper airspace. It has evolved, with other countries, and even the EEC itself, signing up.

Essentially, it deals with ATM (air traffic management), but it has other roles and is particularly concerned with safety and the management of databases.

Evacuation, the 90-Second Rule

The June 2022 Edition of Royal Aeronautical Society magazine *AEROSPACE* had an interesting piece on the EASA/FAA 25.803 rule that if aircraft had a seating capacity of more than 44 passengers that all occupants should be able to evacuate under the following conditions:

1. Only 50 percent of emergency exits are active.
2. The test participants (acting as passengers), as well as the aircrew, do not know which exits will be active.
3. The demonstration is conducted under conditions of reduced lighting.
4. Cabin baggage and blankets are positioned in the cabin aisles.
5. All test participants must evacuate and be on the ground within 90 seconds without assistance from external safety personnel.

This requirement is usually demonstrated by an actual test by the manufacturer.

However, according to the RAES, since the manufacturer chooses reasonably fit people, partly to avoid injuries, this does not replicate real life where one will have children and infants, the elderly, and those with reduced mobility, not forgetting passengers bringing their luggage with them despite exhortations not to do so, and some even stopping to take photos early on.

In recent years there have been several evacuations with the aircraft on fire and passengers—many with luggage—only just getting out in time. There have been cases where it has taken as long as 4 minutes. Recently, we have been lucky, but that may not continue. More people died in the Spantax Flight 995 disaster at Malaga in 1982 because passengers were taking their hand luggage with them and blocking the aisles. (The captain had aborted the takeoff after V_1 because the aircraft was shaking due to the delamination of one of the front wheels, and the heavily loaded DC-10 with fuel to cross the Atlantic overran the runway and caught fire.)

When in 2005 an Air France A340 overran the runway on landing at Toronto and caught fire, a man blocked an aisle as he stood there unpacking his bag despite being ordered not to do so by a flight attendant who had to take an alternative route with the passengers she was guiding. Everyone finally got out but only with a few seconds to spare.

There is a suggestion that the overhead lockers should automatically lock when taxiing and in the event of an emergency evacuation. This would add weight and retrofitting aircraft would be costly, if not difficult. Could bringing luggage in an emergency evacuation be made a criminal offence, or could it be confiscated, or could anyone with luggage be sued by relatives of the dead? The problem is how to ensure passengers are made fully aware—perhaps even being made to sign a form—of the risk of bringing luggage with them in an emergency evacuation.

What those passengers near an exit do not realize is that though they have the time to collect luggage from overhead bins and escape, those much further back and possibly with children may not be able to get out in time with them holding everything up. Furthermore, as firefighters well know, fires can be very treacherous with flashovers suddenly incapacitating everyone in the cabin.

Certification of Aircraft as Regards Evacuation Exits

It is now thought that computer modelling might give a truer result without exposing people participating in tests to injuries as happened in the certification of the supersized Airbus A380 in 2006, where 33 out of the 873 volunteers were hurt. One had a broken leg, and the remaining 32 received slide burns. And that was considered a success. In fact, everyone got out in 78 seconds, though it is a trifle worrying that some of the guinea pigs were from a nearby gym and much fitter than the average passenger.

As mentioned elsewhere in this book, some passengers freeze in case of fire and lose valuable seconds that might enable them to save themselves before losing consciousness due to smoke inhalation.

Some exits over the wing incorporate slides, but where they do not it should be remembered that having the flaps lowered makes it easier for people to slide off onto the ground without harming themselves. Of course, on landing they would normally be lowered to some degree.

Evacuation Dilemma

Captains are faced with a dilemma when an airliner comes down hard in an emergency landing or overruns the runway with no one injured and no obvious fire but with fuel possibly leaking. Should they order a precautionary evacuation or not?

Take the case of Qantas Flight QF32, where Captain Richard de Crespigny carefully brought his wounded Airbus superjumbo A380 back to Singapore and managed to bring it to a stop just short of the end of the Singapore runway in 2010*. Fuel was leaking out onto the ground from about 70 holes in the left wing, which had been pierced by fragments from an exploding engine. There was no fire, but there was the danger that if any fuel met the overheated brakes there would be a conflagration.

Many captains would have ordered an evacuation, but de Crespigny decided that, on balance, the passengers would be safer on board. Any evacuation via the slides inevitably involves some injuries, and especially so when, as on a flight between London and Sydney, Australia, there would be many elderly people, with some wheelchair-bound, and parents grasping infants and babies.

He pointed out in his fascinating book, *QF32*, the primary task of the fire trucks surrounding his aircraft was to save the passengers by preventing the escape chutes catching fire. Extinguishing the fire would be secondary. With no escaping passengers to protect, the fire crews could concentrate on extinguishing any nascent fire and almost certainly put it out.

Aviation fuel for jetliners has a flash point of 38°C (100°F) and an auto ignition temperature of 210°C (410°F). This means that a spark from a passenger's shoe, mobile phone, or camera flash could set it alight, even though it would not instantly go up like a bomb.

Confused passengers meandering about after an evacuation can put themselves in danger by wandering onto an operating runway or as was the case in QF32 risk being sucked into engine number four on the far right, which could not be shut down.

On the other hand, if there is already a fire or the chance of one with no fire trucks it is usually wise to evacuate as the credit card thickness of the fuselage will not keep the fire out for long. As mentioned in the previous entry, to have the aircraft certified the manufacturer must demonstrate that all passengers, in a maximum density configuration, can evacuate in 90 seconds with half the exits blocked.

The decision whether to order an evacuation or not can greatly depend on whether the fire trucks are on hand. In the case of QF32 they were waiting, but on some occasions, they cannot reach the aircraft, in which case a precautionary evacuation may be wise.

Evaluating Enemy Aircraft

Getting hold of an enemy aircraft can provide invaluable information, especially regarding its weak points and strengths. These could even include difficulty in turning left in which case American pilots would be trained to break off in that direction.

In World War II the US had **Technical Air Intelligence (TAI)** specializing in finding and evaluating Japanese aircraft. The British Royal Air Force had a special flight nicknamed the "Rafwaffe" performing a similar task testing German aircraft.

At times the US has offered high financial rewards for defectors bringing with them an aircraft they wished to learn about. In one case a man defected with his aircraft not knowing there was a reward.

eVTOL (Electric Vertical Takeoff and Landing)

A **VTOL (Vertical Takeoff and Landing)** aircraft is one that uses electric power to take off vertically, hover, and land vertically.

The concept is receiving increasing interest as it could, for example, lead to air taxis and the like. See **UAM (Urban Air Mobility), Air Taxis**.

However, until a great stride is made on the battery front, they are not going to be truly successful because so much intense electrical energy is required for takeoff and climb, and then those devices create drag making the vehicle inherently inefficient. Now, for any meaningful range, a *fixed wing* electrically powered aircraft may be a better prospect.

People working in the eVTOL field envisage hundreds of these ferrying people around and avoiding highway traffic jams. Avoiding collisions will be a problem, as will coping with hundreds needing to land in a serious storm over an urban area with nowhere to do so apart from busy roads and even the airport.

Drones carrying urgent packages such as medicines are a good example of efficient civilian use of eVTOL machines. See **Drones** Civilian Use.

Great advances in battery technology, such as solid state batteries, could mean great leaps in eVTOL usage.

Experimental Aircraft

Especially in the United States this can mean two quite different platforms:

1. A **sophisticated hi-tech aircraft**, sometimes referred to as an X-plane, designed to explore new technologies and design concepts such as NASA's X-53.
2. Really a misnomer for a **FAA legal category** for a platform that is almost the opposite of (1) above since it refers to low-tech planes built (often from kits) by the general public that are subject to less strict rules than those applying to even light aircraft.

There is also a **FAA sub-category called *Light Sport***

where the planes are under a certain weight, and have a limited payload, maximum airspeed and low stall speed. For these the pilot does not have be specially qualified, and only needs to be medically fit enough to hold a driving license. These appeal to senior citizens who have the money to purchase one but would not be able to pass a FAA mandated medical.
See Experimental Aircraft Association (EAA) which is for those "interested in building their own airplanes, plus those concerned with antiques, classics, warbirds, aerobatic aircraft, ultralights, helicopters, and contemporary manufactured aircraft."
In the US there are many such planes with people having a lot of fun.

Exponential Effect of Airspeed on Airfoils, 737 MAX

Looking out over the wing, passengers may have noticed that when the aircraft is moving slowly just after takeoff the **Ailerons** over the whole length of the back of wing are moving, sometimes at quite a sharp angle. Then when the aircraft speeds up, they will see only the inner ailerons (taking up perhaps only a third of the length of the wing) being used. In fact, the force required to move an airfoil, and its effect increases exponentially with airspeed. At high speeds the bending force on the wing root produced by the outer aileron would be so great that it could fracture, so only the inner one is used.

In the second 737 MAX crash, the pilots were following the correct procedure in trying to reset the horizontal stabilizer manually, using the trim wheels after the **MCAS** had been pushing them downwards, but because of the exponential effect of their very high airspeed found it physically impossible to move them. They switched on the trim circuit again to help only for the MCAS to butt in again and seal their fates.

Extension (Runway Extension)

A somewhat confusing term used by controllers to mean not the physical runway but an imaginary line projecting beyond its end.

Relevant where a taxiway passes close to the end of the runway and an aircraft on it might be hit by an overrunning or low-flying aircraft.

Eye Tracking

With accidents having been caused by pilots not paying attention to the auto flight mode, eye tracking would seem to be able to play a useful role—besides seeing if they are asleep—in understanding why accidents happen and preventing them.

Unfortunately, precise eye tracking over a wide area in the cockpit without the pilots wearing cumbersome virtual reality (VR) type headsets requires complicated and expensive installations for the infrared emitters and detectors making it not possible now but perhaps possible in the future.

~ F ~

FAA (Federal Aviation Administration) [US]

The US administration that is responsible for supervising airlines, aircraft certification, and air traffic control in the country.

Even at the best of times it had to maintain an awkward balancing act between:

1. The safety of passengers and crew.
2. The commercial interests of the airlines.
3. And those of the aircraft manufacturers.

The FAA's reputation was once, perhaps unjustifiably, such that other countries' aviation authorities would virtually automatically certify any aircraft it had certified. Crashes such as the ValuJet crash and the DC-10 crash in Paris dented its reputation domestically, but it was the two 737 MAX crashes that were the straw that broke the camel's back. They damaged its reputation so greatly that other countries' aviation authorities lost faith in it, and now no longer automatically follow its approvals and decisions. They also send their experts to check on manufacturing facilities in the US, such as Boeing.

However, like nothing else, the 737 MAX scandal did really shake things up with the FAA deter, making it determined to become "great" again. It is being much more rigorous in certifying for new aircraft resulting in tremendous extra costs for Boeing in particular.

Leeham News' Scott Hamilton (see **How to Find Information about Aviation**) pointed out in 2024 that it was Congress that was largely responsible for the ineffectiveness of the FAA—which ultimately led to the two 737 MAX crashes—by not appropriating the funds necessary for it to do it job properly.

Of course, one should not forget the undue influence of Boeing over the FAA gained in part by the "revolving door" with officials knowing that on retiring they would be given roles at Boeing itself or as lobbyists and the like working for them.

Over the years the FAA has been like a political football, particular interests in that it was under pressure from Congress that had been "persuaded" to loosen the rules in the name of making the US aviation industry more "competitive."

From 2011 onwards, the industry and Congress together sought to streamline FAA aircraft certification, reducing red tape, all in the name of strengthening the competitive position of US industry.

The FAA had to knuckle under since its budget re-authorizations by Congress were at stake. Targets were being based on this streamlining with it farming out approvals to Boeing employees under the "Organizational Designation Authorization (ODA) system" using Boeing employees as FAA deputies.

For example, while **Grandfathering** had enabled the Boeing 737 to be evolved into virtually a new aircraft over many years, the Boeing 777X which narrowly failed an ultimate pressurization test was really a new aircraft all in one go with new wings, engines, landing gear, tail, cockpit, and systems. Also, its fuselage employs a different aluminum alloy and has a different length, with different load distribution, door locations, and window structure. In 2024, with the FAA determined to restore its reputation, it was taking its time to certify the 777X.

Boeing has not only lost vast sums over the grounding of the 737 MAX but very much more due to the time taken to get approvals under the stricter criteria.

FAA Regulations

The FAA has numerous regulations ranging from rules for pilots to manufacturing. We mention a few below.

~ TFR (Temporary Flight Restriction)

These are restrictions designed to protect the movement of Very Very Important People (VVIPs) such as the president but are also applied to other things such as major sporting events, natural disasters, air shows, and space launches.

General Aviation pilots can easily find themselves getting into trouble over them.

See **Ramp Freeze, Airport Brought to Standstill for VVIP**.

~ *Part 117* **Flight and Duty-Time Limitations and Rest Requirements**

These are for flight crew.

Part 121 **Regularly Scheduled Air Carriers**

Defined as airlines operating scheduled flights carrying either cargo or more than nine passengers. Includes rules such as that pilots must have 1,500 hours of flight time and must retire by age 65.

See **Pipeline Flying to Build Up Flight Hours**.

FAC (Flight Augmentation Computer) [Airbus]

On Airbus aircraft the FAC controls the inputs to rudder, rudder trim, and yaw damper.

FADEC (Full Authority Digital Engine Control)

Just as on the engines of modern high-performance cars, the control systems on modern aircraft engines must consider so many factors and adjust so many parameters to achieve maximum efficiency that only a digital computer can handle the task.

The point has now been reached where engine control may be completely automatic, with the pilots unable to intervene regarding certain aspects. As a breakdown of the control system could render an engine useless, the systems are usually in triplicate. Though the systems are engineered to allow for extreme power requirements in an emergency, there is a point beyond which they will not go—to protect the engine.

The systems also provide data for maintenance.

False Flag

A term derived from pirates flying the flag of countries with no connection to them. The term today extends to include countries that organize attacks on themselves and make the attacks appear to be by specific enemy nations or terrorists, thus giving the nation that set up the attack on itself a pretext to launch an attack on that enemy or "terrorist" organization.

As used by the Japanese in China in the 1930s it can herald the start of a long-term war.

FAMOUS FEATS OF AIRMANSHIP

Success is a twin-edged sword and was certainly so, in one way or another, for the captains we mention below. All most likely suffered extreme post-traumatic stress, inflicting mental anguish that for some may persist even to this day.

Though others in the cockpit may have played an essential role, the captain gets most of the glory partly because their name gets attached to the event. Also, it should be borne in mind that they shoulder responsibility for the lives of all on board and that is perhaps why they seem to suffer most psychologically.

~ Captain Moody, British Airways Flight BA9, June 24, 1982*

On the extremely long flight from London to New Zealand with many stops, all four engines of the British Airways 747 flamed out high up over the ocean off Indonesia and could not be restarted. This unheard of event was due to ingestion of ash from a volcano that had just erupted.

The aircrew did not know what was happening, only that they were gliding with no thrust at all, knowing only too well that ditching would be difficult in the dark with no engines, and that any survivors would very likely be a late supper for sharks drawn by blood from the injured.

As the 747 sank lower and lower, their captain, Eric Moody made the following "stiff upper lip"** announcement to his distraught passengers.

> *Ladies and gentlemen, this is your captain speaking. We have a small problem. All four engines have stopped. We are doing our damnedest to get it under control. I trust you are not in too much distress.*

[** Even cited by Google as a classic example of the use that British idiom.]

With no engines to maintain the air pressure in the aircraft, it had fallen so much that the pilots had to don their oxygen masks only to find the first officer's would not work. Since it was deemed essential that he remain operational, they sacrificed valuable height to drop down to where he could breathe properly and incidentally would take them unknowingly below the ash cloud.

Little did the aircrew (captain, first officer and flight engineer) know, the ash incrustations gumming up the fan blades had hardened as the engines cooled and were beginning to break off under the shocks received in the course of attempts to restart them. Soon, one engine after the other restarted until all were back working. To their great relief they no longer faced almost certain death.

On learning of this, air traffic control at Jakarta asked them to climb to 15,000 feet to make them visible on their radar. On complying they found themselves back again in the ash cloud and one engine soon failing. Shutting it down, they quickly redescended into clear air and made it over the mountains on three engines to Jakarta Airport.

Once near there, they thought they were safe but could not distinguish the runway lights due to the windscreen having been sandblasted by the ash. Worse, just when they needed it most, they found the glide slope component of the airport's **ILS (Instrument Landing System)** was not working.

The still working localizer (horizontal component) enabled them to be sure they were on the axis of the runway, but not of the height they should be.

Luckily, there remained a narrow slit of clear glass on either side of the windscreen enabling Moody, with the first officer reading out the heights, to see just enough to touch down on the runway and come safely to a stop.

The crew and the passengers, in a state of shock, could hardly believe they were still alive.

The aircrew were feted, showered with attention, and awarded medals in a ceremony grandly orchestrated by the airline. Captain Moody's hair began turning white, and within a year it was to be completely so. Yet, he stayed at the airline only retiring in 1996, with over 17,000 flight hours, to become an aviation pundit and much in demand after-dinner speaker.

It was probably the flight engineer struggling to restart the engines who had the hardest time when they were in the direst straits. Moody's "feat" was perhaps not only managing to land when unable to see ahead out of the windscreen and the ILS glideslope not functioning, but keeping his crew's morale up. He later mentioned that his extraordinary stiff-upper lip announcement to the passengers was really destined for the lead flight attendant.

~ Captain Pearson, Air Canada AC 143, July 23, 1983*

The fuel gauges on the brand-new Air Canada Boeing 767 flying from Montreal to Edmonton, Canada, were not working and the amount of fuel in the tanks was being measured by ground staff before departure using **Drip Sticks**.

The amount of fuel thus determined had to be converted from volume in US Customary units to weight in Metric units for Air Canada had just switched the Metric system to work out if there was the correct amount for the next leg. This would also be entered in the **FMS (Flight Management System)**, which would be able to tell the pilots how much fuel remained simply by deducting the amount of fuel used by the engines from that.

Unfortunately, the technicians (and the pilots) did not intuitively realize they had used the wrong conversion factor as they were only used to the US Customary units system. As a result, the aircraft departed from Montreal with only 45 percent of the fuel the pilots thought they had and, of course entered into the FMS.

Halfway to Edmonton, with the FMS showing they had plenty of fuel left, one engine after the other flamed out for lack of fuel. Air traffic control were unsure of where they were as the aircraft's transponder had stopped working, but finally managed to locate them on primary **Radar** and gave them the vectors for Winnipeg, the nearest airport.

Air traffic control mentioned there was a disused military airfield called Gimli somewhat nearer but could give no information regarding the situation there. In fact, it being a Saturday afternoon meant one of the old runways was being used as a racetrack for drag racing.

Copilot Quintal calculated Winnipeg was too far away, and Gimli Air Station would be their only option. By pure chance, he had done part of his military service there years before and was able to describe the runways to the captain and help guide them in.

Unable to lower the flaps and extend the air brakes to slow their descent, they found themselves coming in much too fast, but not quite high enough to do a circle. Captain Pearson, was an experienced glider pilot, and with their aircraft having become a glider, was able to apply techniques he had learnt, and sideslipping with counter rudder, came in slanted to slow them, before finally straightening out and touching down hard.

Fortunately, with the races over, the spectators, including children, were gathered at the far end except for two boys cycling up and down the runway for fun, not realizing the danger they were in as there was no sound from the idle engines. Everyone was screaming at them, and finally they realized the peril and just managed to get out of the way.

The front landing gear collapsed, but this was a blessing in disguise for the friction of the metal on the bottom of the aircraft against the asphalt together with friction with the rail separating the two halves of the racetrack helping slow them, the enormous aircraft shuddered to a halt just in front of a group of startled people for that was the runway being used for the drag racing. The collapse of the nose gear meant the rear of the aircraft was too high up for the rear door escape chutes to reach right down to the ground resulting in a few people escaping from there sustaining minor injuries, but that was all. With no fuel for a serious fire, the racers' fire extinguishers sufficed to extinguish the small fire caused by friction of the nose against the ground.

No one was able to replicate what Pearson did on the flight simulator.

The Fédération Aéronautique Internationale gave their first Diploma for *Outstanding Airmanship* to Pearson and FO Quintal. Their feat is invariably referred to as the "Gimli Glider" incident, and the aircraft, very soon able to fly again, was given that name.

~ **Captain Dardano, TACA Flight 110, May 24, 1988***

Captain Dardano, on finding he was unable to restart his engines, was about to ditch his Boeing 737 in the water at New Orleans, but at the last moment saw a relatively flat grass levee on the bank.

Despite having only one eye, he touched down so gently that there was no loss of life and for the passengers it seemed like an almost normal landing.

Though little known, he has been called, "The First Sully."

~ **Captains Haines and Fitch, UAL 232, July 19, 1989***

In a feat that again could not be replicated in the simulator, the pilots of a DC-10 tri-jet with no working flight controls managed to bring it down to a runway at Sioux City, USA, by juggling engine power alone.

The reason they could not use the rudder, elevators, trim, ailerons, flaps, or air brakes was that the center engine (at the back of the aircraft) had disintegrated with pieces severing all three hydraulic lines, resulting in the loss of all the fluid.

Providentially, one of the airline's senior captains, called Fitch, was traveling as a passenger and had, following the crash of Japan Airlines 123*, on his own initiative studied how the natural phugoid motion of an airliner with no controls might be controlled using just the engines.

He offered to help, and Captain Haines gave him the task of manipulating the throttles of the two still operating engines, which being very low-slung under the wings could push the nose up.

They, almost unbelievably, managed to make it right to the runway at Sioux City Airport, but with the slow reaction time (following application of throttle) were unable to prevent a wing snagging the edge of the runway and the aircraft breaking up into five pieces.

Incredibly, 185 out of the 296 people on board survived. This included the three pilots found 35 minutes later in the broken-off cockpit lying hidden in the cornfield next to the runway.

~ Captain Piché, Air Transat 236, August 24, 2001*

Halfway across the North Atlantic, an Airbus A330 with Captain Piché in command ran out of fuel due to the severing of an incorrect fuel line installed when the engine was replaced with a slightly different one five flights earlier.

They were 80 miles from the nearest possible landing place, a US military airfield on an island in the Azores called Lajes Field. Reaching it was one thing; adjusting height and speed so as not to overshoot or undershoot was quite another.

Piché came in at 400 miles per hour but could not risk falling out of the sky by making another circuit to lose height and speed. With only the air resistance of the deployed landing gear to slow him, he sharply raised the nose to decrease the sink rate and increase forward resistance. Hitting hard near the threshold, he burst eight tires as the antilocking system was not working.

The aircraft shuddered to a halt three-quarters of the way down the runway, with all on board evacuating in less than 90 seconds and only 12 passengers sustaining minor injuries, though some were vomiting beside the runway after the tension of the long glide.

It was a remarkable feat, though the pilots should have picked up on the fuel problem earlier. They had checked for leaking fuel, but no one was able to see it from the cabin in the dark. [Airbus reprogrammed their systems to make such a problem more evident and revised the relevant checklist.]

On being feted and the center of what he thought was too much attention, Piché claimed it was his tough experience as the only French prisoner in an American jail for allegedly smuggling marijuana in his youth—declared on joining Air Canada—that gave him the fortitude to execute the long drawn out dicey touchdown.

~ Captain Gennotte, DHL Cargo, November 2, 2003*

Outside US Homeland Security circle—believing useful lessons could be learnt—not much attention has been paid to the feat whereby the three-man crew of a European Air Transport/DHL

Airbus A300B4 freighter brought their aircraft safely back to Baghdad Airport by juggling engine power alone.

Shortly after taking off, a surface to air missile (SAM) struck the rear tip of the left wing, setting the fuel tank on fire, but did not cause it to explode as, being full of fuel just after takeoff, there was no vapor. The slipstream meant the leaking fuel was burning along the trailing edge of the wing, with it being gradually eaten away, with the prospect of it in time reaching the main spar, which was on the point of failing when they touched down. Also, a large section of the left wing was missing when they landed, making one wonder how it provided the lift that it did.

Juggling power to maneuver right up to the landing and still doing so on the runway to attempt to steer, they ended up outside the airport boundary in an area where there were mines buried in the sand. The fire tenders had been waiting and one soon reached with the fire tender chief having to tell them how they should tread carefully, with them the three men finally reaching safety.

Altogether, it was a very close-run thing, both in the air and on the ground.

The flight engineer, Mario Rofail, played a major role, showing how the presence of extra people in the cockpit can make all the difference.

~ Captain Sullenberger (Sully), US Airways Flight 1549, January 2009*

Shortly after taking off from New York's LaGuardia Airport Sully's Airbus A320 bound for Charlotte struck a flock of Canada geese at an altitude of only 2,818 feet resulting in the engines winding down and the aircraft reaching a maximum height of 3,060 feet, and the engines ultimately flaming out.

Sullenberger had the presence of mind to immediately start the **APU (Auxiliary Power Unit)**, thus providing electrical power for the computers, and making the aircraft easier to fly.

Analyzing the situation, Sully concluded correctly that he was not high enough to glide back to LaGuardia Airport, or ahead to Teterboro Airport, and ditched the aircraft in the nearby Hudson River with no loss of life. However, with him knowing there had hardly been any cases of the successful ditching of an airliner, it had not been an easy option.

Pilots using flight simulators found after several tries it would have been possible to get back safely to LaGuardia's runway if they had immediately turned back. This was impossible when the inevitable 35 seconds taken to assess the situation was factored in.

Interestingly, Sully always maintains he "landed" (rather than ditched) on the Hudson, and it was perhaps this attitude that helped him do so well.

He was indeed fortunate that there happened to be a wide expanse of smooth water almost below with rescue boats immediately at hand and, not least, having an aircraft that was one at US Airways equipped with extra rafts and life vests for overwater operations, even though its route to Charlotte was over land.

Technically, the Airbus "protections" that he had enabled by immediately starting the APU should have helped him fly without stalling. However, as he asserted, he was trying to reach the maximum capability of the aircraft to soften the touchdown, but it turns out there was a feature of the Airbus **Fly By Wire** system that no pilots knew about called the "phugoid mode" that prevented him from flaring more even though the wing was not at the maximum angle of attack, resulting in the rear of the aircraft hitting the water very hard with some panels damaged and a piece of structure puncturing the aft cabin floor resulting in a flight attendant sitting right at the rear suffering a nasty gash in her leg.

What is often not realized is that Sully's great achievement did not simply involve landing straight down on the Hudson with time for a **Stabilized Approach** as in a normal landing. He had to contend with the enormous George Washington Suspension Bridge, meaning he first had to sacrifice height gliding to it, skirt its high tower and cables on the riverbank, and finally move over to the open river by which time he would not have had enough height to risk a sharp descent to increase his airspeed, level out and prepare to "land" like an A320 normally would.

Also, the checklist for total engine failure and ditching in water written for an aircraft cruising at 30,000 or more feet was so long his copilot could not reach the end with details of the final steps to be taken on coming down to the water.

Without this, Sully was cleverly flying the gliding Airbus to its limit and paradoxically would have been allowed at the last moment to do exactly what he wanted had he been in **Alternate Law**, which would have been the case had he not restarted the **APU**, but then there might have been insufficient hydraulic power to deploy the flaps and other problems, and even a disastrous stall.

Though the aircraft fully responding to Sully's nose-up order would have meant it impacting the water with a higher pitch-up angle and slightly lower vertical speed, one cannot be sure the great result would have been better. Anyway, it was not critical, and the touchdown was remarkable. There was even a baby safely in the arms of a kind man who thought it would be better with him than held by its mother.

The exceptional feat became known as *The Miracle on The Hudson* with Sully idolized worldwide.

Good judgment is the key quality required of a captain, and Sully exemplified it.

~ Captain de Crespigny, Qantas Flight 32, November 4, 2010*

The inner left engine of the Qantas Airbus A380 en route from Singapore to Sydney disintegrated as it was climbing out of Singapore's Changi Airport.

Pieces fell on Indonesia's Batam Island, and though no one on the ground was injured or killed, one piece, still hot, narrowly missed a child in class at primary school.

Its captain, Richard de Crespigny, immediately took over control, adroitly putting the giant aircraft in "altitude hold" to level off and thus place less strain on the aircraft and remaining engines.

In addition to the captain, first officer and second officer, two very senior check captains also happened to be present in the cockpit, since de Crespigny was being checked as well as one of the check captains. While they represented an invaluable reservoir of experience and talent, having such senior people together could have been problematic had not de Crespigny made it clear to his first and second officer that he was the designated captain and in charge. See **Hierarchical Gradient between Pilots**.

Fragments of the engine had pierced the A380's left wing from which a constant stream of fuel was pouring out. They had also severed trunk control wiring embedded in the outer fuselage rendering many systems including fuel transfer and sorely needed fuel dumping inoperable.

With the aircraft sluggish but still flyable, they nurtured it as they flew around for an hour with them troubleshooting the 100 or so faults displayed on the **ECAM (Electronic Centralized Aircraft Monitor)**.

It was vital that they land once they had sorted out the problems the best they could as at any moment something further could go wrong with them totally losing control.

Before landing, de Crespigny nevertheless insisted they perform a "control check," something he previously had experience of in the military, where pilots of a returning aircraft damaged in action had to determine what they could still manage. This consisted of trying out the control settings and airspeeds they intended to use for the landing while still high up with enough height to recover. They were considerably overweight, and with various systems, including those for slowing the aircraft on the ground degraded or inoperative, it would be a close-run thing.

Finally, de Crespigny made a near-perfect touchdown back in Singapore, with them stopping a few 100 feet or so (some 60 meters or so) from the end of the runway, just as one of the check captains had calculated.

Apart from the deftly executed touchdown under considerable mental stress, the great outcome can largely be attributed to de Crespigny systematically making the most of his own talent and that of all the others in the cockpit who were so gifted and experienced.

FANS (Future Air Navigation System)

Traditionally, air traffic control has mostly been exercised by controllers seeing aircraft on their monitors or being told where they are with the possibility of misunderstandings due to language problems or poor radio connections. This is referred to as *positive control*.

In areas over oceans with no radar coverage and only very imperfect high-frequency radio coverage, positive control is not possible, so something called *procedural control* is used where the pilots are not constantly in touch with the controllers but agree on flight plans and timings with, of course, much more separation than with positive control.

This means fewer aircraft in that airspace and the risk of the high-frequency radio connection getting saturated where there is heavy traffic, making it very inefficient.

FAR (Federal Aviation Regulations)

Regulations classified under relevant headings (parts), covering every aspect of aviation, including the operation of flying schools, operation of ultralight vehicles (such as trikes), and certification of pilots and aircraft.

Points to note are that although the regulations are constantly updated, once an aircraft is certified it does not lose its certification and is still allowed to fly, even though it might not meet some new or tighter requirement, say, for more flame-resistant seats. There are limits to this, and something truly dangerous would be the subject of a mandatory airworthiness directive (AD).

In view of the time and cost involved in getting improvements certified, manufacturers can be tempted to forgo them even though they know they would make the aircraft safer.

Faraday Cage

We mention the Faraday cage because we cited it in the context of protecting computers and other devices from **EMP (Electromagnetic Pulse)** attacks brought about by exploding a nuclear device high up in the atmosphere.

The Faraday Cage is an enclosure consisting of a continuous conductive material, or often a mesh of conductive material to protect what is inside from electromagnetic fields.

The domestic microwave oven is an example of the Faraday cage except that it is designed to protect what is *outside* from electromagnetic fields generated *inside*. Though it can theoretically be used to prevent outsiders accessing your mobile phone that should **never be tried** as it would probably not only result in ruination of your phone but worse still explosive difficult to extinguish battery fires and other problems. Buy a cheap Faraday bag to place it in instead.

FAROS (Final Approach Runway Occupancy Signal)

A new system being tried in the US whereby the **PAPI** lights next to the aiming point flash to warn the pilot if an aircraft is on the runway or there is a possible incursion.

"Faster May Be Safer"

There are times when flying faster than normal may be safer. An example would be in the presence of a microburst where rain and wind are coming vertically down onto the runway and splaying out like water from a faucet (tap) into a basin. In that case the aircraft first encounters a headwind that switches to a vertical downdraught followed by a tailwind with its airspeed dropping too low and it possibly stalling. Adding airspeed beforehand gives a greater safety margin.

Pilots are faced with a conundrum in cases where their aircraft has suffered structural damage. On the one hand, the **Standard Operating Procedure (SOP)** is to fly at a reduced or minimum airspeed to prevent the risk of aircraft tearing itself apart. On the other hand, there may be a reason why reducing airspeed to comply with SOP may be fatal.

The incident shown in the photo below is the saddest example of this.

[NTSB]

The left engine of a DC-10 taking off from Chicago O'Hare Airport in 1979 fell off on the runway as it took off leaving the aircraft still able to climb out and fly.

However, in separating from the aircraft the engine spar had damaged the hydraulics lines controlling the slats on that wing. As McDonnell Douglas had not installed a mechanical locking mechanism for the hydraulic slat mechanism, loss of hydraulic pressure allowed the slats on that wing to retract with the result that the wing stalled and the DC-10 tipped over some 120 degrees and crashed. The pilots, not knowing that wing was stalling without the slats, had reduced speed to the minimum takeoff speed (V_2) according to SOP.

Had they been flying just a few knots faster they could have landed safely. As it was, they all perished.

There have been several incidents where pilots have landed structurally damaged aircraft faster than recommended, perhaps saving all on board, apart from those who lost their lives when the original mishap occurred.

An example would be Aloha Airlines Flight 737* in 1988 where passengers in the front area of the cabin were left sitting in the open. See **Decompression (Loss of Cabin Pressure)** with photo of them.

Fasteners

Usually a rivet, bolt, or screw for fastening items together, which sounds very mundane, but which has been highlighted by

Boeing's problems with fasteners for its new 787. First, fasteners were in short supply, and now it seems some, such as those for attaching composites to titanium, were fitted incorrectly, with corrective work perhaps entailing the use of larger fasteners (hence, heavier) to fill the holes already made.

Fastest Air-Breathing Aircraft: NASA's X-43

There has been much talk of the US falling behind in the domain of hypersonic vehicles with it forgotten that at one time it was ahead with NASA's X-43 developed in the late 1990s.

Three were made under the Hyper-X program.

On November 16, 2004, the third version, after being carried up to 40,000 feet attached to a B-52 and released with a Pegasus rocket booster taking it to 110,000 feet and released at Mach 9.8 with the hydrogen powered scramjet flying it for 10 to 12 seconds at Mach 9.65 with it ultimately reaching a speed of Mach 9.68 or 6,755 mph (10,870 km/h) at 109,440 feet.

Though this proved what could be done with certain materials, and was technically a record, the X-43 was very small, had no pilot, and glided down to the ocean with no intention that it be reused.

Fatigue (Human)

As with motoring, fatigue is a major cause of accidents and has been on the NTSB's "Most Wanted" list.

Unlike alcohol levels in the blood, fatigue is almost impossible to measure, which is a pity for being tired can more adversely affect pilot performance than being mildly inebriated.

FBO (Fixed-Base Operator)

This obscure term originated in the early days of general aviation in the US to distinguish people servicing aircraft from a fixed base at an airfield from mechanics traveling around to service a particular machine that could not go very far anyway.

Though not officially recognized internationally, it is now also used outside the US.

FBOs have their own facilities at the airfield or airport and the pilot of a GA aircraft will often look them up beforehand to see their charges for parking (outside on the ramp or in the hanger), fuel, and so on, and let them know in which case they will likely find they are greeted and guided on arrival, with even assistance with luggage, without being a VIP in an executive jet.

FBOs provide a multitude of services, even selling or renting out aircraft, maintenance, aerial photography, and not least facilities for refreshment and transport near the airfield.

FCS (Flight Control System)

As the term implies this is the system consisting of the sidestick/yoke, computers, software that controls how the aircraft flies.

However, it can also mean the controls for unmanned craft, such as drones.

FDR (Flight Data Recorder)

Records the last half hour or more of numerous aircraft parameters, permitting accident investigators to determine more easily what happened. FDRs are getting more and more sophisticated, with modern ones sometimes referred to as DFDR (digital flight data recorders).

One problem is the so-called sampling rate, in that, say, the position of the rudder is recorded at intervals of several seconds, with no indication of how it moved in the meantime. This made it more difficult for investigators of the crash of an Airbus caught up in wake turbulence on taking off from New York's JFK Airport in 2001 to determine precisely how the pilot swished the rudder back and forth, causing it and the vertical stabilizer to break off*.

Fello'fly, Flying Like Migrating Birds to Save Fuel [Airbus]

Migrating birds fly in a V-formation with one riding in the wake of the one slightly ahead. This saves energy and Airbus believes airliners could save 5 percent to 10 percent of their fuel by doing likewise and is studying this in a project it calls *fello'fly*.

Modern technology should make it possible to ensure that two or more airliners flying in formation maintain the requisite position. There are obviously a few problems that need to be ironed out, not least that of air traffic control, especially as regards the joining up and separation of the aircraft. However, having two or more aircraft flying together could free up airspace on congested long-haul routes. Sometimes if an aircraft misses its takeoff slot it must wait a very long time for a new slot in airspace thousands of miles away.

Working out how to share the benefits might be difficult; one way would be to reverse positions midway. One wonders whether birds do that; perhaps the strongest bird always leads.

Fear of Flying, Aerophobia, Why When So Safe

Flying is by far the safest mode of transport but because of people's natural fear of heights, and not least turbulence, some passengers can become so fearful they do not want to fly at all.

Aerophobia can be seen to be either of two elements or a combination of both.

1. Fear of flying itself, though when one looks at the statistics, it is very much safer than almost anything else we do. The drive to the airport may be thousands of times more dangerous.

2. Claustrophobia, the fear of being cooped up in a small container with a feeling of powerlessness and helplessness, which perhaps is more pernicious and difficult to handle. Adding to this, we have turbulence, which at first makes people uneasy and then due to its repetitiveness makes their anxiety grow and grow.

Airlines have courses for those so fearful that they hesitate to fly. At a lower level, some medication prescribed by a doctor can help or, dare one say, a couple of alcoholic drinks. However, there is a school of thought that considers medication ill-advised in the long term as it merely masks the problem and gives no chance for the person to learn to overcome their fear.

A few words with a flight attendant can help. Also, a seat over the wings is likely to be preferable to one right at the back where one is liable to be "thrown around" more.

As mentioned under **Turbulence**, think of it like driving over a road with potholes—something not so terrifying—or better still as a free funfair ride! The fact that the wings bend is a good thing as it reduces the effect of the turbulence on the occupants and means they are not going to break off. In fact, when the wings are tested, they are bent, without breaking off, many times more.

Though turbulence is not going to bring the aircraft down, one should as far as possible always keep one's seat belt fastened as on very rare occasions **Clear Air Turbulence** can lift a person off their seat, even to the extent that they hit their head on the cabin ceiling and then hurt another passenger or themself when they fall back down. The great problem with clear air turbulence is that it may occur with hardly any, or no, warning, at a time just when the very smooth flight at the time has passengers think seatbelts are unnecessary.

Seat belts should be kept attached, even loosely, unless walking away from one's seat.

British Airways has a one-day course for fearful fliers with some too afraid to even try flying, which seems to have a 98 percent success rate. It starts with technical explanations followed by explanations by psychologists and ends with a short flight on an A320 from and back to Heathrow. The person in charge gives a running commentary explaining what the various noises mean. They sometimes bring along a member of their family to show how confident they are that flying is safe.

Other airlines have similar courses.

Feather [Prop Aircraft]

Should an engine on a prop aircraft fail, the angle of the propeller blades for it is adjusted so they cause minimum drag. This is called feathering.

Ferry Flight

A nonrevenue flight to position the aircraft for a subsequent flight or to take it back to home base or other location for maintenance. Sometimes an airline might carry out such a flight to keep its right to a valuable landing/takeoff slot.

Environmentalists criticized BA in 2008 for flying an empty 747 from London to Hong Kong. The airline claimed it was because it could not muster the required cabin crew—if the airline had another flight to Hong Kong that day and few passengers, it might have been cheaper to put all on the one aircraft. Not all is what it seems.

Statistically, nonrevenue flights (as a whole) are more likely to be involved in accidents than revenue passenger-carrying flights, sometimes because pilots may be tempted to experiment.

Ferrying Distance [Military]

With mid-air refueling, a fighter such as the F-12 can stay aloft for 12 hours and travel a considerable distance—just about the distance from the US to Iraq. However, it must then land, not for fuel but so that oil can be replenished. Note, the seat on the F-16 can be pushed very far back during cruise on such long flights.

Fire Trucks

Many of the fire trucks (fire engines) at major airports are very sophisticated and have thermal imaging cameras able to highlight hotspots including overheated brakes which can sometimes be white hot and liable to ignite a fire.

"Fighting at the Controls"—Journalistic Hype

At a time when air crashes were more frequent and aircraft less automated than now, newspapers would often say the heroic pilot had been *fighting at the controls* to save the aircraft, usually unsuccessfully. This often reflected an error of judgment on the part of the pilot.

For in an emergency, it is comprehension, judgment, and decision-making that are critical in saving the situation. Take the famous case where Sullenberger ditched his Airbus A320 in New York City's Hudson River* without any loss of life after an encounter with Canada geese had disabled both engines at only 3,200 feet. It was his quickly switching on the auxiliary power supply and deciding to ditch in the river rather than trying to make the impossible limp to a runway that saved everyone in the aircraft and even many on the ground had he crashed into buildings. It was not his fantastic flying as everyone assumed that saved the day though, of course, professional piloting was essential. Starting the auxiliary power unit provided power to enable the computers to help him fly better.

Fin (Tail Fin); Vertical Stabilizer [US]

Fixed vertical airfoil on tail, usually with the (movable) rudder inset that, as the US name for it implies, keeps the aircraft directionally stable—in other words, stops it yawing.

Final/Final Approach

Final leg of the descent, leading straight to the runway. There is also the term "late final," meaning less than two nautical miles from the runway.

Financing Airliners—Sell and Leaseback on Delivery?

When an airline purchases an aircraft from the manufacturer, it pays in three stages:

1. *Down payment.* Normally 1 percent, paid at the time of placing a firm order.

2. *Pre-delivery payment (PDP).* Normally 49 percent over the two years prior to delivery.

3. *Delivery payment.* Normally 50 percent on the day of delivery.
   An airline with a good credit rating can immediately sell the aircraft to a leasing company or bank on the firm understanding it will lease it back. If the airline

has been able to purchase the aircraft with a large discount because of the size of the order or as a launch customer, it may sell it on for more than it paid and make an immediate profit!

Fire on Board

Nothing strikes fear into the heart of pilots more than fire. Fatal experience has sadly shown that an uncontrollable fire on board may mean they may only have ten minutes to get on the ground, be it at an airport, or anywhere flat, for anyone to survive.

One problem with fire is its unpredictability in that it can smolder in some invisible place and then suddenly turn into a major conflagration taking in the whole space in what is called a "flash fire."

Cargo holds are often sealed so that, in theory, a fire cannot really take hold there due to lack of oxygen. However, there was the disastrous ValuJet crash in the Florida Everglades where the hold contained oxygen generators that fed oxygen to the fire*.

Following the loss of Swissair Flight SR111* in September 1998 where "a fire of unknown origin" broke out in a MD-11 trijet that had just before taken off from New York's JFK bound for Geneva, with the pilots at the beginning thinking it was a PAN, PAN, PAN and not a MAYDAY situation, there was a great rethink regarding how pilots should react, and airliners made safer.

First, it was made clear that pilots should land as soon as possible and not waste time dumping fuel even if seriously overweight—the checklist the pilots had on that MD-11 for dealing with such fires took almost half an hour to complete with the advice to land right at the end!

Second, airliners were made intrinsically safer with smoke alarms in more hidden places and metallized Mylar (thought to be what allowed the fire to spread in the Swissair MD-11 following a short circuit in the wiring for the passenger entertainment system) removed and banned.

Also, it was decided that cabin crew should immediately seek the source of the fire and try to extinguish it. They have fire extinguishers and special bags into which a mobile phone or a battery pack with lithium batteries on fire can be placed. Such items should be carried in the passenger cabin where fires can be dealt with easily.

The fuselage of an airliner is almost as thin as a credit card and can only resist an external fire for a limited time and is partly why an airliner must be certified as allowing all the passengers to escape in 90 seconds with half the exits unusable. See **Evacuation, the 90-Second Rule**.

It is somewhat reassuring that when a light aircraft used to help survivors of an earthquake entered the runway at Haneda Airport in Japan without authority on New Year's Day, 2024, just as an Airbus A350 was touching down, causing it to catch fire, all on board the A350 escaped safely despite the aircraft ending up a burnt out wreck. Though the **Carbon Fiber Composite Polymer (CFCP)** used for the fuselage ultimately burnt, its immediate resistance to fire was seeming better than that of the traditional aluminum.

First Powered Flight by The Wright Brothers

The Wright brothers, Orville and Wilbur Wright, made the first flight in an engine-powered aircraft on December 17, 1903.

First-Person View (FPV)

Term used in connection with drones where the controller (first person), perhaps a very long way away, sees what the drone sees. Also called Remote-Person View (RPV).

Five Eyes

A special, notably intelligence-gathering, relationship between anglophone agencies from the UK, the US, Australia, Canada, and New Zealand. For the US it has the advantage of providing access to "outposts" all over the world.

Flak [Antiaircraft Fire in World War II]

The term comes from the name of a well-known German antiaircraft gun. Though aircraft are nowadays more likely to be downed by SAMs, the term is still used in the name "flak jacket."

In World War II, a Luftwaffe study estimated it took over 3,300 88-mm shells to shoot down a single bomber. Nevertheless, in 1944, flak accounted for 3,501 American planes destroyed, while the fighters shot down about 600 fewer over the same period.

While flak was sometimes dismissed, perhaps out of bravado, it placed great strain on the bomber crews, and as it was often concentrated around targets it put off the bombardiers (bomb aimers) on their bomb runs causing their bombs to fall wide. Sometimes it was so intensive that crews would drop their bombs anyhow rather than go around and risk making a second or third bombing run in the hope they could see the target better.

Flameout

Once started, a jet engine has a flame that burns continuously.

However, if for some reason, such as lack of fuel or ingestion of foreign substances, the flame goes out with the engine no longer producing any thrust, it is called a "flameout." However, the engine will usually continue to rotate due to the windmilling effect, and this is why the term "stopped" is not used.

To prevent the flame from being blown out by a sudden gust of air, engines incorporate a flame holder in the form of a small cusp that behaves like the cusped hand used to light a match in a strong wind. Also, there are electrical ignitors that pilots switch on in conditions such as heavy rain or hail where an engine might flameout. See **Relight**.

People are astounded at how much water (rain) modern engines can ingest without flaming out. Indeed, when badly damaged Qantas Flight 32* stopped just short of the end of the runway at Singapore, the control circuits were so badly damaged one of the engines could not be stopped. Firefighters spraying in as much water as possible could not stop it either and could not leave it because someone could get sucked in. After more than an hour, they had to use foam resulting in damage to the engine.

The reason why engines can ingest such vast amounts of water without stopping is that the efficient high-bypass engines send most of it out without it passing through the combustion chamber—the centrifugal force generated by the rapidly rotating fan blades at the inlet splay outwards into the bypassing air.

Flaperon

Like "elevon," meaning a combination of flap and aileron, a flaperon is simply an airfoil that can act both as a flap and as an aileron. The Boeing 777 is one of the few airliners to use them, and when one was found washed up on a beach in La Reunion on July 29, 2015, it was assumed to be the first item found from Malaysian Airways Flight MH370, which mysteriously disappeared on a flight from Kuala Lumpur to Beijing in March 2014. Serial numbers showed that was so, and that it had most likely drifted some 2,500 miles from the search area off Australia.

Flaps

These are hydraulically powered airfoils set at the trailing edges of the wings that are extended to give the aircraft extra lift both when flying slowly and full of fuel when taking off, and when flying relatively slowly on landing. They usually work in conjunction with **Slats** at the leading edge of the wings which also give extra lift.

The picture below shows the aircraft with the flaps and **Spoilers** configured as they would be just after touching down on the runway. The flaps (at an angle of about 30°) are extended out at the back, and the spoilers (sticking up at about 70°) not only slow the fast-moving aircraft but also push it down on the runway and thus improve braking.

[Canva]

For takeoff, the flaps are usually set at only 5°, which despite the very small angle greatly increases the area of the wing, which together with the **Slats** on the leading edge of the wing greatly increases the lift, permitting the aircraft to lift off and stay airborne at low airspeeds. This is particularly necessary on taking off with the aircraft heavily laden with fuel for a long flight.

For landing, flaps are set at 30° or 40° as shown on the diagram above, allowing the aircraft to land at a low airspeed with a lot of drag to slow it quickly.

There are various types of flaps:

1. Simple flaps.
2. Split flaps.
3. Slotted flaps.
4. Fowler flaps.
5. Slotted Fowler flaps.

There have been crashes where pilots have forgotten to set the flaps on takeoff. However, nowadays such an accident should be impossible for the pilots have to go through perhaps five takeoff checklists with lowering them mentioned, with, as a last resort, a visual and aural configuration warning sounding should they initiate a takeoff without them set properly.

Flare on Landing

Just before touching down and at an appropriate height above the runway, the pilot raises the nose to reduce the rate of descent and at the same time attenuate the airspeed. This softens the landing and means less braking will be required.

If they raise the nose too much there is not only the risk of rear of the fuselage striking the runway (tail strike) but of the aircraft "floating" with the aircraft touching down much too far down the runway. Modern airliners often have software reducing the likelihood of a tailstrike both on taking off and landing.

It should be noted that unlike what can be done in some tiny general aviation aircraft which can virtually stall and plonk down, the flare in an airliner is not a stall but just a way to lessen the sink rate and reduce the airspeed. The aircraft must keep flying so that a last-minute go-around remains feasible.

Interestingly, large military transport aircraft, such as the C-5 Galaxy, cannot perform a significant flare. Instead, they can always augment engine thrust which splaying on the deeply lowered flaps (immediately behind the engines) attenuates the sink rate. That said, with the appropriate thrust setting and good judgment the C-5 can come in very gently to a very smooth landing without the need for major last-minute adjustments.

"Flex Takeoff" [Airbus] and "Assumed Temperature" [Boeing]

With engine power settings calculated automatically by the aircraft for optimum takeoff it is sometimes desirable to "trick" the system to use lower thrust and put less strain on the engines. This reduces wear and saves money on maintenance but can only be done where runway length, altitude, and temperature permit.

Airbus calls this flex takeoff, while Boeing calls it assumed temperature. The pilots simply enter a higher than real ambient air temperature into the flight management system (or flight bag laptop) so the available thrust appears lower than it would normally be—this is because the engine temperature cannot exceed a certain limit.

There are other ways of achieving this.

Flight Controls, Primary and Secondary

1. The primary flight controls are the ailerons, elevator (or stabilator), and rudder system needed to control the aircraft in flight.

2. The secondary flight controls are the flaps, leading edge devices, spoilers, and trim system needed to improve the performance characteristics of the aircraft or in the case of the trim system relieving the pilot of having to apply excessive control forces.

Flight Data Management/Analysis (FDM/FDA)

The analysis of data gathered for specific flights so lessons that can improve safety can be learnt. Obligatory in some countries but not in the US, where it is known as flight operations quality assurance (FOQA), due principally to pilots' fears as to how the information might be exploited. (See **Data Mining**.)

Flight Data Recorder (FDR)

One of the **"Black Boxes" [Cockpit Voice and Flight Data Recorder]** capturing most of the parameters of the flight, including settings of the controls for analysis should there be a crash.

Flight Director (FD)

As its name implies, the flight director is a system that tells both the pilots and the autopilot what they should do according to the **FMS (Flight Management System)**.

The link with the autopilot can be switched on and off, but whether the autopilot is connected or not, the pilots are shown what they (or the autopilot) should do by overlay markings on the attitude indicator right in front of them. If the pilot needs to go left or right, the flight director roll bar superimposed on the attitude indicator tips to the left or right; similarly, if he or she needs to climb or descend, the superimposed pitch bar indicates that.

The system works so perfectly that pilots get into the habit of slavishly following it even when they should not, say, because sensors detecting the airspeed and so on are malfunctioning because of icing, for example. (See **Autopilot**.)

Flight Information Region (FIR)

Airspace throughout the world is divided into Flight Information Regions (FIRs). Each FIR is assigned a controlling authority responsible for providing air traffic services within it.

When **MH370 "The Greatest Aviation Mystery Ever"** disappeared with its transponder switched off, it was in as many as five FIRs (Malaysia, Vietnam, Singapore, Cambodia, and Thailand), meaning was in the ideal place to disappear as that would add to the confusion.

Flight Level (FL)

When aircraft are **high up**, where there is no danger of hitting the ground but still a danger of hitting each other, air traffic controllers assign them "flight levels," with the units being in 100-foot steps. Thus, Flight Level 1, which would never be used, would mean an altitude of 100 feet, while Flight Level 250 would mean 25,000 feet.

In some countries, such as Russia, China, and Mongolia, flight levels are given in full in meters, with no division by a factor such as one hundred.

Aircraft suitably equipped to be able to fly accurately are allowed to fly with a vertical separation of 1,000 feet when above 28,000 feet. In general, flights flying predominantly westwards are given even-number flight levels, say 340 (34,000 feet) or 360 (36,000 feet), and those flying eastwards odd-number ones, say 350 (35,000 feet) or 370 (37,000 feet). There are exceptions to this rule, especially where virtually all the flights are in the same direction, such as first thing in the morning with flights arriving in the UK from the US.

See **Altimeter ~ Barometric Altimeter.**

Flight Time—Pilots Required for Long-Haul **and** Ultra-Long-Haul

1. A flight lasting up to 8 hours of flight time requires only one set of pilots (a captain and first officer).
2. A flight with flight time between 8 hours to 10 hours requires three pilots (two captains and one first officer).
3. A flight of more than 10 hours of flight time requires two sets of pilots divided into teams with each team comprising of a captain and a first officer.

The distribution of tasks for two pilots on duty all the time on flights lasting up to 8 hours needs no explaining.

The three pilots operation is somewhat complicated. Let's make it a 9- hour flight time, with each pilot able to work for 6 hours with 3 hours of rest. Remember, this flight requires two captains and one first officer.

At takeoff and landing it is normal to have all parties present as it is thought the extra pair of eyes adds to safety. After climb out to 10,000 feet the observer captain would go rest and return to take the place of the captain flying.

Three hours after takeoff, the resting captain returns and takes over the control from the active captain. The active captain then goes for rest. After 3 hours of resting, he goes back into the cockpit and takes over control once again. This time it's the first officer's turn to rest, also for 3 hours. As the first officer rests, the flight is flown by two captains, with one acting as PNF (Pilot Not Flying). Three hours later, the first officer returns for duty and relieves the captain who has been there for 6 hours.

Flights longer than 10 hours of flight time with two teams with each having a captain can have a simple distribution of tasks with all usually being present in the cockpit at takeoff and landing.

Flight Training, Becoming an Airline Pilot?

At one time in the US it was alleged the key to becoming an airline pilot was a Mastercard (credit card), though the military was another route for someone without the necessary funds.

With fewer coming from the military, the situation has changed somewhat in the US and has become more similar to that in other countries such as the UK.

Several distinguished pilots caught the flying bug—the desire to fly—when taken up for a jaunt when very young, as was the case for Amelia Earhart!

While many young people may dream of being an airline pilot, it would be a good idea for them to get an instructor at a flying school to take them up to see how they feel. The author knows of one case where a teenager who for years had toyed with the idea of becoming a pilot, only to freak out in turbulence on being taken up and realize it was not for them.

While only high-school qualifications may be essential to become an airline pilot, higher academic qualifications cannot only give more opportunities for employment, but also help the individual find alternative work should he or she be laid off, for instance, on medical grounds. Also, the airline industry is cyclical and there is always the possibility of being laid off.

Flight Mode Annunciator (FMA)

A key cockpit display showing the flight modes being applied. Most important because it indicates the settings to which the flight director and autopilot are subject, such as the parameters for a descent or climb.

An accidental sudden force on the control column can disengage the vertical control component of the autopilot without the pilots noticing as happened in the Everglades crash of Eastern Airlines 401 on December 29, 1972*.

Flight Plan

An official document filed with aviation authorities prior to departure listing details of the flight (names of crew, number of passengers, destination, waypoints, speeds, alternate airports, and so on).

A flight plan may not be legally required in the case of general aviation flights under VFR, but it is usually advisable to file one, so rescuers know where to start looking if the aircraft is reported lost. The GA aircraft of millionaire adventurer Steve Fossett that disappeared in 2007, which was only found by chance months afterwards, is a case in point—he had not submitted a flight plan.

Flight Plan for a Passenger Flying Long-Haul with a Stopover

A passenger flying a very long route with a change of aircraft should consider pacing themselves with a "flight plan." This might simply mean staying awake and eating reasonably on the first leg and then being able to sleep properly on the second to arrive feeling fit.

On ultra-long haul, say 16 hours or the future 20-hour Qantas Sunrise flights from Australia to London or New York, cabin crew will have done the planning for you, adjusting lighting and providing suitable meals at the right times.

Flight Simulator

We need to differentiate between flights simulators used by airlines and those used by individuals, though some pilots, such as the captain of Malaysian Airlines Flight **MH370 "Greatest Aviation Mystery Ever,"** have one destined for individuals at home!

Airline Flight Simulators

Modern flight simulators can simulate most of what pilots will experience except for the g-forces. They are used for training and testing. Interestingly, pilots are not allowed to play purposely crashing the aircraft lest that reinforce negative predispositions.

Simulators are classed according to their degree of realism from Level A to Level D, with D being the most realistic.

Recently, simulators using virtual reality headsets have been tried with the advantage not only that they require less hardware but also can track eye movements. It was found that experienced pilots as opposed to novices were more methodical in their scanning and tended to keep their gaze for longer on the items they were looking at.

Home Flight Simulators

Prices have come down partly because so many people are buying them as computer games. The best for learning about flying are perhaps Microsoft Flight Simulator and X-Plane.

MS Flight Simulator has stunning worldwide scenery even enabling one to go back in time with it showing New York's twin towers prior to **9/11**. Note that it will block any attempt to simulate flying an airliner into them! However, it recently added a tweak enabling one to practice flying an Airbus suffering an engine airstrike.

Because of the enormous market associated with its coupling with Xbox, gaming prices seem to have come down.

X-Plane is also very good for teaching the basics with a Cessna 172, of which more have been manufactured than any airplane in history and are favored by flight schools.

There is even a site https://flightsimcoach.com/ with a selection of instructors to hold your hand when using a home flight simulator.

While these simulators cannot take the place of learning on, say, the Cessna, they can shorten the time required to become proficient. We cannot suggest which versions of the various flight simulator programs to use as updates appear all the time.

For all sorts of examples of flight simulators and sometimes a "short" explaining which is best for a particular application, see "@AaronRheins" on YouTube.

Flight Time and Block Time

Flight time is the time taken from takeoff to landing.

Block time is the time taken from push back at the departure gate to arrival at the gate at the arrival airport and includes taxing time. Since it includes taxiing time, which can be considerable, it gives a truer idea of the time passengers are ensconced in their seats.

It is called "block" because aircraft, for safety and because brakes cool better when not applied, have blocks against the wheels to stop them moving when parked, and the terms "off-block" and "in-block" are used.

Flight Tracking

These sites enable one to see where planes are going all over the world, and also their schedules, real-time estimated times of departure and arrival.

Here are four sites:

1. http://Flightradar24.com
   Perhaps the best known, enabling one to follow flights all over the world. The basic version is free, with payment required to show extra details such as GPS height in addition to barometric height and history of flights over more than seven days.

2. http://FlightAware.com
   More complex but also source of information about many aspects of aviation.

3. http://planeflighttracker.com
   Great display of flights and so on—seems to use FlightRadar24 for tracking but if you look carefully, you can find interesting pages including one showing military airbases worldwide.

4. http://ADSBExchange.com
   This tracking operation is somewhat special in that it started with amateurs working together with simple devices and using triangulation (from three or four) to determine the true location of the aircraft even if it is not broadcasting it. Not looked on favorably by the US military as aircraft having switched off ADS-B and using **S-Mode** are still located.
   To understand what ADSBEX is about and even how to

participate, see their FAQ page:
https://www.adsbexchange.com/faq/

The US Senate is considering trying to limit the ability of enemies but, in fact, also the public to track the movements of aircraft carrying officials. This would also please the very rich travelling in their private jets and allow them to keep their movements and those they invite along secret.

However, flight tracking has enabled the public and press to find out about the waste of public funds, say with high-ranking government officials and military officers enjoying flights on separate "executive jets" to the same destination. It can also reveal an aspect of lobbying.

There is JetSpy, https://jetspy.com calling itself "The #1 Private Jet Tracker" enabling one to track business jets, and the like, from their hull registration number with much historical detail with aircraft in categories such as Individual and Family, Government, Public and Private Corporations, Sports Teams, Churches and Religious Organizations, etc.

Flight tracking even allowed "rendition flights" (the secret moving of terrorist suspects from country to country) to be tracked historically.

Interestingly, when Senator Pelosi flew to Taiwan, so many people were following her, that Flightradar24 sank under the demand and took a little while to recover.

Due Regard

The ICAO convention states that in their management of state aircraft (official aircraft including those of the military, police, and customs services) that while they must sometimes be kept secret they should pay "due regard" to civilian traffic. There are no norms laid down.

It is therefore the responsibility of the authorities and individual pilots to avoid collisions, and it is suggested that as far as possible they should be allotted sole use airspace. There have been some near misses, and should flying without ADS-B become normalized there could be one or more disasters, at a time when there have been so few in the US. The collision between an airliner landing at Washington DC's Ronald Reagan Washington National Airport and a Coast Guard helicopter in 2025 is an example of what can happen in this context.

A truly dangerous potential US enemy would not be much affected since they can track aircraft using satellites, perhaps even picking out those with no transponder switched on as of particular interest. Note that the military can always spoof transponder returns to sow confusion as regards normal tracking. Sometimes they are more than happy to rely on flight tracking to prove their presence in a given theater of operations as a warning.

Contradiction between Safety and (Cyber-) Security

The first consideration of the FAA is safety, and they are loathe to introduce measures such as encryption and long identifying tags that would not only risk overloading their systems but be difficult to realize for all aircraft including old ones and general aviation aircraft. Then there is the rest of the world to consider.

Floating

When landing an airliner, pilots aim to touch down close to the large rectangular markings on the runway denoting the ideal touchdown point. Before reaching there they usually perform a **Flare** which consists of raising the nose to break the descent and slow but not stall the aircraft, so it is still flying and can still perform a go-around if necessary.

Sometimes, say because of coming in too fast or a tailwind, the aircraft remains in the air, eating up runway, and is likely to go well past the touchdown markings. It is then said to be "floating." If the runway is short or contaminated (i.e., water-laden or icy) this can result in the aircraft coming off the end of the runway in what is called an **Overrun**.

If the first officer is the pilot flying (PF) the captain might say, "Get it down! You're floating," as has happened in some overruns.

Flutter

Vibrations that can build up in an aircraft's structure due to flexing and the generation of harmonics, and possibly lead to its destruction, with this sometimes being called "flutter." Adjusting distribution of masses and degrees of rigidity are some of the methods used to prevent its occurrence. Suitable computer software programming for control inputs can often curb nasty tendencies without resorting to structural changes or adding weights, which sometimes consist of depleted uranium, which is 68 percent denser than lead.

London's Millennium Bridge over the River Thames is an example of this phenomenon. Nicknamed the "wobbly bridge," it had to be closed only a few days after its official opening for corrective measures, due to serious wobbling in harmony with pedestrians' footsteps.

Fly America Act

The "Fly America Act" obliges federal employees and contractors to fly on American airlines or book using the US code where a US airline has a code share agreement with a foreign airline. It is a bone of contention in negotiations regarding landing rights and so on.

Fly by Wire (FBW)

The term used to describe aircraft where unlike old ones such as the Boeing 737, where the controls are linked mechanically to the control surfaces, there is an electronic interface. Moving the controls tells the interface or rather a computer what the pilot wants to do, and it does it.

Boeing once laughingly told guests at its facilities how crazy it was that the first fly by wire airliner, the A320, landed perfectly on trees at an air show. In fact, the pilot of that aircraft was engaging in a ridiculously dangerous maneuver at too low an airspeed without realizing the height of the trees*. The A320, in fact, heralded the future and continues to be very successful.

Boeing's first fly by wire airliner was the 777 with control surfaces and other devices not moved directly via, say, cables but via an electronic cum computer interface. However, the Boeing and Airbus aircraft FBW systems differ as regards how the controls show what they are doing. Boeing makes them move as in older aircraft as if they are making, say, the ailerons go up or down, with the position of the control stick showing the position of the device whereas the sidestick position on the Airbus shows what the pilot intends the aircraft to do—in other words, the result the action is meant to produce.

See **Active Sidesticks**.

Flying After Scuba Diving

With people wanting to enjoy their vacation to the full, there is a tendency for scuba divers to continue diving right up to the end,

Though airliners are pressurized, this is normally equivalent to being at about 7,000 feet. If someone has just been diving, when the aircraft climbs, nitrogen dissolved in the blood may revert to bubbles in a phenomenon called the "bends," resulting in all sorts of problems.

To be safe, 24 hours should have elapsed since diving, with 12 hours acceptable in some circumstances.

See recommendations of diving associations regarding decompression sickness.

It goes without saying that other high-altitude activities should be avoided for 24+ hours after diving.

Flying Skills—Good Judgment Most Important

When "Sully" brought his Airbus A320 down safely on a wide stretch of New York City's Hudson River with everyone on board saved by boats three minutes away, he was venerated worldwide for his "flying skills." For the public this meant his flying *technique*.

The NTSB investigators—who were misrepresented in the movie, *Sully*, and received a lot of stick for doing their job—were put in a difficult position dealing with such a hero.

Nevertheless, Sully did demonstrate the most important flying skill of all, namely "good judgment.".

Good judgment can make all the difference in a crisis. That said, it much more often prevents one arising in the first place.

See **Famous Feats of Airmanship ~ Captain Sullenberger (Sully).**

FMGS (Flight Management Guidance System)

The FMGS provides predictions of flight time, mileage, speed, economy profiles, and altitude. It reduces cockpit workload, improves efficiency, and eliminates many routine operations generally performed by the flight crew.

It may include the following units:

1. Two Flight Management Guidance Computers (FMGC).

2. Two Multipurpose Control and Display Units (MCDU).

3. One Flight Control Unit (FCU).

4. Two Flight Augmentation Computers (FAC).

During cockpit preparation for the flight, the crew uses the Multipurpose Control and Display Unit (MCDU) to insert a preplanned route from origin to destination. This route includes SID (Standard Instrument Departure route), EN-ROUTE WAYPOINTS, STAR (Standard Arrival Route), APPROACH, MISSED APPR, and ALTN route as available from the navigation database.

The system then defines a vertical profile and a speed profile, considering ATC requirements and performance criteria.

Each FMGC (Flight Management Guidance Computer) is divided into two main parts:

1. The Flight Management (FM) part controlling the following functions:
   - Navigation and management of navigation radios
   - Management of flight planning
   - Prediction and optimization of performance
   - Display management.

2. The Flight Guidance (FG) part performing the following functions:
   - Autopilot (AP) command
   - Flight Director (FD) command
   - Autothrust (A/THR) command.

FMS (Flight Management System)

When people say pilots have nothing to do these days and that the autopilot can fly an airliner from London to Hong Kong without their help, they really should say the flight management system.

This system is based on two main elements:

1. Databases.

2. Flight plan.

The databases hold navigational data about terrain, air routes, waypoints, navaids, restricted areas, height restrictions, runways, and engine performance data, enabling the FMS to determine takeoff distances and optimum speed and height according to the weight of fuel remaining, together with other data.

The flight plan is loaded into the system prior to departure, with details of the amount of fuel, all-up weight, and temperature, together with all the other necessary information, such as waypoints, destination, and so on.

For regular routes there is often a standard one for the flight in question, with only a few details needing to be changed.

FOD (Foreign Object Damage/Foreign Object Debris)

Damage from foreign objects, such as something on the runway (something that has fallen off another aircraft, for instance) or in the air, such as birds, is always of concern, though engines are tested by having dead chickens thrown into them to ensure they can withstand a certain amount of abuse.

It was a titanium strip that had fallen off an aircraft that brought about the crash of the Concorde supersonic airliner that had departed Paris's Charles de Gaulle Airport in 2000. It caused one of its tires to shatter with the pieces striking the fuel tank above, causing it to burst. The resulting fire caused it to crash. Admittedly, the tires and full-to-the-brim fuel tanks were inherently vulnerable. Though they were redesigned to prevent a recurrence, the aircraft had become unprofitable and remained in service only for a short while.

Folding Wings

One of the problems faced by the Airbus A380 superjumbo was its great wingspan, which prevented it from operating from other than specially prepared airports. Boeing's new 777X has folding wingtips enabling it to use airports that the previous 777s were using.

Interestingly, Airbus is studying the possibility of wings that could "fold," say in the middle, inflight to attenuate turbulence thus making the passengers ride more comfortable and less scary. Of course, getting approval for such a system would not be easy.

FOQA (Flight Operations Quality Assurance)

Routine downloading and analysis of flight recorder data after landing. To make it acceptable to pilots, this is usually sent to a third party for analysis with them anonymously reporting deviation statistics and trends back to the airline.

It is a very useful tool and keeps pilots on their toes and means corner-cutting captains are kept on a leash—easier for the first officer to speak up, with him or her able to say "that will not look good" rather than "that's not a good idea."

There is resistance in the US due to pilots' lack of trust in how the data will be used.

Major deviations from the glideslope would be one of the things picked up on.

In the UK this is called **Operational Flight Data Monitoring (OFDM)**.

See **Aces' Mistakes (Value of Confidential Reporting**) and **OFDM (Operational Flight Data Monitoring)**.

Fractional Orbital Bombardment Systems (FOBS)

See **Missiles**.

Free Fall from 18,000 Feet

In **Survivors Die Waiting for Help ~ Teenage Girl Survived 2-Mile Free Fall (Peru 1971)** we describe how a girl was thrown out of the aircraft still strapped in her seat and falling on trees, survived.

In 1944 in World War II, British airman Nicholas Alkemade was a rear gunner on a burning Avro Lancaster over Germany with his separately stowed parachute burnt and unusable. Given the choice of being burned alive or jumping out, he chose the latter for a nicer death. Falling from 18,000 feet, he first struck young pine trees and then a snow drift, which further cushioned his impact.

When he regained consciousness after perhaps three hours, he was unscathed apart from a twisted knee and, of course, the burns suffered before jumping. The Germans, seeing him with no parachute, thought he was a spy but when he explained in detail, and they found the evidence in the wreckage of the aircraft, they treated him somewhat like a hero. He spent the rest of the war as a prisoner.

Freedoms (Overfly and Landing Rights)

These are commercial aviation rights negotiated between governments. These negotiations often involve brinkmanship and acrimonious horse trading between government officials.

Colonial countries, such as France and the UK, could strike good bargains in negotiating landing rights in the 1960s and 1970s, since many countries' airlines wanted to land both in the home country and in the overseas possessions. Few people know that the one-time Crown colony of Hong Kong was a very valuable bargaining chip for UK negotiators wanting rights for British airlines since so many countries' airlines wanted to fly there.

In 2015 three major US airlines—American Airlines, Delta Air Lines, and United Airlines—with the support of the unions launched an offensive to try to persuade the US government to limit the inroads of the big three Middle Eastern airlines (Emirates, Etihad Airlines, and Qatar Airways) under open skies agreements. They claimed that those airlines receive unfair support from their governments and benefit from cheap oil. It is true that Etihad Airlines and Qatar Airway do not necessarily want to make a profit, and part of their intention is to build up their country's image.

The big three Middle East airlines (M3) counter this by saying the US airlines benefit from hidden subsidies and that the Chapter 11 bankruptcy is one such. Though the US airlines have great lobbying power, the M3 buy a lot of aircraft from another lobbyist with strong unions, Boeing, not to mention the fact that smaller cities have benefited from having direct international flights thanks to the M3.

On the other side of the Atlantic, the fact that Middle East carriers also buy a lot of Airbuses, helps them resist the demands of highly paid Air France pilots that their flights be curtailed.

The rights are as follows:

1. *First Freedom*: Pure overflight rights with no landing except in emergency. Now that spy satellites can anyway see into countries, secretive countries tend to be more willing to let foreign airlines overfly their territory.

2. *Second Freedom*: Right to land for refueling and maintenance without transfer of passengers or freight.

3. *Third and Fourth Freedoms*: Right to take passengers and freight to the other country and back respectively. Usually granted together.

4. *Fifth Freedom*: In addition to the Third and Fourth Freedoms, the right to carry passengers from one's own country onwards to a third country.

5. One subcategory: *Beyond Fifth Freedom*, permits the carriage of passengers from a second country to a

third country, while the other subcategory, *Intermediate Fifth Freedom*, allows passengers from a third country to be transported to the second.

6. *Sixth Freedom*: Right to use one's own country as a transit point or hub to transport passengers or freight from a second country to a third, say between Australia and Europe via Singapore or Bangkok.

7. *Seventh Freedom*: Right to carry passengers or freight between two foreign countries without involving one's own.

8. *Eighth Freedom*: Right to carry passengers or freight within a foreign country in the context of service onwards to one's own. Also called "cabotage."

9. *Ninth Freedom*: Rarely given right to carry passengers or freight within a foreign country without service from or to one's own. Also called "stand-alone cabotage."

FSEU (Flap/Slat Electronics Unit)

The flaps and slats are extended hydraulically by a linked system, which should ensure they extend in the same way on both wings. A discrepancy would push one wing up more than the other, making the aircraft difficult or even impossible to control. If the FSEU detects any such discrepancy it instantly stops the action of the hydraulic pump. Landing faster than usual with little or no flap is better than having them unequal.

Fuel, Kerosene

With a little tinkering jet engines can work on many liquid fuels that burn, but kerosene has virtually the unique qualities that make it ideal.

Fuel Checks

Other than prior to takeoff, fuel checks are made at certain waypoints or every half hour to compare the expected fuel use with the amount left in the tanks, though in some aircraft this is now automatic following the case where an aircraft ran out of fuel due to a leaking hose without this being noticed, and a disaster narrowly averted.

Fuel Dumping

Airliners taking off for a long-haul flight have so much fuel making them exceed their maximum landing weight (MLW), which is much less than their maximum takeoff weight (MTOW). Should a problem, such as an engine failure or engine stall, occur soon after takeoff requiring a return to the airport or one close by, the aircraft may be too heavy to land safely and must get rid of some of the fuel, either by dumping it or flying around in circles.

Such airliners are usually equipped with systems allowing fuel to be jettisoned. Otherwise, the aircraft might have to fly around for an hour or more to use enough fuel up. Normally the pilots will inform air traffic control who then will direct them to an area, often over the sea, where it can be done without impacting people on the ground. However, if there is a fire on board the latest recommendation is to land as soon as possible even if it is overweight, providing the runway is long enough to allow the aircraft to stop in time.

There was a widely publicized case (with dramatic videos) in January 2020 where a Delta 777 taking off from Los Angeles had a problem with one of its engines and dumped fuel at 2,000 or so feet—instead of the recommended 6,000 to 10,000 feet to give it time to vaporize—with it landing on several primary schools. Some 40 children were affected but not badly enough to have to go to hospital. ATC had asked the pilots whether they wanted to dump fuel and they said "No," and still did.

They asked to land on the longest of LA's runways, managing to do so without the brakes heating up enough for the fuse plugs in the tires to melt.

Short-haul aircraft do not have fuel dumping mechanisms because they carry little fuel in the first place, and it can if necessary be quickly burned off. Also, in many cases it does not exceed the maximum landing weight or by so little that a careful delicate landing is possible.

Fuel Gauge

The amount of fuel is determined by using a low-voltage electric current to determine its capacitance, which is related to the quantity.

See **Drip Stick**.

Fuel MAYDAY

Usually, the rule is that pilots declare a "Fuel MAYDAY" if they will be landing with less than 30 minutes' fuel reserve in which case the controllers will give them priority to come straight in.

The controllers will also arrange the landing sequencing to ensure they will not need to perform a go-around due to the presence of an aircraft landing in front of them being still on the runway.

If the pilots have more than 30 minutes' fuel reserve but not enough to divert to an alternate airport, or will soon have to declare a FUEL MAYDAY, say, if ATC continue to delay their landing by keeping them holding, the pilots can declare Minimum Fuel. While, unlike a Fuel MAYDAY, this does not automatically give them priority, it makes the controllers aware of their situation and hopefully avoid them delaying their landing unnecessarily.

Fuel Reserves

Countries and airlines have strict regulations about the fuel an aircraft should plan to have in reserve on reaching its destination to allow for unexpected headwinds or diverting to another airport (again with adequate fuel reserves) should landing there be impossible.

The captain or dispatcher can always demand extra fuel should they feel that circumstances require it for safety. Those circumstances could include bad weather en route with turbulence that might require slower flying or even flying a more circuitous route.

Rules vary from country to country.

> Besides the fuel required to fly the route from the start of engines to the destination the following may be required:
>
> ENOUGH FUEL
>
> 1. For an approach and missed approach
> 2. Diversion to an alternative airport,
> 3. And be able to hold there for 30 minutes at 1,500 feet at planned landing weight after diverting, and land there with 30 minutes fuel remaining.
> 4. In addition, pilots must have an extra reserve equivalent to 10 percent of the estimated flight time. This 10 percent is considerable for a long-haul flight lasting over 10 hours, and unless there is an exceptional headwind or an engine failure, the aircraft will arrive at its destination with a considerable amount of fuel remaining, which is really a great waste. To avoid this, airlines may legally dispatch the aircraft to an airport short of the intended airport, and on nearing it with plenty of fuel in reserve, Redispatch it to the intended one.

Though resorting to the redispatching artifice makes a lot of sense, there are safety implications where the aircraft is operating close to its range limit, with the crew perhaps hesitating to take more circuitous routes to avoid storms or taking off with a light tailwind to avoid taxiing around the airport using up fuel to take off into the headwind.

The supersonic Concorde that crashed in Paris in 2000 was operating near its limit as regards fuel reserves and took off with a (light) tailwind and overweight. Admittedly, other factors, such as the delay incurred, would have also influenced the Concorde captain's decision.

~ G ~

G (Force of Gravity = 1 G)

Force experienced by people or aircraft due to gravity and/or acceleration or deceleration. Expressed as a multiple or fraction of the Earth's gravitational force.

Negative G refers to the situations where the aircraft drops or accelerates downward so suddenly that there is a feeling of weightlessness. (See **Bunt**.)

G-LOCK

Fighter pilots are subject to high g maneuvers reaching 9 g.

1. Pilots first have loss of peripheral vision,
2. then loss of color vision,
3. followed by blackout,
4. and finally G-LOC, with loss of consciousness.

This is because the G forces drain blood away from the brain leading to **Hypoxia**.

Some modern jet fighters incorporate a system where in the event of a pilot suffering G-LOC, control can be transferred to a ground control station (GCS) for as long as necessary. This has apparently saved the lives of a number of pilots.

GADSS (Global Aeronautical Distress and Safety Systems)

In accordance with the ICAO recommendation and European Union mandate, all new aircraft delivered from January 2023 must be able to autonomously report their location anywhere in the world and determine the end of flight location to help rescue teams rapidly locate the aircraft and recover flight recorders.

This has been introduced partly because of the difficulty of finding the wreckage in the ocean in the case of **AF447** and **MH370**.

While traditional automatic **Emergency Locator Transmitter, Distress Tracker (ELTs)** are stand-alone units, the ones for the new GADSS system are linked to the aircraft's avionics system and activate upon detection of a potential distress condition, starting to transmit automatically while the aircraft is still in flight—before any possible crash.

The ELT autonomously acquires the aircraft's location and in real time sends a 406 MHz emergency message including the precise location to the Cospas-Sarsat distress alert organization.

There was always the problem that the volume of data would quickly overload the system. However, this was solved by saying the ELT should transmit every 15 minutes but at least every minute should it detect anything wrong with the aircraft or it behaving abnormal in a way that could indicate a danger. Also it was recommended that the pingers on the cockpit voice and data recorders last ninety days rather than the current thirty days. Also, that cockpit voice recorders should no longer be on a 2-hour loop with the contents overwritten every two hours for privacy reasons, but with present solid state recording be able to retain 25 hours of sound in view of much longe flights and the value of retaining information prior to departure.

The wish of the NTSB (National Transportation Board) to have Cockpit VIDEO Recorders is making no headway in the face of opposition from pilots' professional organizations and unions. One great advantage of being able to show what the cockpit instruments are showing is less relevant now the data recorder can record what they should be showing,

Gait Analysis/Facial Recognition

Everyone has heard about facial recognition, where computer programs are used to try to identify people from still images or frames from CCTV cameras. Not many have heard of gait analysis used to identify people from the way they walk—even though we very often recognize relations or friends in a crowd or at a distance from their gait or posture.

Some thought is being given to recognizing people from above—for example, from satellites or drones—by converting their horizontal shadows to show their gait. Of course, the sun would have to be shining for that, meaning that while it might work well in Pakistan or Afghanistan, it would not be very effective in countries with cloudy weather.

Whether using ground-based cameras or drones, the system would have to be able to recognize those altering their gait by putting a stone in their shoe.

Facial recognition has also progressed. It works somewhat differently from the way humans recognize each other, in that it notes the relative distance between certain facial features.

See **Wide-Angle Motion Imagery (Gorgon Stare)**.

Funneling (Navigation Paradox)

The sky is a big place, and the chances of aircraft hitting each other if flying completely randomly would not be so great in some areas. However, there is the paradox that as navigational systems, such as GPS and even altimeters, place aircraft ever more accurately (plus or minus a few meters) on a flight path, an error by ATC or simply confusion regarding long-distance flight rules could place aircraft at the same spot. It has been estimated that fewer accidents would have occurred over the years had aircraft operated more randomly in certain situations.

While there is the inevitable funneling of aircraft as they are coming in to land, it is not so much of a problem then, since ATC is closely watching separation. However, there may be some unnecessary funneling in other situations. Would the 2002 midair collision in the middle of the night, with hardly anyone around, of a freighter and an airliner full of children over Lake Constance* between Switzerland and Germany have occurred had the Swiss Air Force not appropriated so much airspace, leaving little room for commercial traffic to roam?

"Game-Changing" (Transitional) Years

1. 1969
   Maiden flights of the Boeing 747 jumbo and the Concorde supersonic airliner. (Neil Armstrong took man's first steps on the moon!)

2. 1973
   Oil shock. Oil prices quadrupled before eventually dropping somewhat.

3. 2001
   9/11 (Nine-Eleven*).
   Four aircraft hijacked. Two were flown into New York's World Trade Center, and one into the Pentagon. The fourth crashed in open country after the passengers fought back.
   (See **9/11 Two Unarmed F-16s Scrambled to Down 4th Airliner**.)

4. 2006
   Liquid Bomb Plot* that made security checks at airports a pain in the neck with bothersome rules such as hand luggage not being able to contain liquids or gels of more than 100 cc.
   [The British discovered, monitored, and foiled a plot to bring down flights between the UK and the US by hiding explosives in drink bottles in such a way they would be difficult or impossible to detect.]

We are now seemingly approaching a "game-changing" end of decade as new technologies are on the horizon but not perfected enough to be realized but probably will in the next decade. They include the possibility of switching from "tube and wing" to **Blended Wing** concepts. New types of engine and great advances in battery technology, making eVTOL aircraft and electrically powered high performance drones much more practical. Hydrogen-powered aircraft now suddenly seem a little further off, perhaps towards the end of the next decade.

Gander

Gander is on the great circle route from New York to London and is one reason an extensive airport was built there when aircraft had to refuel.

Although little used at the time, when flights were barred from the US on **9/11,** some 400 flights landed there, and were able to use the vast space to park. Some passengers had to stay for six days, welcomed by the town with limited accommodation.

Gander is also the location of the Gander Area Control Center (ACC) controlling oceanic airspace (over the Atlantic).

Gartner Hype Cycle, Aviation Electric and Hydrogen Prospects

Bjorn Fehrm in his Bjorn's Corner (see Leeham News in **Aviation Knowledge on The Internet**) has, as usual, a very interesting look at the prospects for electric and hydrogen for propulsion in aerospace based on the Gartner Hype Cycle.

According to Gartner, Inc., Gartner Hype Cycle methodology gives you a view of how a technology or application will evolve over time, providing a sound source of insight to manage its deployment within the context of your specific business goals.

Bjorn Fehrm suggests we are now (2025) near the Peak of Inflated Expectations and the bottom of the Trough of Disillusionment will be in 2026 or 2027. Disillusionment in the aviation sector is apparent from the recent abandonment of several of the much hyped projects. Furthermore, he believes that both for electrically powered craft (eVTOL, etc.) and hydrogen-powered craft, it will be in the automotive field and in particular large trucks (lorries) that the necessary technical advances will first be made.

Early in the next decade we should be gradually climbing to the Plateau of Productivity.

GARTNER HYPE CYCLE

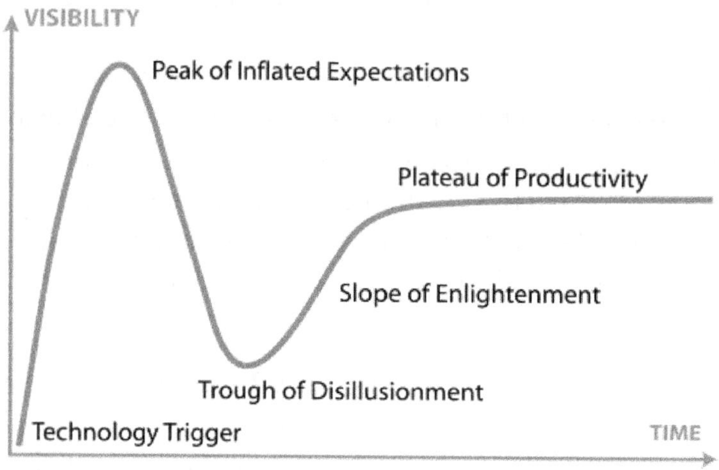

[GARTNER, INC.]

Geared Turbofan (GTF)

Pratt & Whitney, part of United Technologies of the US, has invested considerable funds in the development of the so-called geared turbofan engine. Sophisticated planetary gearing enables the front fan to operate at a different speed from that of turbines inside. This is theoretically advantageous, because fans, such as the large-diameter rotating blades one sees at the front of aircraft engines, are more efficient at low speeds, whereas turbines are more efficient at high speeds. The fan can have a larger diameter without the tips of the blades exceeding the speed of sound. Having both operating at optimum speeds should allow higher bypass ratios, a more efficient, cleaner fuel burn, and much quieter operation.

Only time will tell whether the great risk Pratt & Whitney has taken with this more complex technology will pay off. Surprisingly, many of the problems that cropped up with the engines did not concern the gearing mechanism.

Gender Neutral Language, Inclusive Language [NASA]

Originally the term "unmanned" once just seemed to mean no one, either male or female, piloting a vehicle. A "manned" craft could presumably have a woman aboard.

Recently, some companies have started referring to their UAVs as "Uncrewed" aerial vehicles. With thousands of articles using the traditional terms this poses problems for reference books such as this one, with us using both terms, and mainly the old ones in keeping with the original texts.

NASA first accepted women as potential astronauts in 1978, notwithstanding the fact that Valentina Tereshkova had been the first Soviet woman in space in 1963.

Now, to keep up with the times, NASA has come up with the following guidelines:

"MANNED" should be

HUMAN or CREWED or PILOTED

and

"UNMANNED" should be

ROBOTIC or UNCREWED or UNPILOTED

The author feels that saying "manned" or "unmanned" pulls more on the heartstrings when wondering whether lives may or may not be at stake.

General Atomics Aeronautical Systems

A subsidiary of general atomics with headquarters in Poway, California, and using Grey Butte Field near El Mirage as a testing facility for its drones.

Designs unmanned aerial vehicles and radar systems for the US military and for ostensibly commercial use abroad. Big operation.

General Aviation (GA)

Especially in the US, General Aviation (GA) refers to aircraft operations and aircraft that are neither military nor commercial (scheduled airline or cargo flights).

This sector thus covers anything from the flying club Piper Cub, through business jets, to unscheduled cargo and unscheduled airliner operations. In practice, the business jets and unscheduled cargo and airliner operations flying into and out of major airports are supervised very much like commercial flights, in contrast to most GA operations in and out of small airfields, where regulations are less strict.

Generation (e.g., Fifth, Sixth Generation)

Both in the field of computers and aircraft, and notably fighters, the Fifth Generation seems to indicate the most modern generation as regards the technology. In the case of fighters, there was a "Generation 4.5," where there was no great technological leap on the aerodynamics front but where great advances were made in electronics and avionics. Now there is talk of a 5.5 Generation and Sixth Generation.

"Get-There-Itis"

Pilots can be under a great deal of pressure to land at their destination and not divert. Accidents have resulted from this.

Where the weather is marginal pilots can be lured into making multiple attempts to land in the expectation conditions will improve until there will not be enough fuel to divert. To avoid this, LCC Ryanair has a rule that diversion is mandatory after two failed attempts unless visibility at least doubles.

Glass Cockpit

Modern cockpits with liquid crystal displays and the like replacing the traditional analog instruments and gauges. More information can be displayed in a concentrated space, with it being possible to add head-up displays.

See **Primary Flight Display**.

Glide Path/Slope

Imaginary path, aligned with the runway and at an angle of about three degrees, leading to the touchdown point on the runway. Some airfields surrounded by high terrain may have a steeper glideslope of five degrees.

On his YouTube Blancolirio channel, Juan Browne has mentioned, when describing the far too high and rushed approach of the Pakistan airliner that touched down without the undercarriage lowered, a useful rule of thumb he calls the *three-in-one rule* for reckoning where an aircraft should be when following a normal three-degree glideslope: *At three miles out, the aircraft should be at 1,000 feet.*

See **ILS (Instrument Landing System)** and **VASIS.**

Most airliner landings are made using **ILS**, even when the visibility is not poor enough for them to need it to get almost on the runway.

Glide Ratio—How Far Can Aircraft Glide with No Fuel?

Indicates how far an aircraft can fly with no thrust from the engines and is simply the distance gained forwards divided by the height lost or the forward speed divided by the sink rate. It is also called the lift-to-drag ratio (L/D), which is the term most used when one wants to show the efficiency of the airframe for powered aircraft (not intended to glide hundreds of miles, though a plus if they can).

Gliders (sailplanes) can have glide ratios approaching 60:1, whereas modern airliners have glide ratios (L/Ds) of around 16:1 to 18:1. Thus, an airliner can perhaps glide two nautical miles for every 1,000 feet of height and therefore some 80 miles from 40,000 feet. Of course, it is not so simple because some height will invariably be lost while the pilots try to work out a solution if the engine failure is not obviously simply due to fuel exhaustion. Headwinds and tailwinds have a great impact, too.

There have been remarkable cases where airliners have been able to glide considerable distances from cruising height after running out of fuel and effect safe but hard landings, such as *A330 Glides Eighty Miles Over Atlantic to an Island**.

Airliners frequently glide with engines set at idle when they approach their destinations. The only difference is that the pilots know that power can be applied whenever needed, although this was not true in the case of a British Airways 777 flight* from Beijing to London's Heathrow.

In that incident, when the pilots sought some thrust to complete the final part of the glide and get them over the airport perimeter fence, it was not forthcoming. However, by reducing flap, they were just able to touch down—or rather plonk down—just short of the runway. This was due to ice in the fuel.

Global Airspace Management—FIRs

GPS (Global Positioning System)

See **Global Navigation Satellite System**.

Global Combat Air Program (GCAP)

A program involving BAE Systems, Leonardo, and Mitsubishi Heavy Industries to produce a next generation fighter aircraft, often referred to as "Tempost."

Global Warming Impact: 2.5 percent, or 4 percent if Contrails Included

Aircraft engines have become much more efficient, aerodynamic improvements have been made, and more passengers per flight and higher load factors mean that the amount of fuel burnt per passenger per mile is much less than it used to be. However, the great increase in the number of people flying mean that the effect on global warming is getting worse rather than improving.

Furthermore, as steps are taken to reduce global warming through nuclear power, wind, solar, and using new technics in industries such as steel and cement, the proportion of global warming attributed to aviation will increase.

According to Our World in Data https://ourworldindata.org/global-aviation-emissions

aviation accounts for 2.5 percent of global CO_2 emissions, but 4 percent to global warming because other emissions, and notably contrails consisting of water vapor, add a further 1.5 percent.

Unfortunately, reducing global warming due to aviation in the sky is far more difficult than on the ground. At the moment batteries are far too heavy to have a meaningful impact, and even in the future, hydrogen will produce water-laden vapor trails and not be so wonderful as it might seem.

Sustainable Aviation Fuel (SAF) is expensive and only available in limited amounts.

The French have tried to get people to use their excellent trains for short journeys with limited impact since the regulations are not strict. Airlines not having to pay for the intermediate infrastructure (the air) and its maintenance can often offer ridiculously low fares.

GNSS (Global Navigation Satellite System)

A navigation system, now well known to the public as *satnav*, determines location, speed, and direction of travel of the aircraft by tracking three or more satellites with a special receiver.

Originally developed for the US military, GPS was offered free to the public following the shooting down of a Korean Airlines 747 in 1983 that had supposedly lost its way. It used to have potential inbuilt errors to stop enemies using it to guide missiles and so on, but that faculty is not now employed—at least, in times of world peace. New techniques enable GPS destined for the public to be locally degraded in war zones.

To increase accuracy, and not least to confirm readings are accurate at a time when users are increasingly dependent on the system, supplementary systems are being introduced. These work by having ground stations monitor the satellite signals and broadcast corrections to make the positioning accurate enough, say, to land an aircraft.

In the US there is the Wide Area Augmentation System (WAAS), designed to cover most of the country, and the Local Area Augmentation System (LAAS), covering particular areas, such as those in the vicinity of airports. Other countries have similar systems.

Interestingly, a GPS-based system called positive train control (PTC) is being developed in the US for preventing train crashes by plotting trains' positions, noting their speeds, warning them of the risk of collisions, and applying the brakes should drivers fail to take appropriate action. It incorporates a WAAS-type system, called Nationwide Differential Global Positioning System, to increase accuracy to about a meter. With train crashes less feared than air crashes, railroad companies are reluctant to install it widely, despite the fact that it would raise average speeds and allow trains to pace themselves, thus saving fuel.

The Europeans have their own GPS system, called Galileo.

Russia and China also have their own systems, Glonass and BeiDou, with limited coverage.

The US military, or rather the Air Force, still runs—inefficiently—the GPS system with it falling technology-wise behind the systems just mentioned. It has been suggested that managing and developing the GPS system should be passed to a civilian body and a separate system be created for the US military with the less vulnerable M receivers and so on.

Interestingly, "Mike" Holmes, head of Air Combat Command in the US said, "My U-2 guys fly with a watch that now ties into GPS, but also China's BeiDou, Russia's Glonass, and the European's Galileo systems so that if somebody jams (our) GPS, they still get the others."

Go-Around

The term *go-around* simply means the pilots decide to abandon the landing for some reason, such as a poor approach due to gusty winds, wind shear, bad visibility, or because the aircraft that landed before them is still on the runway having taken longer than expected by the controllers to exit, or one taking off has delayed their takeoff run. At busy times with many aircraft waiting to land and take off, ATC may try to squeeze in as many as possible that are landing knowing they can always order a go-around if necessary.

Following the collision at night between a military helicopter and an airliner landing at Ronald Reagan Washington National Airport in 2026, the media began shaking passengers' confidence by reported go-arounds all over the US as if they were near misses and disaster was narrowly averted when usually they were nothing of the sort. One key point in making them benign was that right up to the landing the incoming aircraft were flying (i.e., not stalling) and TOGA, described below, could be engaged effectively.

Flight controls include a special **TOGA (Takeoff/Go-Around)** button that when pressed automatically reconfigures the aircraft and resets the throttles for an immediate climb away. Pilots put settings for the go-around into the flight management computer as there are specific rules regarding the way a go-around for a given runway has to be executed to avoid collision with other aircraft and even obstacles.

Since the aircraft's computer is programmed to treat a TOGA input as a possible emergency, it makes the aircraft react abruptly, which can scare the passengers. Consequently, some pilots perform the go-around manually when they think the situation permits. Even so, the passengers often come back telling stories of their frightening experience, even when it was just a precautionary action taken well before landing. To passengers, banks, turns, and climb-outs can seem steeper and more dangerous than they are, partly because unlike on the road they cannot anticipate them.

Short-haul pilots may experience several go-arounds in a year, whereas depending on the airports flown into, long-haul pilots may not experience any and so might not be ready for one.

GPWS (Ground Proximity Warning System)

With hindsight, the ground proximity warning system is a no-brainer for warning pilots when they are inadvertently too near the ground, and one that has done more than any other to prevent controlled flight into terrain (CFIT). It was made mandatory for large airliners in the US in 1974.

Using the data fed to the radio altimeter giving the actual height above ground, if the aircraft is dangerously close to it or sinking so fast it might hit it, it gives a verbal warning such as "Pull up! Pull up! Terrain!"

Early versions would sometimes give too many spurious warnings, resulting in it being switched off in countries where at the time it was not mandatory. A case in point was the French Air Inter Airbus A320 flight into the side of Mont Sainte-Odile in 1992*, with only eight passengers and one crew member surviving, where the pilots claimed there were too many false alerts, and they knew the terrain so well it was unnecessary.

The great drawback of the original TAWS was that it could only warn pilots if the ground immediately below was dangerously close and not if they were about to fly into a mountain or cliff directly ahead. However, there are now enhanced versions called EGPWS where radio altimeter data is complemented by data held in the computer's database about the terrain ahead, and the pilots can be warned in consequence. Also, instead of just considering a straight down and back signal, modern GPWS systems emit and receive over a wide-angle window and cleverly take the one received back first as the distance to the terrain thus enabling them to work properly even when the aircraft is banking or pitching steeply.

Graveyard Spiral and Spatial Disorientation (SD), Death Spiral

Mention of the term "graveyard spiral" evokes the fatal disaster that befell the light aircraft President Kennedy's son, John F. Kennedy Jr., was piloting over water at night on July 16, 1999, with his wife and sister-in-law as passengers.

Investigators found nothing wrong with the aircraft, engines, or equipment. It was thus presumed that Kennedy became disorientated because he could not see the horizon and allowed the aircraft to bank, until it entered a graveyard spiral, where, with the aircraft already tipped on its side, instinctively pulling back on the controls would only further tighten the spiral and cause the aircraft to drop even more precipitously. When disorientated and unable to see the horizon, a pilot should rely on his, or remove his hands and feet from the controls and rely on the autopilot to extricate him from his predicament.

In Australia's the *Sunday Age* in 2007, aviation medicine expert Dr. David Newman is quoted as saying, "80 percent of a person's sense of what is up or down is determined by sight and the remaining 20 percent determined in equal share by vestibular [ear] and seat-of-the-pants sensations."

The trouble is that the latter sensations can be confusing when combined with G-forces, while vestibular sensations are subject to drift, like gyroscopes, but over a period of seconds rather than hours in their case. Thus, spatial disorientation easily comes about when pilots cannot see the horizon, with it being claimed it is more likely to occur when pilots do not have much to do.

Pilots are trained to cope with it by having an instructor put a mask over their eyes so they can only see the instruments, then confuse them by executing various maneuvers, getting them to redress the situation using only the instruments. Even so, it is said that quite a number of the general aviation accidents that do occur are due to spatial disorientation, which can even make pilots feel they are sitting on the wing!

General aviation pilots are more often subject to the consequences of SD because they are used to flying under visual flight rules rather than instrument flight rules depending on instruments.

See **Somatogravic Illusion**.

Grandfather Rights

New designs for airliners and even minor modifications have to be approved and certified by, say, the FAA in what is a costly and time-consuming procedure. Though well intentioned, this can lead to delays in bringing an airliner into service and result in manufacturers delaying the implementation of minor improvements. Even after crashes due to a latent danger, the manufacturer might find it easier to put a warning in the operating manual rather than modify the system.

As technology advances and as authorities become more stringent, the norms for certifying new aircraft become tighter. However, by continuing to produce the same series of aircraft, such as the Boeing 737 or Airbus 320, the manufacturer—within limits—can benefit not only from not having to devote so much time to get approvals but also from only being subject to the norms applicable to the original type certificate, under what are called "grandfather rights."

Perhaps an indication of how "grandfathering" can be abused is that the term originally referred to the system whereby, when some Southern US states had introduced literacy and residence rules in the late-nineteenth century, illiterate whites in the US could be allowed to vote under grandfather rights, while African Americans with no grandfathers in the US were not allowed. As years went by, these grandfather rights were judged unconstitutional.

Great Circle Route

A straight line in a two-dimensional map is not the same as a straight line on a three-dimensional globe. This means that airliner routes taking the shortest route may seem odd and sometimes pass over the poles. The great circle route is the shortest.

Ground-Based Strategic Deterrent (GBSD)

The US ground-based strategic deterrent (GBSD) consisting of Minuteman III missiles in underground silos is now going to be replaced by new and, it is hoped, even more reliable ones. See **Missileer**.

Ground Effect

Depending on the type of wing—the supersonic Concorde, with its delta wing, perhaps being a prime example—an aircraft can get extra lift and show greater efficiency when close to the ground. Some believe it is rather like a cushion of air pushing the aircraft up, when, in fact, the aerodynamics are much more complex than that.

Ground Speed

True speed over the ground, as opposed to the speed through the air.

~ H ~

HALE (High-Altitude Long-Endurance), MALE (Medium-Altitude LE)

In the field of **UAV**s (unmanned aerial vehicles), the acronym HALE is much easier to remember and cite in training than "high-altitude long-endurance vehicle," for, say, the **RQ-4** drone. There is also MALE (Medium-Altitude Long-Endurance) for drones such as the MQ-9A.

HAPS (High-Altitude Pseudo Satellite)

These are not orbiting satellites but platforms that can stay aloft at a height of more than 20 km for months or years and can be airplanes, airships, or balloons.

Perhaps the most interesting is the recently developed ultralight electrically driven vehicle covered with solar panels that has stayed up for several months and is expected to be able to stay up even longer. It has batteries that are recharged in daylight for power at night. Applications include military ones. Being very delicate, it must be launched in calm weather conditions and, of course, cannot function near the poles where there is little sunlight.

Advantages over satellites:

1. Cheaper, and since they can land and take off again, they can be continuously reprogramed and upgraded with the latest technology.

2. Operate lower, which means they are better at filming, can more easily pick up radio transmissions and radiation from radars. Furthermore, being lower there is less latency, say when used for 5G.

3. Unlike satellites they are easily and cheaply repositioned.

HAPS is also taken to mean High Altitude Platform Systems used to provide broadband and telecommunications in isolated areas with little latency.

HARM (High-Speed Anti-Radiation Missile)

These are missiles usually launched from an aircraft that home-in on the radiation from enemy radar, or more rarely a radio transmission, to destroy its "eyes"—sometimes aircraft will also deploy, say, cluster bombs to make ancillary equipment and facilities unusable. The main use was to knock out enemy's radar at the start of combat so following aircraft are less vulnerable.

The US has supplied some to Ukraine. However, since they cannot integrate with their aircraft systems, the general target coordinates have to be inputted beforehand, with them seeking the radiation source on arrival there.

Handoff, to Hand Off

A term used in air traffic control, where aircraft are handed off from one ATC facility to another.

Hard Landing

A hard landing is defined as one where the g-force is equal to or greater than 2.1 and requires a check by technicians to ensure the aircraft has not suffered damage requiring repairs. If the pilots believe a landing falls into that category they report it, though there is onboard equipment that does so automatically.

For various reasons, pilots may make a *firm* landing that passengers might consider hard but was not so.

The reasons might be:

1. For safety, where a crosswind may make the pilot decide it is better to put the aircraft down very deliberately rather than let it float and risk them getting blown to the side.

2. Again, for safety, where the runway is short or even covered with water and the aircraft might aquaplane—

the sharp touchdown makes the tires cut through the water.

3. Wind shear or when on "Autoland" in fog where the autopilot performs the "flare" higher than usual.

4. Where the captain feels it is better to allow a pilot being trained to get the feel of a landing even though it might well not be perfect.

A soft landing is now the norm, with it being usually easier to execute on large aircraft with bogies hanging down where the rear tires of a pair touch first, and shock absorbers attenuate some of the shock.

Though accelerometers can detect an overall hard landing, one wheel or bogie can come down particularly hard without it showing up. For this reason, some landing gear struts have pop-up indicators at the top that trigger in the event of it having sustained a pressure exceeding a certain limit. These would be apparent on routine inspection after the flight.

Hardpoint

A hardpoint is a fixture on the aircraft structure used for mounting weapons or optional extra equipment like fuel tanks or sensors. There can be as many as a dozen.

Modern stealth aircraft either do away with them or can fly in a mode with nothing affixed to them as they substantially increase the radar cross-section. Such aircraft carry their weapons in an internal weapons bay which has limited dimensions. See **Beast Mode**.

Heading

The heading is the direction in which the aircraft is pointing, as opposed to the aircraft's track over the ground, which is affected by the wind direction/velocity. The drift angle is the angle between the heading and track over the ground. (See **Crabbing**, where the aircraft is pointing to one side but tracking in line with the runway to land in a crosswind.)

Head-Up Display (HUD)

A head-up display permits the pilot to see pertinent information on the windscreen or helmet while looking straight out of it with his or her head up. HUDs were originally developed as sophisticated gun sights for fighter aircraft.

HUD systems can be in increasing degrees of sophistication:

1. **Basic Systems**
   with information such as height, heading, airspeed, and so on, displayed using special symbols.

2. **Enhanced Flight Vision Systems** (EFVS)
   with also an enhanced view of the terrain ahead, obtained via an infrared camera situated close to the pilot's line of sight so that on coming out of the clouds the view will be the same.

3. **Synthetic Vision Systems** (SVS)
   with a computer-generated view of the terrain ahead derived from databases and radar.

The helmet for the pilot of the F-35 Lightning is "tailor made" because it must fit perfectly to work properly. It is said to cost $400,000.

According to Arthur De Ruiter of Collins Aerospace:

The F-35 helmet, and visor system is truly an engineering marvel that represents today's fifth-generation warfighter. The key difference between the F-35 helmet and a traditional fourth-generation warfighter helmet used on the F-16 is that all the flight and mission data is presented to the pilot on a display, mounted to the helmet.

In addition, the helmet is connected to six infrared cameras located around the aircraft, providing a 360-degree view and unprecedented situational awareness. If the pilot looks around in certain conditions, they don't see the cockpit or even their legs. All they see are their surroundings like terrain, with the flight critical and mission data projected inside the helmet.

Head-Wearable Display (HWD)

These are sometimes called **Heads Up Displays (HUDs**) but range from attachments to glasses to sophisticated, and very expensive, helmets. Essentially they are devices that project information directly into the user's field of vision,

They can be used in many domains besides aviation. At the simplest level they can show your smart phone screen or be used for gaming.

Headwind/Tailwind

In most of the northern hemisphere the prevailing winds are from west to east, and hence around-the-world balloon attempts are made in that direction. What is sometimes forgotten is that a headwind when flying in one direction will be a tailwind for when flying in the opposite direction, thus at least doubling the difference between the flight times for the two directions. Also, if the wind is extremely strong and the aircraft is very slow in comparison, the aircraft can in theory seem to be hardly moving when flying against it and must land to refuel.

"Heavy"

"Heavy" is a call sign suffix (for instance, "United three-two-one heavy") that indicates to air traffic controllers the presence of a large aircraft requiring more space (separation minima) than lighter craft. Heavy is defined in the US as an aircraft that has a gross takeoff weight greater than 255,000 pounds, and greater than 300,000 pounds in some other parts of the world. This includes the larger aircraft in use today. However, some aircraft are on the borderline, as their takeoff weight can depend on their seating configuration and other factors. One aircraft that can have it both ways is the 757; however, it is invariably called heavy, as it is notorious for producing nasty wake turbulence.

The Airbus A380 is referred to as "Super."

Height above Mean Sea Level (MSL) as Opposed to AGL

AGL is above ground level.

Helicopter

Helicopter rotors incorporate a mechanism allowing the blades, but not the engine, to freewheel. In other words, the engine drives the blades, but the blades do not drive the engine and as a result, if the engine fails the blades will continue to rotate and prevent the helicopter dropping like a stone. However, to keep the freewheeling blades rotating and enable the helicopter to make a controlled landing without engine power, the helicopter should have some forward speed, which is one reason some news organizations instruct their pilots to move around rather than hover for extended periods when filming incidents.

That said, helicopters, with their expensive spare parts, are nowadays very reliable and incidents of engine failure rare. Accidents are more likely where the great convenience of travel by helicopter tempts people to take risks in bad visibility. See **Tilt Rotor Helicopter**.

Hellfire Missile [includes R9X: Knife Bomb," Flying Ginsu"]

Since there are many iterations and generations, the Hellfire is a genre rather than a particular missile.

Although the name looks as though it was chosen to make the missile seem like a terrible weapon from Hell, the origin of the name was an abbreviation derived from "Heliborne Laser FIRE (and forget)" missile.

It started off as an antitank missile in the 1980s. Initially launched from helicopters such as the Apache, it has since been developed for deployment on many different platforms.

Essentially, they are precision-guided missiles that can be mounted on helicopters and drones and are used widely. The US alone has had more than 100,000.

One version, the Hellfire R9X, known by various nicknames, including the "knife bomb" or the "flying Ginsu" has the explosive charge replaced with six 18-inch blades that deploy on reaching the target.

[Ginsu is the brand name for a kitchen knife used by a US company in the 1970s. It is made with special Japanese steel with more than 2 million sold.]

The aim is to avoid collateral damage, though that may not be always be feasible

Lockheed Martin's AGM-179 Joint Air-to-Ground Missile (JAGM) is the latest Hellfire missile combining laser and radar targeting in the one missile. This greatly simplifies inventory and means the operator will not have to decide which version to launch.

Helmet-Mounted Display (HMD)

The helmet and associated hardware and software on, say, an F-35 is highly sophisticated and expensive. The F-35 distributed aperture system consists of six infrared cameras mounted below the fuselage looking in all directions enabling the pilot to feel they are seeing through their aircraft and even look at and shoot down an aircraft behind them.

Hellscape, Thousands of Drones to Protect Taiwan

One possible US plan would involve launching thousands of drones, from surface vessels and submarines to fight Chinese invading forces as soon as they begin to cross the Taiwan Strait, effectively acting as a kind of first line buffer, valid until the arrival of possible help.

See **Replicator**.

Herbst Maneuver

Dogfighting or airshow maneuver whereby 4.5- or fifth-generation fighters with thrust vectoring can put their aircraft in a steep angle of attack and turn to put the aircraft flying in the opposite direction. Also, called the "J-turn."

Hierarchical Gradient between Pilots, CRM

Gone are the days when it was thought acceptable and even desirable for the airliner captain or head surgeon to act like God, with the hierarchical gradient between them and the underlings being almost vertical.

In aviation, a prime example of an extreme vertical gradient would be the Korean Air Cargo Flight 8509 that took off from London's Stansted Airport in the early evening of December 22, 1999*. The captain was a former colonel in the Korean Air Force with 13,490 flying hours, of which 8,495 were on the Boeing 747. As was the custom, senior military officers such as him were catapulted to the rank of captain and had little understanding of flying as part of a team. His copilot had only 195 flight hours on the 747 out of a total of 1,406. The flight engineer, however, was very experienced but was not listened to.

Just after lift-off the captain banked to make a turn, but with no bank angle showing on his faulty artificial horizon he could not see he was doing so. Even though the bank angle was increasing dangerously and evident on the first officer's side, the latter was afraid to speak up or take any action to save them. In the end, the bank angle was so extreme they crashed.

Korean Air, under pressure from the country's president, had a complete shake up and with the help of Delta made a cultural change, becoming one of the safest airlines.

While the dangers of a vertical gradient are obvious, a gradient too close to the horizontal can be dangerous too. Two management captains flying together, both thinking they know it all and not taking orders, have led to crashes.

An extreme case of a combination of very senior pilots that could have led to trouble was Qantas QF32*, where there were two very senior pilots present in the cockpit *in addition to the captain* when an engine on the superjumbo A380 disintegrated, damaging numerous systems. Fortunately, before departure from Singapore, the nominated captain, Richard de Crespigny, had made it clear to the first and second officers that though he was less senior, he was ultimately in charge. In the end, the pilots all worked extremely well together, solving the problems, and worked out the best course of action.

The tragic case of Air France **AF447*** is an example of *no gradient*—for Air France did not have a protocol regarding who should take charge in the event of major problems occurring in the absence of the captain.

HIJACKINGS AND BOMBINGS from 1933 to 2001

One forgets that hijackings were once, and especially in the 1970s, 1980s, and 1990, a regular occurrence.

29 incidents:

1. 1933: The first proven case of an airliner in the US brought down by an explosive device was in 1933. However, neither the perpetrator nor a possible motive was found. It was United Air Lines Trip Flight 23, a Boeing 247, with four passengers and three crew, who were all killed. It had stopped off in Cleveland, Ohio, and was on its way to Chicago.

2. 1955: The second known case of an airliner brought down by an explosive device was United Air Lines Flight 629 with 5 crew and 39 passengers in 1955. The son of one of the passengers had placed the device in her case, doing it out of revenge for putting him into foster care and for the insurance. He claimed to have been inspired by the downing of Canadian Pacific Air Lines Flight 108 in Canada in 1949. In that case there had been three conspirators who had done it for the insurance and so one of them could eliminate his wife and marry his mistress.

3. 1970: A New York-bound Pan Am 747 was hijacked after takeoff from Amsterdam and diverted to Cairo where the aircraft was blown up.

4. 1970: One of the most publicized hijackings where as many as five airliners (TWA, Pan Am, and El Al) were hijacked over three days by the Popular Front for the Liberation of Palestine (PFLP). Three of the five planes

were diverted to a remote airfield in Jordan, called Dawson's Field, and finally blown up, fortunately with no one aboard.

5. 1971: D. B. Cooper*
   Parachuted with ransom from a Boeing 727 using the trijet's rear airstair in a storm which prevented the following fighters seeing where he came down. He was never found or identified, though a little of the money was much later found on the ground by a boy.
   See **D. B. Cooper Mystery**

6. 1972: A JAT (Yugoslav Airlines) DC-9 en route from Copenhagen to Zagreb exploded at cruise with no survivors.

7. 1972: To murder his fiancée, a Thai police lieutenant supposedly placed a bomb on a Cathay Pacific jet flying from Bangkok to Hong Kong, killing 81 people.

8. 1973: Eighty-one people were killed during the attempted hijacking of an Aeroflot airliner over Siberia.

9. 1974: A TWA 707 flying from Athens to Rome crashed into the sea near Greece, due to an explosive device.

10. 1976: Cubana de Aviación Flight 455 DC-8 crashed near Barbados after the detonation of two devices killing all 73 people on board. Cuban exiles were responsible.

11. 1976: Famous Israeli commando operation at Entebbe to rescue 105 hostages on Air France Flight 139 diverted there. The flight had originated in Tel Aviv and had many Jewish passengers. The rescue was considered a daring success, though three passengers, one Israeli, and some 40 Ugandans, besides 7 hijackers, were killed. The perpetrators were the PFLP and Revolutionäre Zellen (RZ).

12. 1977: The hijacker of Malaysian Airline System domestic Flight 653 wanted to divert to Singapore. With the pilots saying they did not have enough fuel, shots rang out, killing the pilots. The aircraft then crashed killing all 100 occupants.

13. 1977: Lufthansa Flight 181 was hijacked by four members of the PFLP and over six days the plane was taken to Rome, then Larnaca, then Dubai, then Bahrain, then Aden, and finally Mogadishu. At Aden the captain was assassinated. A rescue mission by German commandos finally freed the remaining hostages. The case was famous in Germany with flight attendant Gabriele von Lutzau hailed as the "Angel of Mogadishu," because she sacrifice herself for her passengers.

14. 1985: TWA Flight 847 was hijacked on a flight from Athens to Rome. Negotiations lasted two weeks, with all hostages (except for a US Navy diver who had been shot and thrown on the tarmac) released, and the Israeli government agreeing to free some 700 Shiite prisoners.

15. 1985: The world's worst single-aircraft bomb disaster that could have been far worse for the Sikh militants based in Canada were simultaneously proceeding to bring down an Air India 747 taking off from Tokyo.
    That failed because the perpetrators did not know Japan did not have summer time—the bomb exploded prematurely killing two Japanese baggage handlers and injuring four others at Tokyo's Narita Airport as an item of luggage was being transferred to Air India Flight 301. However, the Air India 747 flight between Toronto and Bombay was blown up over the North Atlantic near Ireland with 329 fatalities.

16. 1986: As TWA Flight 840, a Boeing 727, passed through 10,000 feet on its approach to Athens a bomb detonated in the cabin causing a rip in the fuselage through which four people were ejected.

17. 1987: Korean Air Flight 858, a Boeing 707, was blown up on a flight from Baghdad to Seoul by two North Korean agents (a man and a woman). They both took suicide pills when about to be arrested, with the woman who had placed the device on board the aircraft surviving. Though condemned to death, she was pardoned in 1990 on the grounds that she had been brainwashed and put under close guard in a secret location for her own safety. She had switched sides, explained how she had been trained, and revealed who had told her to carry out the terrible deed.

18. 1987: At Los Angeles a ticket agent working for USAir, the parent of Pacific Southwest Airlines (PSA), who had been fired, but had kept his pass, boarded PSA Flight 1771 about to take off for San Francisco. When the flight reached cruise level, he shot both pilots and allowed the plane to crash, killing all 44 on board.

19. 1988: The Lockerbie (Scotland) disaster that partly contributed to the ultimate demise of Pan Am because people feared that its fame as the number one US carrier meant it would very likely be targeted again.
Pan Am Flight 103, a Boeing 747 bound for the US with 259 people on board, broke up half an hour after takeoff from London's Heathrow due to a bomb in an item of luggage. All on board were killed plus 11 more at Lockerbie below. Astonishingly, part of the device for detonating the explosive was found on the ground and that coupled with evidence from a shopkeeper in Malta resulted in two Libyan secret service men, al-Amin Khalifa Fhimah and Abdel Baset Ali al-Megrahi, being brought to trial in the Netherlands with three Scottish judges. Only Al-Megrahi was found guilty, with his defense not being given information that could have helped his case, and he was sentenced to life imprisonment. Although saying he did not know anything about it, the Libyan leader Gaddafi paid damages to the families, perhaps because there were

much greater pressures on him.
Al-Megrahi never admitted guilt, sought several retrials without success and, suffering from cancer, was released by the British government in 2009, to die in Libya three years later. It was a very murky affair with the US paying the Maltese shopkeeper and offering money to others, and evidence favorable to the defense withheld. It was thought that following the shooting down of an Iranian airliner by a US warship* they might have been directly or indirectly via the Palestinians more likely responsible. Also, around that time the US government had been signaling embassies and others about warnings and intel about the likelihood of such an incident.
However, years later, in December 2022, the US announced it had taken custody of the alleged bomb maker in the 1988 Lockerbie attack, a Libyan, Abu Agila Mohammad Mas'ud Kheir Al-Marimi. Could this eventually clarify the affair and show how or by whom the bomb was placed on board and prove the US might have been, to some extent, right all along?

20. 1989: Nine months after the Lockerbie disaster, Libya was indeed responsible for the downing of France's UTA Flight 772, with 172 people on board. An explosive device went off in the forward luggage hold of the McDonnell Douglas DC-10, flying from Brazzaville.

21. 1989: Avianca Flight 203 flying from Bogotá to Cali was blown up by a bomb five minutes after takeoff. It was planted near an empty fuel tank full of fumes by Pablo Escobar's cocaine cartel gang to eliminate an inimical presidential candidate who was not on board as expected. In total, 107 people on board and 3 people on the ground were killed.

22. 1994: A Fed-Ex flight engineer about to lose his job arranged to travel as a deadhead passenger on Federal Express Flight 705 with the intention of overcoming the aircrew (two pilots and a flight engineer) and crashing

the aircraft to allow his family to benefit from his life insurance if he died in an "accident" at work.
So the crew's injuries would look like that they'd been sustained in an accident rather than a hijacking, he brought a speargun and a hammer on board a cargo flight to disable the pilots, thinking such weapons would not raise suspicions on being found in the wreckage and invalidate his insurance claim.
On coming to their senses after being hit by the hammer, one pilot struggled with him, leaving the dazed captain at the controls. After grappling with the perpetrator for some time, he had to call for assistance, whereupon the captain put the plane on autopilot and came to help. Together they managed to restrain the perpetrator and land the plane safely despite their injuries.

23. 1994: Philippines Airlines Flight 434 flying from Cebu in the Philippines to Tokyo suffered an explosion under the seat of a passenger, blowing off his legs and making a hole underneath him. The liquid explosive device had been prepared and planted by Ramzi Yousef who had been one of the main perpetrators of the World Trade Center bombing in 1993. It was a test run for the so-called Bojinka Plot planned for the following year. The unusual layout of the seating on that particular Boeing 747 meant the bomb was not fatally placed above the fuel tanks as intended. The aircraft was able to divert to Okinawa with ten injured and the dead passenger who had been above the device.

24. 1994: In what might have been the precursor to **9/11**, the hijackers of the Air France Airbus A300 in Algeria intended to crash it into a Paris landmark such as the Eiffel Tower.
After negotiations with the Algerian authorities with a passenger killed from time to time to show the hijackers meant business, the Airbus was allowed to take off for Marseille with the hijackers expecting it to be a refueling stop—they wanted so much fuel it was obvious what

their intentions were. French commandos stormed the aircraft in a remarkable action, killing the hijackers with one holding out and firing for 20 minutes. No hostage was killed, though the copilot jumping out of a cockpit window suffered quite serious injuries due to the height*.

25. 1995: Failed Bojinka plot to blow up several airliners over the Pacific, conceived by Ramzi Yousef (see Philippines Airlines Flight 434 above). The plot was discovered because the bomb makers working in an apartment kitchen in Manila carelessly discarded chemicals down a sink in their kitchen resulting in fumes that forced them to suddenly go out into open air leaving behind evidence including a laptop found by the police, which even showed flights they were thinking of targeting*.

26. 1996: An Ethiopian Air Lines 767 was commandeered over the Indian Ocean by hijackers who had read the inflight magazine mentioning the long range of the aircraft and wanted to fly to Australia. They would not believe the pilots who explained they had only loaded enough fuel for a short hop. The jet ran out of fuel. The captain who had already experienced a hijack pilot ditched aircraft off the Comoros Islands. But wrestling at the controls with the hijackers he could not level out with the result the left wing snagged the sea, and the aircraft broke apart. In total, 50 survived, but 125 including the 3 hijackers lost their lives—some because they prematurely inflated their lifejackets, and the water pushed them upwards trapping them inside the cabin*.

27. 1999: All Nippon Airways Flight 61 was hijacked by a 28-year-old unemployed man who gained access to the cockpit by seizing a female flight attendant and threatening her with an 8-inch knife he had smuggled aboard having discovered a flaw in airport security. He stabbed the captain and took control, with disaster narrowly averted as the aircraft once dropped below

1,000 feet. Finally, the hijacker was restrained, allowing the copilot to land the 747 safely with its 503 people, including the dead captain, on board.
The hijacker who had Asperger's syndrome had practiced flying on simulators. He said he hijacked the aircraft to fly under the Rainbow Bridge over Tokyo Bay.

28. 2001
9/11, September 11: Four aircraft, taking off for long-haul flights in the US, and hence carrying large quantities of fuel were hijacked almost simultaneously. Two were flown into New York's World Trade Center's North and South tower, which as the fires burned ultimately collapsed with some 3,000 people killed.
A third one was flown into the Pentagon near Washington DC killing some 125 people on the ground and 64 in the aircraft.
The fourth hijacked airliner was delayed, enabling its passengers to learn of their probable fate by cell phone. They rose up and confronted the hijackers but unfortunately could not prevent the plane crashing in open country—probably saving many lives on the ground as it was headed for Washington DC, and either the White House or the Capitol building*.
See **9/11: Two Unarmed F-16s Scrambled to Down 4th Airliner**

29. 2001: Al-Qaeda operative Richard Reid was seen by a flight attendant trying to detonate plastic explosive concealed in his shoe on American Airlines Flight 63, flying from Paris to Miami. It failed to detonate because the day before departure he had been held at security at the airport in Paris as suspicious and after spending the hot night in his clothes, he had sweated so much that the explosive material had become damp. His failed attempt led to passengers having to remove their shoes on going through security.

30. 2026:As an "afterthought" we must mention a novel attempt at a hijack where one of trusted pilots was the hijacker bringing into question the advisability of having impenetrable cockpit doors as did Germanwings suicide case where the copilot locked out the captain who had gone to the toilet and crashed the aircraft into a mountain killing all on board.

    In this case the Omani first officer of an Israel-bound Flydubai flight said he was getting up to pray and struck the captain with the axe kept in the cockpit. A woman in the toilet next to the cockpit heard the altercation, and shouted out to the cabin crew and others. The half-conscious captain just managed to unlock the cockpit door enabling passengers to intervene.

    The first officer had apparently cut off fuel to the engines had put the aircraft in an earthwards plunge at a speed exceeding its maximum allowed speed. A plumber allegedly helped put the aircraft out of its dive, but fortunately not sharply enough for the wings to snap off, though most of the rudder broke off.

    Two off-duty Flydubai pilots who had been seated at the rear were able to take over and save the situation—no easy task.

    Before joining Flydubai the criminal first officer had been taken off flying duties and given an office job? He could not bear to be in the presence of women.

Above, we have listed what are some of the most notorious incidents with many omitted. Those marked with an asterisk * are described in full in our classic *Air Crashes and Miracle Landings*.

"High-Speed Climb" Request to ATC to Climb Without Flaps

In the US, airspeed up to a height of 10,000 feet is limited to 250 knots for safety reasons. This avoids the air traffic control system getting overwhelmed and gives more time for the resolution of problems. Sometimes an aircraft such as a "heavy" needing flaps to fly so slowly will ask permission for a "high speed climb" to save fuel by flying without flaps. Permission is normally granted.

High Bypass Turbofan Engines

Modern commercial airlines' jet engines are much quieter than the screaming versions prevalent in the 1960s and 1970s. This is partly achieved by having the turbines at the back of the engine drive an enormous fan at the front that blows air to the back, with much of it bypassing the combustion section. Hence, the use of the term "high bypass."

High-Speed Trains, Maglev Train in Partial-Vacuum Tube

Airlines have one great advantage over other modes of transport in that they do not have to pay for infrastructure such as roads or railway lines between the departure point and destination. Of course, there is the cost of the fuel, the airport infrastructure, air traffic control, and overfly charges on international flights, not to mention taxes and expensive pilots.

High-speed trains can compete over limited distances but often with higher fares because of the cost of the infrastructure, often allowing **Low Cost Carriers** to compete with give-away fares. Also, highspeed rail, unlike commercial aviation, is only theoretically possible between points with very heavy traffic.

Interestingly, there is the prospect of "rail" being in some situations even significantly faster than subsonic flights. Already, we have the maglev train running between Shanghai and its main airport at a maintained maximum speed of 299 mph (481 km/h) and the Japanese maglev project already under construction linking Tokyo to Nagoya and, eventually, Osaka.

Looking further into the future there is the prospect of maglev "capsules" running in partial-vacuum tubes at far higher speeds, with this sometimes referred to as "hyperloop." Just as aircraft fly with less drag high up with lower air pressure, the same would apply to maglev trains in a tube under a partial vacuum.

Probably, if feasible, such hyperloop systems will be first deployed on any scale in China—they have the technology and are not held up getting permissions. They already have something experimental called the "T-Flight." Chinese engineers from CASIC (China Aerospace Science and Industry Corporation) have reported that their T-Flight maglev train recently reached a record speed of 387 mph (623 km/h) on a test track apparently in a tube, with the engineers hoping to reach 621 mph (1,000 km/h).

CASIC is China's state-owned major missile maker and not only has considerable resources but is also involved in the development of all sorts of technologies. As regards the hyperloop project, CASIC envisages it developing in stages with the first phase attaining speeds up to 621 mph (1,000 km/h) for domestic intra-city commuting; the second phase up to 1,243 mph (2,000 km/h) for domestic intra-region traveling to linking conurbations such Beijing, Shanghai, Chengdu, Guangzhou, and Wuhan. Double those speeds are envisaged for the third phase.

This would make it more than competitive with air travel. However, it is doubtful whether the capsules in a tube would have sufficient capacity to make it a general mode of transport. It would be mainly for VIPs.

Highest Salaries for Airline Pilots

Apart from the nominal salary, the amount of money a pilot receives depends on their tax status. Some Middle-East airlines not only offer high salaries but also have low or negligible income tax.

Whatever the airline, salaries can be quite low for first officers at around $60,000, with the really big jump coming on a move to captain. At Delta a captain can earn $350,000, and at Lufthansa $180,000 for short-haul, and $340,000 for long-haul.

Hijack Protocols

Before **9/11*** where the hijackers' intention was to use airliners full of fuel as flying bombs, the protocol for dealing with a hijacked aircraft was to acquiesce to the hijackers' demands and allow them to land and engage in negotiations.

Now cockpit doors have been reinforced and pilots in general told not to let anyone in regardless of the threat, say, to a flight attendant. Passengers, aware that they may themselves be weaponized, are more likely to take some action. They also know the airliner might be shot down if heading for a major city or nuclear reactor.

The precise instructions airlines give to their pilots regarding what to do in a hijacking are confidential, but whatever the airline, the first thing pilots would normally do is to enter the "hijack" emergency code 7500 into the transponder. Note that failure of the radio can lead air traffic control to suspect a hijack, in which case the pilots enter the 7600 "radio failure" code into the transponder.

Because of the precautions taken, and the fact that pilots will no longer open the cockpit door whatever the menaces, hijackers know that succeeding is nigh impossible. Hijackings are, as a result, rare.

See **Behavior Detection Officers**.

HIL (Hold Item List)

Noncritical maintenance items such as switches that need fixing or replacing but because they have backups do not prevent the aircraft from flying. However, they must be noted and fixed within a certain time. Normally items that should not be used are highlighted with an "INOP" sticker. Also see **MEL (Minimum Equipment List)**.

HIMARS (High Mobility Artillery Rocket System)

With the conflict in Ukraine constantly in the news one often sees reference to HIMARS. It is essentially a pod launcher on a 5-ton army truck able to launch either six missiles or one missile. They can launch all sorts of missiles ranging from the very simple ones to long-range ones with sophisticated guidance systems.

Holding Pattern (Stack)

When aircraft cannot land at an airport because others are already waiting to land, or due to bad weather and so on, they are put in a queue, but as they must keep moving to stay in the air they are stacked up in rectangular patterns, usually in the vicinity of a VOR beacon. Each aircraft will be at a different flight level and will drop down to the next level when told to do so by ATC.

Hollow Charge, Shaped Charge

Sculpting the forward end of an explosive cartridge to form a cone or hemisphere concentrates the effect, with the metal lining fusing and carried with it, enabling it to penetrate, say, metal or concrete, easily.

The Bazooka, which took its name due to its resemblance to a musical instrument with that name, was a tube firing a rocket propelled hollow charge projectile developed in 1942. Although it was created to be a portable weapon to destroy tanks it was mainly used to attack emplacements and fixed fortifications, not as anti-tank weapons. General Dwight Eisenhower is said to have called it one of the four "tools of World War II victory."

There are suggestions that at the end of World War II the Germans were contemplating using hollow charges to make a nuclear fusion bomb. How far their atomic research had progressed is uncertain, but the allies were worried, and the development of revenge weapons of mass destruction was one reason to end the war as soon as possible. There was talk of a V-3 and V-4 rocket with a very lethal warhead able to reach New York.

Home-Built Aircraft 2½ Times More Dangerous than GA

Having one's own aircraft and flying oneself is much sought after in the US. But with them so expensive and subject, as well as their pilots, to strict FAA regulations including medical, the prospect of making one's own—usually from a kit—is appealing.

The FAA regards such aircraft as experimental and subject to virtually no regulations—in fact, the FAA could not be everywhere. It is not only the "aircraft" that can have problems but also the pilots—perhaps seniors with spare time and money—who know they would be unable to pass GA medicals.

As a result, the accident rate for home-built aircraft is said to be two-and-half times that for GA aircraft, and perhaps more. That is not to say it cannot be worthwhile when done with care and with truly capable people at the controls.

See **Ultralight Aviation (Microlight)**.

Horizontal Stabilizer (Tailplane)

The horizontal stabilizer is a large horizontal fin in the tail **Empennage** that is in a fixed but adjustable position that helps keep the aircraft steady in the longitudinal plane, which compensates for the perhaps inherent tendency of the aircraft to pitch down, and can be tweaked by what is called trimming so the pilot or autopilot does not have use the elevators to do what they want to do.

Because of its large area, it is very powerful and can override the elevators, as happened in the two Boeing 737 MAX crashes where the MCAS prevented the pilots saving the aircraft.

The vertical stabilizer sticking up vertically on the tail and often with the logo of the airline painted on it similarly has a large area but is not usually adjustable. It helps stop the aircraft yawing, but when struck by a crosswind it tends to pull the nose of the aircraft into the wind.

There are V-tailed aircraft where the functions of horizontal and vertical stabilizer are combined. Their operation is complex.

Canards (mini wings right at the front of the aircraft) are sometimes referred to as horizontal stabilizers.

HSI (Horizontal Situation Indicator)

"Horizontal situation" means location with horizontal reference to points on the ground. Can be set in ILS/VOR/NAV indication modes.

Hub Buster

New term for ultra-long-range versions of an airliner, making it possible to fly ultra-long routes with a full load without refueling at the customary intermediate hub.

HUD (Head Up Display)

See **Head Up Display (HUD)**.

Human Error

With the great improvements in technology and materials and the greater reliability of turbine engines as opposed to piston engines, human error now represents an increasingly large proportion of accident causes.

One problem is that with automation doing so much, pilots are losing their flying skills—on a 12-hour flight it may just be a few minutes at takeoff and landing with pilots able to claim thousands of hours of flying experience, which looks good but does not mean much.

Hydrogen

Hydrogen was expected to be the next big thing in aviation, but enthusiasm has cooled even at Airbus who were once announcing plans for it. Limited availability and the time taken to solve the problems of storage aboard mean plans to use it will be postponed.

Hypersonic Missiles, Ramjet, Scramjet, Rocket

As "hypersonic" is the new buzzword with it being thought that the US is lagging behind Russia and China in this domain, it may be a good opportunity to cover the whole topic of missiles under this heading.

Aircraft, drones, and missiles are referred to as hypersonic if they fly at airspeeds above Mach 5, or five times the speed of sound. One may ask why Mach 5? The answer lies in the type of engine used for cruise missiles, and the fact that at Mach 5 and above air resistance becomes problematic.

Different types of engine listed according to operating speed:

1. **Turbofan engines.**
   These engines found on airliners use large-diameter

fans to draw in air compressed by turbines, mixed with fuel and ignited with the greatly expanding gasses expelled from the rear to create thrust. These work well up to transonic speeds with the performance thereafter dropping off with them almost useless after Mach 3. This meant the Anglo-French Concorde flying at Mach 2 was feasible and the cancelled US SST flying at Mach 3 was a much more difficult proposition, not to mention lack of materials to cope with the heat generated. The great advantage of these engines is that they can be started from a stationary position.

2. **Ramjet engines.**
   These engines are in some ways the simplest of all in that they do not have fans or moving main parts. The forward velocity of the vehicle rams the air into the mouth of the engine where it is slowed to subsonic speeds and compressed in a chamber where it is mixed with fuel and ignited. The thrust is produced by the expanding air ejected from the back. They work most efficiently at Mach 3, and theoretically have a maximum speed of Mach 6 due to shockwaves. The problem with the ramjet is that unlike the turbofan it must already be moving, normally above Mach 0.5. They can be started by dropping them from say, a Boeing B-52.

3. **Scramjet engines.**
   "Scram jet" stands for Supersonic Combustion Ramjet. As the name implies, combustion takes place with the air/fuel mixture moving at supersonic speed, and while this appears the simplest engine of all, mixing in the appropriate amount of fuel at such hypersonic speeds is extremely complex and delicate. They can operate right up to Mach 24 but must be started with them moving at around Mach 5. To achieve such speeds a rocket is usually necessary

though that is also possible with them dropped from a B-52.

4. **Rockets.**
   Rockets differ from the air-breathing engines just described in that they carry all the elements, including oxygen, needed for combustion. They are used for ballistic missiles or short-range missiles where fuel efficiency is not of prime importance.
   Their great disadvantage is that once the fuel is exhausted and they reenter thicker air, they glide at progressively reducing speed on a predictable path, and though using fins slight deviations are possible, they become easier to track and destroy. Nevertheless, ballistic missiles are still moving very fast when they strike making them difficult to counter at that stage. However, the Chinese have developed the two-stage hypersonic glide vehicles (HGVs), DF-17 and DF-27, combining a rocket and a glide vehicle able to twist and turn.

Hypersonic cruise missiles launched from B-52s with a range of 1,000 miles and on part of the trajectory speeds up to almost Mach 20 are envisioned. The greatest problem with hypersonic missiles is that their great speed makes it difficult for them to significantly change course when seeking moving targets such as ships. That said, the location of a seaborne invasion is fixed, and once known could have hypersonic missiles rained down on it from nearby or even from afar with no need for them to change course.

Hypersonic missiles, for instance, are not new, but what makes the new iterations so lethal is their ability to maneuver in flight and confuse a defender uncertain of where they are going. With thrust vectoring some can even make a 180-degree turn.

Hypoxia

Oxygen deprivation.

For the pilots there can be several causes including flying above 10,000 feet with the aircraft unpressurized without supplemental oxygen, or carbon monoxide poisoning where the blood cannot carry the oxygen.

It is especially dangerous because of its insidious onset. The signs and symptoms often develop so imperceptibly that the victim becomes virtually incapable before it can be recognized. Also, the indications and symptoms vary from person to person.

Crashes have occurred due to hypoxia. One stands out partly as it might have had a happy ending like a Hollywood movie with someone coming from the passenger cabin to fly the plane and save everyone.

It was Helios Airways Flight 522* in 2005, a 737 flight from Cypress to Athens.

Taking off from Cypress with the cabin pressurization unusually still set to manual and climbing through 10,000 feet, the cabin altitude warning horn sounded, but as the same horn was used to alert them to a mistaken configuration at takeoff, they misconstrued its significance. As the aircraft climbed further, the two pilots lost consciousness due to hypoxia with the flight management system (FMS) which they had earlier programmed taking them on up to their cruising altitude and on to Athens, their destination.

On arriving there with no instructions to descend, the FMS put them into a holding pattern near the airport. The Greek controllers, alarmed at not being able to make radio contact, and wondering what was happening, sent two of their Air Force fighters to investigate.

The pilots reported seeing two people in the cockpit. It later transpired it was a steward with some flying club experience and a female flight attendant. They had been trying to get help over the radio, but had been unable to make contact as it was still set to the original Cypress frequency. Finally, fuel ran out, and the 737 lost height, crashing into hilly terrain leaving no survivors.

In general, should the aircraft be flying high enough for the pilot to need oxygen in the event of depressurization and one pilot being left alone in the cockpit, for example with one of them having to go to the bathroom, the remaining pilot should don their oxygen mask with supplemental oxygen working. However, this does not apply if they have a "quick donning" mask at hand.

~ I ~

Iannello, Victor, Radiant Physics [MH370]

A highly qualified engineer hosting a group of technical experts, each with their own expertise cleverly analyzing the disappearance of MH370. Very technical.

See https://mh370.radiantphysics.com/.

IATA (International Air Transport Association)

Airline trade organization based in Montreal, Canada.

At one time, with many carriers being national airlines (flag carriers), governments gave them dispensations under anticartel regulations to fix fares through IATA and even colluded themselves. However, now, with deregulation, the body no longer has the role of fixing airfares. Although the international airfares were usually set too high, the system had some advantages in that you could use your IATA ticket on any airline and could switch even at the last minute, even though many passengers were not aware of that.

Sometimes IATA comes up with what would seem to be useful proposals that it cannot enforce. For instance, it suggested setting an international standard for the size of carry-on luggage that would not only end the confusion but, by making it 20 percent smaller than the norm in the US, create more room in the overhead bins on board. Air China, Azul, Pacific, China Southern, Emirates, Lufthansa, and Qatar were in agreement, but American, Delta, and United in the US opposed it, with Delta saying it had invested in larger bins, which were what the US public wanted.

ICAO (International Civil Aviation Organization)

United Nations agency responsible for aviation matters, also based in Montreal, Canada, but not related to IATA, which is based there too. It has seven regional offices. It sets many international norms.

ICBM (Intercontinental Ballistic Missile)

See **Missiles**.

IED (Improvised Explosive Device)

Because the abbreviation is short and can mean anything from a hollow-shaped charge able to blow up an armored vehicle to a big firecracker destined to sow panic, it is widely used by official bodies, such as the US's TSA, to describe any improvised (homemade?) explosive device—in other words, anything other than standard munitions.

IFE (In-Flight Entertainment)

In-flight entertainment systems allowing passengers to watch movies have become very important as a way to make journeys less disagreeable and seem shorter.

In view of the amount of wiring required, the use of Wi-Fi to transmit the data needed for flight entertainment systems has been very seriously considered for the latest aircraft but has encountered two obstacles. The first is getting approval from the official international body controlling such matters for a special intra-aircraft waveband. However, it should finally be given. The second is that with more and more demands being made on the systems, Wi-Fi may finally not be able to provide the necessary bandwidth.

That said, some say that other than in first and perhaps business class, IFE systems, with their heavy screens and wiring, should be done away with, since more and more passengers in economy have their laptops and iPads, which they can use to see their own videos or videos transmitted by onboard Wi-Fi.

IFF (Identification of Friend or Foe)

The first transponders were IFF devices to tell radar operators in World War II whether aircraft were friend or foe.

IFR (Instrument Flight Rules)

The situation where poor visibility or air traffic control requires the pilots to fly according to air traffic controllers' instructions and their instruments rather than visual observations of landmarks. Pilots must be suitably qualified. See **VFR (Visual Flight Rules)**.

In the US, general aviation aircraft flying under VFR in uncontrolled airspace set their transponder to squark 1200.

Illegal Intelligence Operative

A term used for spies who don't have diplomatic protection.

ILS (Instrument Landing System)

A system whereby two radio beams (one in a vertical plane for the glide slope and the other in a horizontal plane as the horizontal localizer) create an artificial glide path and tell pilots, or the aircraft autopilot, whether they are above or below the glide slope, or to the left or right of the projected centerline of the runway.

In good weather the pilots can make a visual approach, with the **VASIS** lights near the touchdown point changing color to provide similar but less critical information.

The procedure for joining the ILS glideslope is to first find the localizer (the horizontal beam showing one's position) and then come onto the glide slope (vertical beam) flying almost level meaning one approaches it from below as it is sloping. Coming onto it from above is trickier and can lead to problems, not least of which is sinking up well below it. If way off the localizer, the ILS system can give false information regarding the situation above or below the glidepath.

IMC (Instrument Meteorological Conditions)

This is the visibility below that stipulated for flight under **IFR (Instrument Flight Rules)** and incidentally means that the approach must be stabilized by the time the aircraft descends to 1,000 feet on its approach.

Incapacitation of Pilot

There have been cases of pilots falling ill with cabin crew "carting" them away and helping the remaining pilot. There was even a case where undersize bolts were used when a windscreen panel was replaced, and it blew out with the captain sucked halfway out and only saved by a member of the cabin crew grasping him. The very inexperienced first officer landed the aircraft safely.

See **Subtle Incapacitation of Pilot.**

Inertial Reference System (IRS)

This is one of the most sophisticated systems on the aircraft and is notable for working independently unlike GPS, which depends on satellites. For this reason, it is used for the targeting of cruise missiles and more notably Intercontinental Ballistic Missiles (ICBMs).

In its already very effective original version developed in the 1960s, mechanical gyroscopes and accelerometers calculated how far the aircraft has moved from a given place such as the coordinates of the departure airport entered prior to takeoff. Though the indicated "position" tended to drift over time the error was small with the possibility of the system being reset at a known location. Before the advent of GPS, aircraft very much depended on it in isolated places such as over oceans with it being called the inertial navigation system.

The Korean airliner (KL007*) shot down by Russian fighters in 1983 would have been using inertial navigation, and it is said that incident led to President Reagan giving permission for the civilian use of GPS. There was also the Air New Zealand sightseeing flight to the South Pole where the programmed Inertial Navigation course took them directly into a high mountain*.

Mechanical gyros can easily break down and require expensive maintenance, and a much better system has now been perfected using ring lasers instead of mechanical gyros.

See **ADIRU (Air Data Inertial Reference Unit).**

Infra-Red Search and Track (IRST) vs. Radar [military]

Infrared detection has, unlike radar, the advantage of being passive, in that it does not emit any signal that could alert the aircraft or ground unit being tracked. However, it has the disadvantage of being greatly affected by cloud, rain, and other precipitation.

Another disadvantage is that it cannot tell how far away an object is, and while radar can determine distance the object of attention would be alerted.

One way to avoid using radar is to couple the tracking with a similar infrared sensor in another aircraft, or in a drone, some distance to the side and determine the distance by triangulation. Another way, assuming the object tracked is flying a straight course is for the tracking aircraft to weave from side to side and simulate the above triangulation method taking the bearing and position at each extremity.

Insulation (Electrical, Thermal, Acoustic, Etc.)

Insulation used in the construction of airliners sometimes means insulation used to prevent the propagation of heat/cold or noise rather than electrical insulation.

Intelligent Flight Control System (IFCS)

Aircraft computers are programmed to fly the aircraft even when several systems fail, since key systems can be in triplicate on the belt-and-braces principle, or an alternative backup can logically be found. However, there can be extreme situations in war or in normal flying where pilots can find themselves in potentially disastrous situations, flying severely damaged or malfunctioning aircraft, and NASA is developing new “smart” software that will enable aviators to control and safely land disabled airplanes.

In essence, the system would use “self-learning neural network software” or AI to try to resolve the situation. In mentioning the worst-ever single aircraft crash, Japan Airlines 123, we described how the pilots, having lost the tailplane and hydraulics, tried to maneuver the aircraft by varying the power of the engines. They succeeded for half an hour, finding it difficult to anticipate and minimize the phugoid motion. The computer using this new software *might* have been able to do a better job.

Recent studies have tried to establish whether such an AI approach could be applied to situations where a pilot is disorientated by being upside down or for other reasons unable to cope.

Intersection (Navigation Using Radio Aids/VOR)

The points where radials from VOR beacons intersect are known points and have identifying codes, with those pertaining to an approach to a specific runway incorporating that runway's number.

ISR (Intelligence, Surveillance, and Reconnaissance)

This is an often used acronym, and makers of sophisticated drones often claim they can perform it (Intelligence, Surveillance, and Reconnaissance). Besides actually firing missiles and dropping ordinance, drones can perform target acquisition in what is called ISTAR (intelligence, surveillance, target acquisition, and reconnaissance).

These drones can fly for 40 hours and reach as high as 40,000 feet and be controlled almost anywhere in the world from its base at RAF Waddington in the UK.

ISTAR (Intelligence, Surveillance, Target Acquisition, and Reconnaissance)

There are now drones like the British Protector drone that besides the usual ISTAR role can themselves carry weapons such as the powerful 500-lb Paveway IV laser-guided bombs and Brimstone 3 missiles to carry out strike operations.

ITCZ (Intertropical Convergence Zone)

A zone at the equator, renowned for its stormy and fickle weather.

See **AF447**.

~J & K ~

Jackscrew/Jack Screw

These are relatively long threaded bolts that can be rotated to set the position of a threaded attachment and set the angle of, say, the horizontal stabilizers or flaps. Their great advantage is that they are strong and maintain the item in a fixed position despite strong forces.

There have been accidents due to corrosion of badly lubricated jackscrews. If the jackscrew for the horizontal stabilizers is failing pilots should not take actions to force it and risk it failing completely—with the aircraft inverting and crashing as happened in the case of Alaska Airlines Flight 261, a McDonnell Douglas MD-80, on January 31, 2000.

JADO (Joint All-Domain Operations)

Jane's

Jane's Information Group describes itself as *a world-leading provider of intelligence and analysis on national and international defense, security, and risk developments.*

Initially famous as a British publisher of books on fighting ships and aircraft, it has expanded into a major group. One can sign up to receive (free!) weekly emails with tempting extracts from its various publications. It is a valuable source for the latest information on air transport and airports, in addition to military and security matters.

JADOC2 (Joint All-Domain Command and Control)

"Janet" Call Sign for Secret Airline Flying to Area 51

Aircraft, mainly 737s with the call sign *Janet* and no markings apart from the US aircraft N- registration number and a red stripe running the length of the fuselage, could be seen mostly in the early morning taking off from Las Vegas's (now-called Harry Reid) International Airport in the early morning and returning in the evening.

Some said *Janet* stood for Joint Air Network for Employee Transportation, for it is a US Air Force commuter airline for civilians, ferrying employees. It even has its own AECOM terminal with the large car park packed during the day and almost empty overnight.

The aircraft take off with their transponders working but on reaching restricted D airspace some 70 miles (112 km) north of Las Vegas their transponders are turned off so no one can see exactly where they are going, for in this general area there are many ultra-secret facilities including Area 51 where the civilians obviously work. There are other ultra-secret areas where the same obfuscation happens.

We are not revealing anything really secret as details are available on the Internet and even YouTube. In fact, Janet represents a clever way of combining easy commuting with security, for it would be much simpler for a potential enemy to track employees traveling on land in their own vehicles or by bus from which the occupants themselves could see out of the windows.

Whether it applies to all Janet aircraft is uncertain, but those flying to Area 51 may have their windows covered, even though passengers might be working there and have a high-level clearance. This might be to prevent them catching sight of, say, secret aircraft and drones under development or being tested.

In 2023, the US Air Force for some reason—perhaps because the present operator was more interested in other things—started looking for a new operator for Janet flights.

JAPAN

As early as 1918, Tokyo University's Engineering Faculty opened the Aviation Research Institute, with an aeronautical course in the Department of Shipbuilding. This course expanded to become the Department of Aeronautics in 1920. Graduates included the legendary designer of the Zero fighter, Jiro Horikoshi, born in 1903. Thus, aviation design in Japan had a strong academic base.

Early in World War II, the US believed the Japanese had many more Zeros than they, in fact, had, because their unbelievable range enabled them to be everywhere. In fact, production of the Zero at Nagoya was almost artisanal with it being impossible to scale up as required, unlike the US that had numerous factories producing civilian goods that could be switched to production for military purposes.

For details of how the US learnt about the performance of Japanese aircraft in World War II see **TAI (Technical Air Intelligence)**.

Interestingly, just before the war Japan had begun building DC-3s under license, producing some 500 in all, mostly used as military transports.

Postwar

While German designers were spirited away to the US in **Operation Paperclip** and the like at the end of hostilities, most of the Japanese ones stayed on in Japan where under the occupation the country had been forbidden from engaging in any aviation production work for seven years. The Korean War broke out in 1950, with the US needing Japan to supply goods, giving the Japanese economy and industry a great boost.

This led to Japan's recovery, highlighted by three outstanding undertakings:

1. The creation of a world-beating high-speed train, benefitting from the fact that Japan was a long, thin country with most of the population having to travel long distance on one single line.

2. The creation of an airliner, the YS-11.

3. The Olympic Games to be held in 1964.

The YS-11 turboprop airliner and the "Bullet Train" were ready by the time of the Olympics, with the Olympic torch being ferried round the country in the airliner of which the Japanese were so proud.

Interestingly, the Bullet Train was found to have aerodynamic problems due to its high speed and the World War II designers of the Mitsubishi Zero fighter and the Ohka (a rocket-powered suicide bomber) were brought in to solve the problems. Not many people know the bulbous front of the original bullet trains looked somewhat like that of the Ohka suicide bomber!

Japan's expertise in new materials and engineering means that it now supplies composite wings and other aviation related items to the US and notably wings for the Boeing 787 Dreamliner.

Jeppesen

A US company, now owned by Boeing, that over the years has developed into a key supplier of navigational charts, particularly ones required for a flight into and out of airports. With worldwide coverage, Jeppesen charts and systems have become something of a standard. The British have their similar *Aerad* flight guide and navigational chart system.

Jet Lag

Sleep disturbances and problems encountered when traveling between time zones are normally only significant when it is more than three. It is said to be worse when traveling from West to East apparently because it is easier to stay up late. Children usually adapt more easily to the new time zone.

In considering it, specialists refer to the "circadian rhythm" meaning the body's internal clock.

There are claims that passengers traveling in cabins such as those of the Airbus A 350 and Boeing 787 or top of the range business jets where the cabin altitude is lower are less than subject to jet lag, or perhaps the discomfort of air travel.

Jet Stream Highway Across the Atlantic

One of the busiest, and for some airlines the most profitable, routes is across the Atlantic from the US to Europe and vice versa with all the aircraft flying carefully spaced out in parallel lines in a corridor which can vary according to the jet stream. It is the busiest oceanic airspace in the world.

The jet stream runs from East to West with the result that the difference in time taken for flights across the Atlantic is usually one hour or more, with Eastbound flight times shorter.

With so much traffic and mostly no radar coverage, there is the need to have the aircraft relatively close together and only have aircraft that can accurately maintain their positions. This is achieved by applying **Reduced Vertical Separation Minima (RVSM)** with pilots and the aircraft having the appropriate Approval(s) issued by the State of Registry or the State of the Operator.

Jetpacks

Though there are already civilian applications for jetpacks where a person is taken aloft by thrusters attached to their body, the greatest advances have been made for ones destined for the military, and notably by JetPack Aviation in the US.

JetPack has one with two turbojet engines that can take a soldier up to 10,000 feet and is able to reach speeds of up to 125 mph. Though it can for now only stay aloft for 30 minutes, that is already enough for an assault by a small band of soldiers on an objective with mines and other ordinance. Rendering a ship inoperative would be one possibility.

As jetpacks improve, their value in civilian operations will become apparent, with these including firefighting in awkward places, rescues, and so forth.

JFK (John F. Kennedy International Airport)

New York's major international airport, formerly known as Idlewild, is so well known that the public generally refers to it by its code, JFK.

It has almost the longest commercial runway in the US. The somewhat longer one at Dulles Fort Worth is at an altitude of 185 m and subject to higher temperatures so the one at JFK being at sea level in colder climes is theoretically better if one ignores the noise-abatement limitations.

Job Opportunities at Airlines

There are many sources of information, including approaching an airline directly.

However, one site (https://latestpilotjobs.com) lists not only jobs for pilots but also for cabin crew, mechanics, and ground crew, including pilots for sophisticated business jets. In addition, they provide aptitude tests and interview preparation.

Salaries for pilots at the major airlines are highest, with much depending on seniority. As regards **Business Jets**, salaries for corporate pilots are considerably higher than those for charter pilots. A corporate pilot is one working for an individual or company, whereas a charter pilot works for private jet charter company or one providing an air taxi service. Sometimes a charter pilot is working thus to build up enough flying hours to join an airline.

Joint Special Operations Command (JSOC) [US]

As its name implies, the JSOC carries out—as well as the CIA—special operations including deployment of the **Hellfire Ninja Bomb**.

John Doe Immunity

In November 2006, a group of imams returning from a US imam conference were removed from a US Airways domestic flight after a passenger passed a note to the crew reporting their anti-American remarks. Flight attendants had also noted they were not sitting together but dispersed, while two of the party had asked for seat belt extensions, although one did not require one. The following November a judge allowed the imams in question to pursue their discrimination case, but the "John Does" (the passengers who had cast aspersions) were no longer cited, because the US Congress had meanwhile granted immunity to passengers on aircraft reporting suspicions. In the US, the term John Doe is used in legal proceedings when there is no specific name.

Joystick

Despite some assertions, the origin of the term for the "stick" in various shapes between the pilot's knees used to control planes is uncertain. Its use predates World War I.

While many airliners now have control columns rather than a joystick, it is notable that Airbuses have sidesticks, consisting of a short control stick to the left and right of the captain and copilot respectively.

Judgment Errors

In their book *A Human Error Approach to Aviation Accident Analysis*, Wiegmann and Shappell point out that analysis of over 4,500 pilot-causal factors associated with nearly 2,000 US naval aviation incidents, showed that:

1. Judgment errors (for example, decision-making, goal setting, and strategy selection errors) were associated more often with major accidents.

2. Procedural and response execution errors are more likely to lead to minor accidents.

Putting it in a motoring context, they compare the former to deciding to run a red light, which could result in deaths, and the latter to mistiming one's braking and causing what in the US is called a fender bender. Pushing the argument further, they say this dispels the widely held belief that the (causal) difference between a major accident and a fender bender is little more than luck and timing. One should perhaps add the caveat that navy pilots encounter more situations where decisions are needed compared with airline pilots flying set routes.

Just Culture as Opposed to a Blame Culture

Professor J. Reason, famous for his Swiss Cheese accident theory where accidents occur where the holes happen to line up, foresaw an organizational culture where people trust each other and exchange safety-related information as important for safety.

Kapton

A type of electrical insulation material used for many years for electric wiring in aircraft because of its many excellent qualities. However, recently there have been concerns about brittleness with aging in the case of aircraft over 25 years old.

It is now being phased out and replaced by other materials, such as TKT, which is Kapton sandwiched between two thin layers of Teflon enhancing resistance to mechanical abrasion and impeding water intrusion.

The turning point was the crash in September 1998 of Swissair flight 111* en route from New York to Geneva killing all 229 on board. Shortly after leaving New York, a fire behind the cockpit paneling started filling the cockpit ceiling with smoke but was not thought to be a MAYDAY emergency.

They took their time diverting to Halifax Airport in Canada only to find the fire so intense that it burnt through wiring harnesses including those carrying essential wiring for the flight controls. The MD-11 crashed into the ocean over which they had wanted to jettison fuel.

It was thought that Kapton used in the wiring for a new entertainment system was responsible for arcing and the outbreak of fire. Another lesson was that aircraft should land as soon as possible in the event of fire and not waste time jettisoning fuel even if they are overweight.

A new (expensive) form called TKT, which is Kapton with two thin layers of Teflon to make it more resistant to mechanical abrasion and better withstand water intrusion is being introduced.

Knot (kt) (Nautical Miles Per Hour)

Aircraft speeds are usually measured in nautical miles per hour or knots.

1 knot = 1.853 km per hour or 1.16 statute miles per hour.

~ L ~

Ladkin, Peter Bernard

Ladkin, born in 1951, is professor of computer networks and distributed systems at the Bielefeld University in Germany. He and his team have done much work on air accidents.

Lady Grace Drummond-Hay, "Around the World by Zeppelin"

This courageous woman, born in Liverpool in 1895, had gained her title of "Lady" by marriage to a member of the nobility older than she.

After the demise of her husband and under the sponsorship of the newspaper proprietor Randolph Hearst, and with little experience as a journalist, she joined an around-the-world voyage in the German airship the *Graf Zeppelin* in 1929.

Departing from a location near New York, the airship flew to its home to Germany with little incident. Then it flew on to Tokyo, crossing the Soviet Union in cold conditions with the risk of colliding with high peaks. It then flew on to Los Angeles, encountering a severe storm and believed lost, and finally on to the point of departure near New York, to a rapturous reception.

Her reports, transmitted from the airship—one short note was sent by carrier pigeon on leaving Japan—and during stopovers, made her famous, and she used that fame to further the cause of flying and women's roles in subsequent years. Interned during World War II by the Japanese in the Philippines with her journalist companion, Karl von Wiegand, she fell sick after her release and died in 1946.

A TV documentary broadcast on BBC Four in the UK on February 7, 2010, as *Around the World by Zeppelin* movingly portrays that around-the-world trip, with remarkable footage from the actual times. See link in "Other Links" section on our https://flyingknowitall.com website.

"Laker's Revenge"

At London's Gatwick Airport there is often some turbulence as the wind swirls around buildings near the runway. BA crew used to call this "Laker's Revenge," referring to the time the airline had done all possible to sabotage Freddie Laker's low-cost airline.

Laminar Flow vs. Turbulent Flow

When water flows slowly out of a faucet (tap) it is likely to be clear, but when turned fully on it becomes cloudy since when the water is very fast flowing it gets mixed up. In the first instance, it is experiencing so-called laminar flow and in the latter turbulent flow.

Laminar flow over an aircraft wing reduces drag.

Lasers

Laser-based systems are seen as the future of air defense. One great advantage is that they travel at the speed of light. A great disadvantage is that cloud, fog, and rain limit their effectiveness, something that also applies when they are used to "light up" a target so a weapon can lock onto it.

"Last Class"

An unofficial term for a cheap no-frills class, even below Economy or Coach, where *everything* is extra, changes cannot be made or are extremely costly, and with no help with rebooking in the event of cancellation.

Launch Customer

When an aircraft manufacturer is conceiving a completely new design aircraft, it hopes to find a respected airline willing to be a "launch customer," and one preferably willing to place a significant order. This confirms to it that there is a serious market, and above all encourages others to follow suit.

For the airline, being a launch customer can add to its prestige, not to mention the fact that it almost invariably benefits from the keenest prices. However, it can be saddled with early production models subject to teething troubles or even some serious defect that means it later must take the aircraft out of service for a month or more to correct it.

Pan Am was famously the launch customer for both the Boeing 707 and 747. However, the company's capital expenditure on the 747 weakened its balance sheet just when it was losing passengers worried that it had become a prime target for terrorists following the loss of a Pan Am 747 over Lockerbie in Scotland. With no US domestic network to compensate, the situation was so bad that the company went out of business.

Lufthansa was the launch customer for the 747-8, and for a time it seemed it might be the only customer. No doubt it was pleased with its launch customer prices but perhaps not quite so much as Qantas was with the 787 Dreamliner. As one of the launch customers for that aircraft, Qantas was granted options at extremely low prices.

An option being a firm price for a new plane for delivery at a specific time, Boeing had to pay Qantas hundreds of millions of dollars in compensation when the 787 program suffered a three-year delay. Qantas, with problems of its own to sort out, waited ten years before feeling it was in a position to exercise those options before they expired. Its claim that it found the 787 to be better for its requirements than the A350 has to be taken with a pinch of salt considering how much it would have to pay for properly priced A350s. Interestingly, it is going to use the A350 for its **Sunrise** project.

Learning Curve (Aircraft Production Costs)

Since World War II, aircraft and other manufacturers have been aware of what they called the "learning curve." It is the notion that doubling production reduces production costs by 20 percent because staff learn how best to perform their tasks. Since it is easier to double production from a small number, this results in a curve that is very steep at the beginning. In fact, aircraft that come off the production line in the very early years are often sold at a considerable loss, even if development costs are averaged over the expected longer run.

This is in addition to extra manufacturing costs associated with certain out-of-the-ordinary projects, such as the Boeing 787 Dreamliner, employing composites for the fuselage.

This also applies to a lesser extent to variants, and manufacturers acceding to too many special demands from clients can fall foul of this and lose money, even where the total for a given aircraft seems quite high.

LAWS (Lethal Autonomous Weapons System)

The term means weapons that can identify and kill targets (people) autonomously using AI without human intervention.

Such weapons can identify and engage what they think are military targets without input from human operators. The development and deployment of AI-based LAWS represent a significant shift in military technology, with profound implications for the future of warfare.

Such systems, operating with virtual autonomy, not only work in actual combat but integrate with intelligence gathering, surveillance, and reconnaissance (ISR). In fact, AI in may be applicable in all military decision-making processes..

Lawyers and Insurance Underwriters

Although lawyers are much maligned, the enormous claims they can make in the case of litigation in the United States has perhaps helped make flying safer, and especially with regards to maintenance. Insurance companies also have a big role in making flying safer.

LEO (Low Earth Orbit)

Low Earth Orbit (LEO) is a region of space at altitudes ranging from about 100 miles (160 km) to1,423 miles (2,000 km) used by many satellites and even the International Space Station.

It has the advantage of being high enough to be virtually out of the atmosphere yet low enough for high resolution images and communications with Earth without too much.

There, satellites travel at about 17,500 mph (7.8 km per second) meaning one orbit around the Earth takes only 90 minutes.

One problem is that it is getting rather congested.

Leading Edge

The front edge of something, usually of the wing or tailplane, which meets the airflow first.

Lease (Wet/Dry, and So On)

Airlines often lease aircraft rather than buy them outright.

Leasing can take the following forms:

1. **Wet lease**, where the lessor provides—and pays for—almost everything, namely the aircraft, the aircrew, and cabin crew, maintenance, and insurance (hull and third-party risk).

2. **Damp lease**, which is the same, except that the lessee (the airline) provides the cabin crew.

3. **Dry lease**, which is purely a financial arrangement, with the aircraft operated by the lessee (the airline) as if it were its own.

A dry lease can either be an operating lease, where the lessee employs the aircraft for a relatively short period, usually from two to about seven years, and the aircraft does not appear on the lessee's (the airline's) balance sheet. This means the airline's balance sheet looks healthier. Or a financial lease, where the aircraft appears on the lessee's (the airline's) balance sheet, and the length of the lease may be three-quarters the usable life of the aircraft, after which the lessee (the airline) would have the option of purchasing the aircraft at an agreed price, and so on.

A wet lease of less than a month would be called an *ad hoc charter*.

Not only does leasing free up cash but an airline, having purchased an aircraft from a manufacturer at a substantial discount, might sell it to a bank at true value, thus recording an instant profit, and lease it back with payments split over many years. Leasing companies themselves often place large orders, thus obtaining discounts and a good place in the queue for coveted aircraft.

Legacy Carrier

Shorthand for airlines that became bloated in the heyday of commercial aviation, where extortionate regulated ticket prices enabled them to pay pilots and cabin crew sky-high salaries and set up expensive facilities for hub-and-spoke operations.

By declaring bankruptcy under so-called Chapter 11, which allows US corporations to continue trading while releasing them from the burden of collective bargaining agreements with employees, including those relating to pension schemes and established salary scales, US legacy carriers were able to "reinvent" themselves.

European airlines, such as Air France and Lufthansa, where the pilots had got used to being pampered and paid enormous sums, were not able to do that and experienced a series of debilitating strikes as they struggled to move their less profitable operations to low-cost subsidiaries.

Lethal Autonomous Weapons System (LAWS)

Lidar (Light Detection and Ranging)

Term coined in a similar fashion to the way "**Radar" (Radio Detection And Ranging)** was but where light waves, instead of radio waves, are involved.

Lidar is a new method, using light waves from lasers instead of radio waves, to detect turbulence, wind shear, and wake turbulence. Lidar also refers to state-of-the-art technology in the aerial mapping domain, with accuracy down to 10 to 15 cm possible.

Life Vest—Do Not Inflate Prematurely!

If the aircraft has to ditch in water, passengers will normally be wearing life vests but should not inflate them until "safely" outside the aircraft. When inflated, their bulk impedes exit. Worse still, should the cabin start filling with water the wearer of an inflated vest is liable to float upwards and become completely trapped. This happened to many passengers when an Ethiopian Airlines 767 ran out of fuel after being hijacked, forcing the pilot to ditch it in the Indian Ocean*.

LIFO (Last In, First Out)

Term used to represent how elements in a queue are treated with the last to arrive being the first to be removed. When an airline suffers a downturn in business, it is usually the most junior pilots (the last to arrive) who are laid off first as, partly due to the unions, much at an airline depends on seniority rather than ability. This means low-cost carriers (LCCs) and commuter airlines have some particularly good pilots joining them after a downturn.

FIFO (first in, first out) would be the reverse. (See **Age of Pilots**.)

Lift

Lift is the term used to describe the upward forces generated by the forward motion of the aircraft through the air. An aircraft can suddenly lose lift (stall) if it flies too slowly or its configuration changes so that the behavior of the air passing over and under the wings suddenly changes.

The phenomena involved are much more complex than claimed in some books. If it were that simple, manufacturers would have already designed the perfect wing, and if it were merely the Bernoulli effect, aircraft could not fly upside down.

Lithium Batteries, Danger Thereof

Because they have involved fatal disasters to cargo aircraft, fires caused by lithium batteries have not received much public attention.

Passengers are told to have items with lithium batteries and battery packs with them in the cabin where any fires can be dealt with. The cabin crew have extinguishers and special bags into which they can be put. However, accessing a burning lithium battery in the overhead bins can be difficult with some airlines insisting they be kept at the passenger's feet.

Should a phone or battery pack fall into the mechanism of a passenger's seat, they should ask the cabin crew to deal with it as if the passenger adjusts the seat, it could be crushed and catch fire.

There have been at least two harrowing cases of cargo aircraft carrying lithium batteries having fires breaking out in their main holds and flying on, trying to make an emergency landing with smoke and flames spreading, and crashing.

One was a UPS 747-400 that had flown from Hong Kong to Dubai and took off the next day, September 3, 2010, for Cologne. The fire in the main hold generated so much smoke that, because the air conditioning unit meant to blow clean air into the cockpit was not working, it meant the first officer—the captain had collapsed on going back for oxygen—could not see his instruments even to change the radio frequency to that of Dubai and crashed on trying to land there.

The other was Asiana Flight 991 that took off from Incheon International Airport on July 28, 2011, for Shanghai. As they got nearer to Shanghai, a fire broke out in the main cargo hold containing lithium batteries in pallets close to paint and other flammable material. Shanghai ATC recommended an airport closer to them for their emergency landing, but the conflagration was so intense that the aircraft started breaking up and crashed into the sea on the way there.

LMF (Lacking Moral Fiber)

Shaming of Royal Air Force aircrew refusing to fly further MISSIONS.

See WORLD WAR II.

LO (Low Observable)

Low observable technology reduces an aerial vehicle's radar signature so that it may penetrate enemy airspace undetected. In other words it is **Stealth** technology.

Lobbyists

According to *Open Secrets* (Following the Money in Politics):

> In 2021, 926 lobbyists of whom 65.87 percent were former government employees spent $108,604,062 on air transport for a total of 239 clients.

Boeing was the highest spender at $13,450,000.

Surprisingly, FedEx and UPS were next on the list at $9,590,000 and $8,181,434.

The three US majors (airlines) all spent similar amounts in the region of just over $4,000,000 each. However, it should be remembered that the airlines lobby "in kind" by granting extremely favorable travel perks to senators and the like.

Sometimes a company will reward a government official by making them lobbyists after they retire.

One should not forget lobbying related to defense industries.

Boeing had great influence on the FAA, including using revolving door tactics whereby retiring staff could be given jobs. Before the MAX debacle it could get the FAA to do much that it wanted.

Localizer (LOC)

The horizontal guidance component of the **ILS (Instrument Landing System)** tells the pilots where they are relative to the centerline leading to the runway—in other words, right on it, or to the left or right of it.

The **Glide Path/Slope** is the vertical component, telling them whether they are on it, or above or below.

LOCKHEED (LOCKHEED MARTIN)

In terms of revenue from defense-related work, Lockheed Martin is the world's largest defense contractor. The present incarnation is the result of the merger between Lockheed and Martin Marietta in 1995.

When the original company first started manufacturing commercial aircraft in 1912, it was called Loughead Aircraft Manufacturing Company. The name was so difficult to read that there were posters telling people to pronounce it "Lock-Heed."

Around the same time the Glenn L. Martin Company was founded in Los Angeles. In 1916 it merged with Wright Company, but in 1917, and with outside backing, Glenn Martin pulled out and reestablished the Glenn L. Martin Company in Ohio.

The two companies, Lockheed and Martin, went on to great things, with the former producing flying boats, airliners, and fighters. In 1961 Martin merged with the American Marietta company to form Martin Marietta, and although the latter's forte was aggregates, cement, and chemicals, the stronger financial footing no doubt helped it become a force with which to be reckoned. In 1993 Martin Marietta acquired General Electric Aerospace for $3 billion, thus giving it a footing in the satellite domain, in addition to its already strong presence in the rockets and missiles domain.

One highlight—or rather, lowlight—came in 1982 with a hostile takeover bid by the Bendix Corporation. Bendix managed to acquire a majority of Martin Marietta shares, but Marietta's management in turn managed to use the interim period between ownership and cession of effective control to sell off noncore businesses and to take over Bendix itself using the so-called "Pac-Man" defense. Ultimately, Bendix was sold off, leaving a somewhat indebted Martin-Marietta intact.

Despite its glittering history as a maker of civilian airliners, such as the Electra used by Amelia Earhart, and the elegant piston-engine Constellation, Lockheed later pulled out of the civilian airliner business, deeming it too risky.

This was because the company had lost tremendous sums, more through bad luck than through any failure of its own, when its superior wide-body Lockheed TriStar (L-1011) missed the boat in the face of competition from the ill-fated, hastily produced McDonnell Douglas DC-10.

Lockheed had missed out on initial sales due to problems Britain's Rolls-Royce had in perfecting the new, more powerful engine required. However, in those days there were three suppliers in contention in the wide-body market, so it was easier for one to fall by the wayside. Now there are only two, Airbus and Boeing.

With Lockheed brought to its knees financially, Lockheed had to obtain a government loan subject to its accounts being opened up to investigators. This revealed a long history of corruption to obtain foreign sales, the most discreditable being the sale of the Lockheed F104 Starfighter—called the "widow maker"—to West Germany where 61 pilots lost their lives. Though a good aircraft in some conditions such as over dry deserts in the US, it was totally unsuited for German conditions and the role it was destined to play there. There were also commissions paid to middlemen in Saudi Arabia, Denmark (Royal family) and Japan. The State Department at some point said the foreigners' names should not be revealed as it might harm relations but was overruled. In the case of Japan, it implicated the prime minister (Tanaka). Of course, companies in other countries were also paying kickbacks to intermediaries in the pursuit of sales of aircraft. Ironically, in the case of Japan and Hong Kong (Cathay Pacific), kicks were helping the sales of the best aircraft, the ~ **Lockheed TriStar (L-1011)**.

Lockheed cleaned up its act and like other US defense contractors spends money on lobbyists in the US.

See **Defense Contractors**.

~ Lockheed Constellation Military 1943, Civil 1945/856

A fast for its time and beautiful four-engine piston-engine airliner conceived just before World War II. Nicknamed "The Connie."

It partly owes its elegant appearance to the fact that its fuselage has a continuously variable profile, with no two bulkheads having the same dimensions. More modern airliners are mostly tube shaped because a tube, besides being cheaper, is better able to withstand pressure changes. The legendary Howard Hughes apparently greatly influenced the design.

In the 1950s it became a frequent sight at major airports all around the world as the first serious long-distance pressurized airliner. The entry into commercial service of jet-engine airliners, such as the Boeing 707 in 1978 and subsequently the DC-8, quickly rendered the Constellation obsolete for such routes.

While the Constellation depended on its powerful Wright R-3350 engines for its success, these were something of a liability. Indeed, the FAA ordered that the aircraft be taken out of civilian service until modified after incidents in the early months.

Cruising speed was 340 mph (295 kt, 547 km/h) at 22,600 ft (6,890 m). It could carry more than 100 passengers but would usually carry rather fewer, in configurations that were more comfortable.

Military versions, some for electronics-related missions, continued to operate for many years, even in the Vietnam War.

~ Lockheed Electra (Electra)

Name given to two distinct airliners:

1. **Electra 10 1934/149**
   Ten seats, with slightly smaller or larger variants. Used by many airlines and by the military in World War II. A specially adapted version with extra fuel tanks in the place of seats was used by Amelia Earhart for her ill-fated round-the-world attempt in 1937*.

2. **L-188 Electra 1958/170**
   A very promising fuel-efficient turboprop airliner seating up to 92 passengers in an economy configuration. It was largely due to bad luck that Lockheed got its fingers burnt with this airliner, since the two most dramatic accidents were due to vibratory harmonics (flutter) at high speed that no one could have predicted at the time, and which were easily dealt with once recognized.
   Those early accidents harmed sales prospects. Once it

was rendered very safe, it was too late because the public had come to believe that pure jets were intrinsically better than turboprops.
Some were later used in South America.

~ Lockheed TriStar (L-1011) 1972/250

Outstanding tri-jet airliner developed by Lockheed in competition with the initially ill-fated McDonnell Douglas DC-10. It was destined for the medium- to long-haul wide-body market just below the high-volume long-haul market being captured by the Boeing 747.

Unfortunately, as mentioned, problems that Britain's Rolls-Royce had in developing the new RB211 engine meant it arrived late on the market, with the result that McDonnell Douglas—already well established as a supplier to US airlines—took the lead order-wise. This was despite Lockheed being persuaded to offer sweeteners (bribes via agents), which ultimately led to the resignation of the Japan's prime minister, Tanaka.

Thus, Lockheed never sold enough to make the venture profitable, and indeed it lost so much money that it decided to abandon civilian airliner production altogether.

Loiter Time (LT), Dwell Time

A rather appropriate military term for the length of time an aircraft on a mission can stay at the scene, this is simply the endurance minus the commuting time (from and back to base).

Particularly relevant to maritime patrol aircraft, ground/air support aircraft, surveillance aircraft, and AWACS-type aircraft, not forgetting drones. There is even now the notion of *loiter munitions*, in the form of tiny missiles that would lurk overhead for hours to be directed at the enemy at the opportune moment. Their precision—say with laser targeting directed from the ground—means only a tiny warhead would be required.

Lock-On-After-Launch (LOAL)

Case where the missile launched from an aircraft acquires the target after launching. This is what is done with the F-22 Raptor, where the missiles have to be kept within the body of the aircraft so that stealth can be maintained.

Lock-On-Before-Launch (LOBL)

Case where the missile launched from an aircraft acquires the target after launching.

Longest Scheduled Flight (Ultra-Long-Haul)

In 2024, the longest scheduled flight was Singapore Airlines' SQ24/SQ23 between Singapore and New York's JFK Airport. It must be doing well enough for there is a daily flight and another daily one, to New York's Newark Airport.

The **Flight Time** is roughly 17 hours 40 minutes and the **Block Time** (gate to gate) just over an hour longer. The return journey is some 40 minutes longer on account of the headwind.

The distance is almost 9,000 nautical miles, or over 10,000 statute miles.

SIA is upgrading its A350s with seats originally destined for the Boeing 777-9, which is being constantly delayed. Though this will apply to most A350s the seat configuration on the A350ULR used on what has hitherto been the long flights between Singapore and New York will be changed with four first-class seats, 70 in business and 58 premium economy seats. This represents an increase of three business-class seats and a decrease of 36 in premium economy to allow for the first-class cabin.

"LongShot"

As part of a broad program called *LongShot*, the US Defense Advanced Research Projects Agency (**DARPA**) in 2021 engaged General Atomics, Lockheed Martin, and Northrop Grumman, to develop designs for an *air-launched missile-toting* air-to-air combat drone.

This unmanned vehicle would be launched from a larger manned standoff platform (aircraft) and then fly to a designated area to engage multiple aerial threats with its own weapons. This would have many benefits, not only as regards protecting the primary launch platform but also in terms of the **Loiter Time** of the secondary platform.

Look-down Radar

Radar looking down from above more likely able to detect stealth aircraft designed to be invisible from the side or front.

Low-Cost Carrier (LCC)

Southwest Airlines of the US is said to be the pioneer with other airlines such as the UK's Ryanair and easyJet copying its techniques with lower fares and few frills. Many have made an art of quoting a very basic fare with no significant baggage and adding extras for baggage, choice of seat, food and drink.

In the US, there are even ULCCs (ultra-low-cost carriers), such as Allegiant and Spirit Airlines. However, with Spirit Airlines losing money, and declaring bankruptcy, the ultra-low-cost model no longer seems viable in the US, for the majors can pull the carpet from under them by offering a few cheap seats at little cost with profit coming from the expensive seats with many passengers willing to pay extra for comfort. Allegiant used to have very old MD-80 aircraft which, being cheap, did not have to be used as intensively as at other airlines. Also, they concentrate on holiday traffic from minor airports. Also, a third of their revenue is from commission on hotel bookings.

Southwest Airlines is having some difficulties now that the majors are competing with them, offering some cheap seats on flights with the cost of the flight already more than covered by "regular" passengers. Also, their fleet management has become more efficient thus lowering their costs.

Southwest is now abandoning its policy of not allotting seats and letting passengers board and seat where they like or can, so they can charge extra.

Loss of Control In-Flight (LOC-I)

Loss of control in-flight (LOC-I) is something that is much more prevalent in **General Aviation (GA)** than in commercial aviation. In GA it often results from failure to prevent or recover from a stall or upset.

In commercial aviation LOC-I is rare thanks to all the protections and warnings, but since there are many fatalities in such crashes, it must be taken very seriously.

LOC-I can cover all sorts of situations ranging from cases where something critically wrong with the airliner (such as on Japan Airlines Flight JL123*) makes it completely uncontrollable to cases (such as Air France Flight **AF447***) where the pilot stalled the aircraft and did not know how to recover at high altitude. Indeed, it was **AF447 ONLY A MINUTE TO SAVE THEM** in particular that made **IATA**, the **FAA**, and **EASA** stress the need to train pilots to cope with LOC-I, or better still avoid it.

Loss of Situational Awareness

There have been disasters due to pilots not following their instruments and feeling they are in a dangerous climb due to **Somatogravic Illusion** and putting their aircraft into a steep dive.

Only recently, a Qatar Airways flight taking off from Doha for Copenhagen was put into a dive by the first officer—perhaps due to somatogravic illusion—with the captain making a recovery with a fair amount of height to spare but little time.

Perhaps the oddest case of a pilot making an untoward maneuver was that of a pilot who had nodded off and on waking mistook the bright planet Venus for another aircraft and took extreme action to avoid it!

Low-Altitude Fuel Consumption

Airliners' jet engines are very much more efficient at high altitudes and that is why they climb as quickly as they reasonably can. Furthermore, even on an exceedingly short flight they will climb to high altitude even if they must soon come down again.

One should note that the regulations for calculating fuel reserves required for a flight must include the fuel required to hold for 30 minutes at 1,500 feet above that airport, at planned landing weight after diverting, and land there with 30 minutes' fuel remaining.

Loyal Wingman

Refers to unmanned (often cheap) aircraft working with an (expensive) fighter or bomber, and able, if necessary, to take a hit instead.

In 2024 the US Air Force picked General Atomics and Anduril—a defense startup—for its "loyal-wingman" Collaborative Combat Aircraft (CCA) drone program.

See **Attritable** and **Collaborative Combat Aircraft (CCA)**.

LRASM (Long-Range Anti-Ship Missile)

Twenty-four of these can be carried by B-1B bombers to attack ships, say, in the Black or South China Sea. See Missiles.

LSA (Lowest Safe Altitude)

Lowest safe altitude for a given air route, with a safety margin over the highest obstacle. Allows for the aircraft straying, say, 2,000 feet off course. Over congested areas, clearance must be 1,000 feet, while for populated areas it might be 500 feet. Over open water or sparsely populated areas, it might be less but requires a minimum horizontal distance of at least 500 feet from any person, vessel, vehicle, or structure.

Called **MSA (Minimum Safe Altitude)** in the US.

~ M ~

Mach

Mach is the ratio of an aircraft's airspeed to the speed of sound in the ambient air around it. Thus, an aircraft flying at Mach 1 is flying at the speed of sound, and one flying at Mach 2 is flying at twice the speed of sound.

An interesting point is that the speed of sound varies with temperature and not pressure, with Mach 1 being about 15 percent less in terms of airspeed at a cruising height of some 35,000 feet, where it is extremely cold.

The typical cruise speed for the Boeing 747 at 35,000 feet would have been (Mach 0.855 = 567 mph (912 km/h).

Airliners have Mach indicators, enabling the pilots and autopilot to stay aware of the Mach speed. (See **Subsonic, Supersonic, and Hypersonic**.)

Maintenance Checks

One reason flying in airliners is so safe is the mandatory check system whereby an aircraft is examined at set intervals, with the infrequent final D-check involving virtually taking it apart being very costly and time-consuming.

The C and D checks include taking the aircraft out of service thereby reducing revenue among other things. However, the time out of service can be shortened somewhat by dividing those checks into segmented blocks or segments.

The checks are as follows:

1. A-Check.
2. B-Check.
3. C-Check.
4. D-Check.

MANPADS (Man-Portable Air Defense Systems)

Shoulder-launched antiaircraft missiles, such as the ones used to attack the DHL Airbus freighter taking off in 2003 from Baghdad

Airport. In that remarkable case, which received little attention because there were no passengers, the missile—with two embedded French journalists filming the launching—hit the left wing, damaging the hydraulic lines. Controlling the aircraft by adjusting engine power alone, the pilots were able, in a remarkable feat, to bring it back to the airport and touch down without killing themselves.

The Israeli airline El Al equips its airliners with mechanisms to defend them against MANPADS. Other airlines do not believe they are worth the cost since they try to avoid dangerous areas. Malaysian Airlines MH17 was shot down over Ukraine in July 2015, but that was with a much more sophisticated Russian-made Buk missile, able to reach 70,000 feet. Other airlines were flying through the area, which should have been off-limits. However, the Ukrainians were earning a considerable amount of money by allowing airlines to take that shortcut.

Markings on Weapons Indicating Whether "Live" or "Inert" [US]

Fighters very often carry dummy weapons when training.

The dummies, often filled with concrete or just metal, have a blue stripe to indicate they are inert, while weapons usually have a yellow or green stripe to warn they are live.

Manual Flying

With computers flying airliners more accurately and usually more safely than pilots, some airlines are loath to follow the FAA's recommendation in 2023 that they make their pilots fly part of their route manually—apart from takeoff, which is always done manually—for fear that something will go wrong or that they would use more fuel by flying less accurately. The truth is that especially when it comes to long-haul routes the number of flying hours a pilot has on a given aircraft is meaningless since it is only a few minutes at takeoff and landing. It does not cover manual flying at cruise altitude, such as was the case for **AF447**.

A number of crashes have, to a great extent, been due to pilot error without investigators daring to acknowledge it for fear of upsetting pilots' unions and the like.

Massive Ordinance Air Blast (MOAB), GBU-43/B

Nicknamed the "Mother of All Bombs," it is the US's largest non-nuclear bomb. Weighing 22,600 pounds, and filled with 18,700 pounds of explosives, it has a blast radius of one mile and a yield equivalent to 11 tons of TNT.

It is detonated above ground, so its energy is neither absorbed by the ground nor reflected upwards. Like any high-explosive ordinance, its role is taking out surface targets while acting as an anti-personnel weapon.

MAX Crashes—Southwest Airlines Insistence on No Extra Training

Boeing's infantile mistake in the design of the software for the 737 MAX where they incorporated the **MCAS** system depending on a single angle of attack sensor was unconscionable.

However, it was the original customer for the 737 MAX, Southwest Airlines, that was unknowingly partly responsible due to its rigid demands and insistence on no extra pilot training. It even ordered a large number with Boeing liable for unbelievable financial penalties per aircraft should that not be the case.

Maximum Height Airliner Can Fly

Flying high saves fuel because there is less drag. Also, it has the advantage that it gives more time to sort things out or glide to a diversion airfield in an emergency. Also, there tends to be less turbulence and better weather. Flying very high up can help one avoid other traffic and mean less headwind when flying westwards.

The fuel savings are so great that even on a short hop it can be worth climbing high up and almost immediately gliding down to one's destination.

The **FMC (Flight Management Computer)** tells the pilots the optimum maximum height they can fly depending on various factors and notably the all-up weight and air temperature. If the flight is long haul, it may have to perform a **Step Climb**, increasing its altitude in steps as it burns off fuel.

Though aircraft such as the Boeing 757 can fly as high as 42,000 feet, and a 767 at 43,000 feet, they do not usually do so. It seems that airliners such as the Boeing 787 and Airbus A350 with fuselages made of **Composites** tend to fly more often at 40,000 feet once they have consumed enough fuel to be light enough. This is perhaps because their fuselages are less subject to failure due to repetitive expansion and contraction.

Private jets, such as Learjets, Gulfstreams, and Falcons fly at around 45,000 feet or higher, and sometimes up to 51,000 feet.

One of the purported advantages of business jets is that their cabins are also more pressured than those on commercial airliners, meaning their occupants suffer less from jetlag.

The discontinued supersonic Concorde flew at Mach 2 at 60,000 feet!

Maximum Take Off Weight (MTOW)

Several factors, including the wind speed/direction, the temperature, and altitude of the airport—the hotter and higher the airport, the more difficult it is for the aircraft to take off—determine how heavy the aircraft can be at takeoff.

The overall weight of the aircraft (see **AUW/All-Up Weight**) increases according to the amount of fuel carried and the combined weight of freight, including baggage, and passengers.

Apart from the maximum permissible takeoff weight, determined according to the above conditions, there is another limiting factor, which might come into play in the case of very powerful engines able to get the aircraft into the air regardless of the conditions. That is the limit imposed by the structural strength of the aircraft, and notably that of the landing gear.

See **Maximum Zero Fuel Weight** below.

Maximum Zero Fuel Weight

Airliners carry fuel in central fuel tanks in the fuselage and, perhaps most importantly of all, in the wings. The advantage of having fuel in the wings is that the wings lift the fuel at the same time as they lift the aircraft and extra-strength wing-fuselage mountings to support it are not needed.

It sometimes happens that a long-range aircraft designed to carry a massive amount of fuel in the wings is assigned to a short leg with hardly any fuel but with very heavy freight in addition to a full load of passengers—all concentrated in the fuselage. This can mean that although the all-up weight including fuel is well within the MTOW limits, the stress on the wing roots would be excessive. To guard against this, airliners also have a maximum zero fuel weight.

"MAYDAY ... MAYDAY ... MAYDAY"

Prefix to distress call, equivalent to SOS in the days of Morse code. Derived from the French *M'aidez!* Meaning *Help me!* The word is repeated three times.

When an emergency arises, pilots also change their transponder squawk code to the emergency code, which is 7700 in the US and other places. If they are very busy dealing with the emergency, they may not have time to issue a Mayday call immediately and might only reset the transponder—and they might not even have time for that, as in the case of the Air France Airbus A330 lost between Brazil and France in 2009.

Much less known is the tentative distress call *Pan-pan*, repeated three times, transmitted when the pilots have a (worrying) problem requiring their urgent attention but do not believe there is an immediate threat to life or to the aircraft that would require air traffic control and emergency/rescue services to take immediate action as they would in response to a Mayday call.

Like Mayday, it derives from a French word, *panne*, meaning a fault.

In the case of the fatal Swissair flight from New York's Kennedy Airport in 1998, the pilots smelled smoke but initially did not think the situation was too serious and made a Pan-pan call to air traffic control. Unfortunately, the intense fire had already gained a hold, making the situation hopeless.

In the case of a Mayday emergency landing with the need for emergency services to be present, the controller will invariably ask the *amount of fuel* and *number of souls on board.* They say "souls" since it could be a flight with bodies, say from a war zone, who do not need rescuing.

MCAS (Maneuvering Characteristics Augmentation System)

According to Boeing:

The Maneuvering Characteristics Augmentation System (MCAS) flight control law was designed and certified for the 737 MAX to enhance the pitch stability of the airplane—so that it feels and flies like other 737s. It is designed to activate in manual flight, with the airplane's flaps up, at an elevated Angle of Attack (AOA).

It was responsible for the 737 MAX crashes.

See **MAX CRASHES** and **Boeing 737 MAX, Crashes**.

McDonnell Douglas (MD)

In 1967 Douglas accepted an offer to merge with McDonnell Aircraft to become McDonnell Douglas, to be subsequently merged with Boeing in August 1997. Though the entity was called Boeing, the commercial way of thinking at McDonnell Douglas took precedence.

See **DOUGLAS** and **BOEING**.

McDonnell Douglas DC-10 1971/386

~ McDonnell Douglas KC-10 1981/64 [Tanker]

The DC-10 was a three-engine airliner destined for the medium- to long-haul wide-body market in competition with Lockheed's TriStar in the 1970s. It was developed in haste, and a gentleman's agreement with the FAA allowed it to fly in the US despite some safety concerns. While the conditions attached to that agreement, regarding the installation of a special window in the hold doors for checking, might have been sufficient in the US, they were not for operators overseas with less well-trained staff. This resulted in a crash near Paris in 1974 with no survivors. See **Decompression (Loss of Cabin Pressure)**.

Early on a super-DC-10 had been envisaged but never realized, due to the accidents and various downturns in the aviation business. Finally, a derivative was developed and called MD-11.

The KC-10 is an airborne refueling platform derived from the DC-10-300. The KDC-10 is one derived from modified Japan Airlines DC-10-40s.

~ MD-11 1990/200

This larger version of the ill-fated DC-10 (which when improved proved a very safe aircraft but could never shake off the accidents in people's minds) never lived up to expectations.

~ MD-80 Series 1980/1,191

This second-generation DC-9 medium-size medium-range airliner has been very successful, and many are still operating.

~ MD-90 1995/114

A stretched version of the MD-80 with other enhancements but less successful, in part because production ceased in 2000, since no further orders were taken following the purchase of McDonnell Douglas by Boeing in1997—it would have competed with other Boeing aircraft, notably the 737.

Medical Tourism

Thousands of people are traveling abroad for medical treatment that can be 70 percent cheaper in Turkey than in England even factoring the travel and hotel cost. Treatment can range from dentistry through cosmetic surgery to joint replacement.

To save money and to avoid being too long away from work or family, many post-operative passengers board their return flights much too early with them even dying, say due to blood clots. Flights may have to divert at considerable cost to the airline.

At certain airports such as Istanbul airline staff question the bandage-covered passengers to try to ensure they are fit to fly. Some airlines even demand a medical certificate that it is safe for them to fly.

Memory Items

These are facts and procedures a pilot must remember so as to be able to take immediate action without having to refer to documentation.

When something is not right pilots will usually refer to a quick reference card, and then the quick reference handbook, and finally a checklist covering that specific problem. This is because there are so many possibilities, and it is important to do the right thing and most importantly not overlook something.

However, there are situations where immediate action is required, leaving no time for thumbing through documents. How to initially deal with them must be memorized.

Mental Health of Pilots

With memories fading of the Germanwings disaster in March 2015, where a pilot with known mental problems (that under German law the doctors could not reveal to the airline) purposely crashed his airliner, another shocking event in 2023 refocused attention on the problem of pilots' mental health.

Case Where Intention Not Suicide

Though, as in the Germanwings case the perpetrator—Joseph David Emerson—had been suffering from undeclared depression, it was not that in itself that was responsible for the near disaster, but a full-blown psychotic episode triggered by psychedelic mushrooms. He was a pilot on a week's leave.

To commemorate the premature death of a dear friend, Emerson together with two mutual friends took advantage of this to go had been on a trip a long way from home for their friend's death had impacted them all deeply.

Although strictly speaking against airline regulations, Emerson acceded to his friends suggestion he take some magic mushrooms to help him cope—after all, it was six days before he was to resume flying duties, and any effect would surely have worn off by then.

That night he had a dreadful reaction and was still hallucinating as his friends drove him back to the airport. The Alaska Airlines flight he was to take back home happened to be full but as a company pilot he was allowed to sit as a deadhead in the jump seat in the cockpit of the Alaska Airlines Flight 175, an Embraer with 81 people on board. This was all above board and airline staff had not noticed anything strange in his demeanor.

The aircraft climbed to cruise altitude, and he and the two pilots chatted normally. There followed a pause during which Emersone muttered something like he was not feeling right, to which, being busy, the operating pilots did not immediately respond. Shortly after that, Emerson leant over and pulled the engine fire extinguishing handles for both engines, switching off the fuel supply to the engines, which in turn would be followed by the firing of the fire extinguishers on depletion of the fuel.

He was pulled away, and his action was cancelled before the fuel remaining in the lines ran out and the engines flamed out. It had been a close call, for had the engines flamed out and the fire extinguishers fired, restarting them would have been nigh impossible. They landed safely with Emerson detained and subsequently charged with the attempted murder of 80 people.

Like many other pilots with mental problems, he had not sought help for his depression despite the urging of his wife for fear he would lose his high-paying job, and at the very least fall on hard times without pay for months even should he ultimately be reinstated.

The case of Emerson was unusual in that he was hallucinating due to the mushrooms and not essentially bent on suicide.

Humiliation and/or Revenge as Motive for Suicide

The leitmotif in some cases of pilots committing suicide with all on board their aircraft seems to be "humiliation cum revenge," as in the case of EgyptAir Flight 990 that took off from New York's JFK on October 31, 1999*, where the soon-to-retire pilot obviously crashed it, fighting at the controls against the captain who had somehow managed to get back to the cockpit.

He had been humiliated, having been told by the company's head pilot (also on board) that he could no longer be allowed to fly to the US following allegations of sexual misconduct, including exposing himself to teenage girls, propositioning hotel maids, and stalking female hotel guests. Flights to the US came with extra pay besides prestige, so his downgrading and loss of face would be obvious.

Another case was possibly in China where a very senior captain was seemingly humiliated by being demoted to first officer, and the aircraft on which he was rostered crashed in a vertical high-speed dive in what seemed to be suicide. The Chinese authorities have erased all mention of it and although meant to keep everyone informed of their investigative conclusions merely say it is complicated, and at the time of writing have not been forthcoming. However, following the crash more attention to the mental state of pilots has been paid in China.

The instances of pilots using their airliner to commit suicide are fortunately extremely few and far between and will hopefully become less likely with more attention paid to mental health, and perhaps care to avoid someone with a grudge or humiliated having access to the cockpit, even after taken off actual flying duties.

See **Psychometric Evaluation of Pilots**.

Impact of Mental Health on Pilots' Performance

Of far greater concern must be the effect of mental problems such as depression on the performance of pilots. They are human, and like anyone can have problems in their marriages and life in general. Crashes have surely occurred that could have been prevented had those pilots not been afraid to seek help. It seems the FAA is beginning to recognize this and not be so rigid in applying rules for pilots in difficulty.

It is an intractable problem as to how to convince pilots that seeking help for mental problems will not put their careers at risk, for what is the physician to do if they conclude the individual represents a danger? Perhaps the answer would be for every pilot to be "insured" so they would continue to receive an excellent salary on being taken off flying duties. However, it would be difficult to prevent it being abused and becoming too costly.

The degree to which pilots' mental problems such as depression leading to sub-par performance have featured in crashes may have been overlooked because investigators are afraid to publish personal medical and psychiatric details for fear of flak from pilots and their unions. For instance, the French investigators examining the crash of **AF447 ONLY A MINUTE TO SAVE THEM** even hid the simple fact that the captain said, "he had only had an hour's sleep and that it was not enough," on the grounds that it was a private matter.

Unfortunately, lawyers whose probes have contributed greatly to safety have little incentive to study relatively poor pilots, other than where it was obvious the airline should not have employed them.

While serious aptitude, intelligence, and personality checks must be made before pilots are accepted even for training courses, they then must be looked after, with it possible for them to be open about problems.

Metamaterial

These are unusual artificial materials able to have effects on the transmission of electromagnetic waves, light, sound, and heat among others. Applications include super lenses, optical filters, medical devices, sensors, infrastructure monitoring, lasers, lenses for high-gain antennas, ultrasonic sensors, and aerospace.

The subject is so complicated and applications so diverse that they are impossible to explain properly here, and we recommend consulting Wikipedia.

METAR (Aviation Routine Weather Report)

Standardized weather reports are issued hourly, except in the case of radical changes, by airfields and permanent meteorological stations.

Pilots and aircraft dispatchers use a METAR for flight planning and en route updates. Said to be derived from the French *Météorologique Aviation Régulière*.

A METAR is generated automatically, whereas a TAF (terminal aerodrome/area forecast) is prepared by a human, though not necessarily by someone stationed at the site in question.

Metric System

Cockpit instruments such as altimeters can usually be switched to metric unit for use, say, over Russia. However, with so many foreign carriers transiting that country they started allotting flight levels (in hundreds of feet) above the transition altitude and were even introducing QNH in feet below it.

Scientific work is usually based on the metric system with the US ostensibly resisting change to it. NASA uses a combination of both, say, with the basic design work done in metric, and manufacturing and usage by astronauts done in imperial, or US standard.

NASA lost an unmanned vehicle due to part of its trajectory being calculated in metric and the people controlling the vehicle not realizing it.

The **Gimli Glider** incident* arose because Canada was converting to metric, and the amount of fuel was miscalculated.

MCDU (Multi-Function Control and Display Unit) [Airbus]

A way of saving space in the cockpit by allowing a single screen to be used for multiple purposes. So-called soft keys located on the periphery are pressed to make the selection.

The **DCDU (Datalink Control and Display Unit)** is in front of both pilots.

Missile Defense

Were it possible to create a 100 percent effective antimissile system a country might risk a preemptive nuclear strike in the knowledge they would not be wiped out by retaliatory strikes, meaning **MAD (Mutually Assured Destruction)** preventing nuclear war would have failed.

Destroying enemy missiles in the boost phase would seem optimal as the heat given out makes them easily detectable and with them massive and loaded with fuel very vulnerable. Furthermore, destroying them then takes out not only all the warheads but also the decoys and other countermeasures.

However, as the Russians point out in the case of the **Sarmat**, with the missile launched over the enemy's territory where quick access is difficult and numerous defensive measures in place this is not easy.

MH370 "The Greatest Aviation Mystery Ever" *

Just after midnight local time, on March 8, 2014, the MH370 Malaysia Airlines 777 flight to Beijing took off from Kuala Lumpur with 239 souls on board.

Forty minutes later, just at the point where it was leaving Malaysian air space—and signed off saying "goodnight"—its transponder stopped working, meaning the aircraft did not show up on the Vietnamese controllers' screens. Since they had not called the Vietnamese to sign on, the absence of an echo was not noticed.

Only quite some time later when the Vietnamese asked their Malaysian counterparts what had happened to the Beijing flight did questions start to be asked. However, a statement from Malaysian Airlines management wrongly asserting the aircraft was over Cambodia delayed matters. After that was checked out and found not be the case, a search of the sea off Vietnam on the route to Beijing was launched.

MH370's ACARS was not working either, possibly because the 777's electrical power supply had been cut off leaving just enough power from the ram air turbine (RAT) for very basic instruments. However, an hour or so later a handshaking function of the 777's ACARS came to life (perhaps because normal electrical power had been restored) and from then on made routine hourly handshakes with an Inmarsat satellite.

The air and sea search off Vietnam was proving fruitless, and then, to everyone's surprise, some days later experts at the UK satellite company Inmarsat analyzing the handshakes were able to calculate that the aircraft had come down on an arc in the Indian Ocean somewhere off Perth, Australia! Review of Malaysian military radar recordings showed the aircraft had doubled back and crisscrossed frontiers to the other side of Malaysia to Penang (where the captain had been brought up—was he having a "goodbye look")—before turning north and going out of range.

The location, a long way off Perth, Australia, determined by Inmarsat was very imprecise and by the time the search switched to there, any wreckage floating on the surface would have sunk or been dispersed in the notoriously rough seas there. Also, the batteries for the emergency locator transmitters attached to the cockpit voice and flight data recorders only had a life of 30 days, assuming the perhaps old batteries were in good condition.

A year later, a flaperon from the 777 was found washed up thousands of miles away on the shore of La Réunion, off Africa, with analysis of the currents showing it could well have come from the general area determined by ACARS handshake analysis. That, and other debris found in that general area, extinguished all hope that the families of the passengers, many of whom were Chinese, had held that they were still alive.

The fact that communication with the aircraft was lost at precisely the point between air traffic control zones, where the disappearance would not be noticed, immediately raised suspicions. Despite expensive and extensive searches, the aircraft has never been found, leaving no definitive answer to what is said to be the greatest ever aviation mystery.

The most plausible explanation is that the captain was responsible. See **Victor Iannello, Radiant Physics [MH370]**.

Microbursts and Wind Shear

A microburst is a weather phenomenon occurring near the ground which is particularly tricky and dangerous because the aircraft first encounters a headwind that can make the pilots think they are flying too fast and reduce thrust, then a downdraft (downdraught) pushing them down, and finally a tailwind meaning a sudden further drop in vital airspeed making it difficult not only to combat the effect of the downdraft pushing them down, but also causing them to stall due to insufficient airspeed, with them possibly crashing in either event.

To visualize the phenomenon, imagine water flowing into a sink from a faucet (tap) and tiny toy boat coming towards the splaying out water. Before entering it, the boat is in virtually still water and moving at its normal speed. On entering the splaying out water its speed relative to the water its speed) increases greatly. On reaching the center its speed suddenly drops to what it was before with it being rammed downwards. Then as it enters the water splaying out in the opposite direction its speed relative to the water falls in what is equivalent to a tailwind for an aircraft. In the case of an aircraft, not only may be it stalling, or virtually so, without enough height to recover since the engines take time to spool up.This is why microbursts can be so treacherous.

Wind shear without the downdraft but a sudden reversal of wind direction is similarly dangerous.

On August 2, 1985, Delta Flight 191* was landing at Dallas/Fort Worth in stormy weather when it encountered a 26-knot headwind, which switched to a 46-knot tailwind—a 72-knot drop in airspeed! Though they started spooling up the engines to compensate for such a formidable drop, at such a low height it was impossible to recover in time, and the aircraft came down short of the runway, hitting a large water tank and catching fire.

This disaster induced the FAA to finally order the installation of Doppler radar at various airports to detect wind shear and microbursts of which were little understood at the time. Airliners also have radar able to detect wind shear provided the air is not too dry.

Other than avoiding microbursts altogether, the only solution is to fly faster than one normally would.

Microsoft Flight Simulator 2020 (August 2020)

Some pilots have said it was Microsoft Flight Sim played as a child that made them want to become a pilot. The original represented the nuts and bolts of flying, and millions were in use. Then, with versions attempting to apply modern games technology, there followed a decade and a half with it languishing forgotten.

Six years in the making, and based on Asobo's game engine, we now have this new version with Microsoft Azure cloud computing at its core. From its hyper realistic flight handling to gorgeous visuals, Microsoft's cloud computing elevates the game into a technical marvel. In incredible detail it depicts every inch of the entire globe using data from Bing Maps to generate 3D models including terrain, roads, vegetation, buildings, and even a herd of cattle. Two petabytes (!) worth of data is available in real time via an Internet connection. Though there are some errors in the depiction it is pretty accurate. A selection of views is possible including those from outside the aircraft and even that for a passenger looking out of the cabin window over the wing.

The downside is that it is difficult for beginners, and some complain about the size of the downloads required. Also, some reactions to controls are not always correct and involve delays making it not necessarily a perfect guide to actual flying. However, it is a great experience. Travel the world!

Largest (Planned) Military Transport Called "Windrunner"

The American company Radia Aerospace has designed the world's largest transport aircraft called the "Windrunner." Originally conceived to carry wind turbine blades, it is expected to have other military and commercial applications.

It would be 108 meters long, 24 meters high and have a wingspan of 80 meters.

Minimum IFR Altitudes

MEA (Minimum Enroute Altitude)
MOCA (Minimum Obstruction Clearance Altitude)

Minimum Safe Altitude (MSA)

Altitude in each area below which aircraft under IFR are not allowed to go without specific permission from ATC. Shown on charts and especially important where there are mountains near an airport.

Missed Approach

When an aircraft is performing a landing, the pilots may decide, or be obliged, to abandon the landing for any number of reasons.

Including:

1. Being unable to see the runway when down to the decision height.
2. Being badly lined up, or too high or too low.
3. Aircraft not properly stabilized or traveling too fast.
4. Presence of wind shear or crosswind exceeding company limits.
5. The presence of another aircraft or a vehicle on the runway.

If continuing with the landing is impossible, the pilot initiates the go-around and missed-approach procedure, which the pilots always have in their mind, or in the flight director, as they come in.

See **Go-Around**.

While experience has shown diverting to an alternate airport could have averted many an accident or overrun, there can be great pressure on a captain not to do so. First, there is the cost to the airline of a diversion, involving the booking of hotel rooms and the aircraft being out of position for the next flight. Second, but not to be entirely discounted, are the captain's personal plans not to mention their delayed return to their family.

MIRV (Multiple Independently Targetable Reentry Vehicle)

A multiple independently targetable reentry vehicle (MIRV) is a ballistic missile with multiple warheads, where each one targets a different site with them being as much as several hundred miles apart. A less sophisticated version is the MRV (Multiple Reentry Vehicle) where the warheads are not individually targetable but somewhat spread out for greater effect.

MIRV gives a greater advantage to the attacker.

Missileer

The term "missileer" was coined in the 1950s for personnel responsible for launching missiles. In the widest sense, it can cover personnel in the air, on land, and undersea in submarines.

In the narrow sense, it covers (in the US) a special breed of person in the ICBM underground silos. In the US at least, selection is extremely rigorous, and perfection is sought. There are triple checks, so no mistakes. After satisfactorily completing their training enrollees work in pairs and incidentally do not know where their missiles—which can quickly be retargeted from afar anyway—are targeted.

An "engagement" can last four years and interestingly be a good career move both in the service and civilian life, for being accepted demonstrates certain qualities and notably reliability.

See **Nuclear Triad** and the **Two-Person Rule**.

Missiles [ICBMs, Hypersonic, Glide Missiles, etc.]

For decades now, the Intercontinental Ballistic Missile (ICBM) has maintained peace under the **MAD (Mutually Assured Destruction)** principle whereby a country would respond to any nuclear strike by launching an annihilating counter strike.

The redundant US **Nuclear Triad** consists of a combination of (1) land-based ICBMs, (2) submarine-launched ballistic missiles, and (3) bombers with nuclear bombs and missiles removing any doubt about the credibility of the response.

The launching of what might have been a **Fractional Orbital missile (FOB)** by China caused concern.

The FOB with a hypersonic glide vehicle as the final stage comes down into the atmosphere and unpredictably makes its way to the target. This is worrying for the US because it can come from an unexpected direction such as the South Pole with scant radar coverage rather than, say, over Canada.

The trajectory of a traditional ballistic missile shot up into space and hurtling down to the targets is predictable and can be detected by satellites, which means it in theory can be destroyed in mid-trajectory. Even modern versions of ballistic missiles, as their name implies, work somewhat like a shell or bullet on the principle that after they have been launched, they follow a predictable trajectory, though some slight mid-course corrections when high up can be made. During this relatively lengthy mid-course period they can be easily tracked by radar and satellites, and in theory destroyed or thrown off course, though this is by no means easy. And some would inevitably get through.

Alarmingly for those living near land-based ICBMs sites located in isolated places in the US mainland, they would therefore be vulnerable and seem pointless. However, they could serve to make an enemy use up its stock of missiles in attempting to destroy the well-known silos, with the missiles on impossible-to-locate US submarines untouched.

Even so the US is planning to install new ICBMs and their associated silos to replace the present Minuteman III missiles at great expense.

With America having some success in antiballistic missiles (ABMs) there is the fear in some quarters that they could risk a first strike in the expectation that they could defend themselves against any retaliation. In practice this would be virtually impossible now were they to be subjected to a whole swarm of missiles. However, it might already be enough to significantly defend against a limited attack, say from North Korea.

See **Circular Error Probable (CEP).**

Taking all this into account, the Russians and Chinese have been developing much more difficult to intercept hypersonic missiles.

There are two categories:

1. Hypersonic glide vehicles
   Launched somewhat like an ICBM by a rocket into space, but unlike an ICBM they drop down to the atmosphere and skate across it using a phenomenon called compression lift. Once near the target they glide down steeply under control at hypersonic speed rendering them difficult to counter.

2. Hypersonic cruise missiles
   The great difference from traditional cruise missiles is that they fly at hypersonic speeds making them almost impossible to intercept. To do this they require a scramjet engine which only works at speeds above Mach 3, which means they must speeded up to this speed using a rocket for instance before they can be started.
   The high speeds around Mach 7 at which they may ultimately fly with the scramjet engine means special exotic materials must be used to withstand the intense heat generated. Furthermore, the heat and shockwaves generated result in a plasma enveloping the missile which interferes with GPS-based systems and incoming signals ordering course corrections such as when required for targeting an aircraft carrier. Also, the great heat makes them visible to satellites.
   Since there is no way to distinguish a nuclear-armed hypersonic missile from one with a conventional warhead and hardly any time to decide how to react, the targeted country is placed in an invidious position and might overreact.
   Directed Energy Weapons still in their infancy are one way to combat hypersonic missiles. In time satellite detection systems will be able to track them through their heat signature over their course from launch to impact, giving a certain amount of time, provided they are not launched close to their target.

Mode

The state or setting of a program or device.

For instance:

1. It can be in the cockpit with it essential that the pilots are aware of what mode is set as certain systems such as the Autothrust can work differently. Failure to understand the implications of a mode, or even selecting the wrong mode by mistake, has led to catastrophic accidents.
2. The type of transponder is referred to by its Mode. For example, there is Mode-S for the most sophisticated one as those transponders can handle a great amount of data and include cockpit instrument settings enabling the controller to see what the pilots intend to do.

Mode C Intruder

General Aviation Aircraft (GA) wandering into controlled airspace near airports risks collisions with traffic arriving and departing. Following some incidents due to this, Congress and the FAA have mandated that such GA aircraft must have Mode C transponders transmitting their altitude with there being a new system called Mode C Intruder Alert that warns controllers visually and aurally of the intrusion so they can immediately take action to prevent risk of a disaster.

MORA (Minimum Off-Route Altitude)

To ensure that aircraft do not hit the ground or a mountain if they stray a little from their assigned route, they are given minimum altitudes to allow for that.

Most Profitable Airlines

According to AeroTime (https://aerotime.aero) in 2024 the top ten most profitable airlines as regards profit margin (profit ÷ revenue) were:

1. Emirates 14.24%

2. Singapore Airlines 14.06%
3. Ryanair** 12.65%
4. Turkish Airlines 10.57%
5. IAG 8.67%
6. Qatar Airways 7.70%
7. Delta Air Lines 5.61%
8. United Airlines 5.52%
9. Air France-KLM 5.10%
10. LUFTHANSA 3.73%

** Ryanair under pressure from competition has had to lower prices so the profit margin for 2025 is somewhat lower, though revenue is still increasing as more routes are added.

The revenues for the US's Delta at $61 billion and United at $57 billion are almost double the approximately $30 billion for Emirates, IAG, and Air France-KLM, with LUFTHANSA higher at $40 billion.

Most Profitable Seats for Airlines

It's not First Class and not Economy, often called "Coach" in the US.

First Class seats of which there are fewer and fewer is often kept merely to add prestige to the airline itself and to its other offerings, and notably Business. Not only the vast amount space allotted to First Class makes it very costly for the airline, there are also all the extra services that must be provided, not to mention the cuisine with perhaps unlimited caviar and selected wines. Psychologically, having First Class can make those in Business think they are being reasonable.

It is Business Class and surprisingly to a lesser extent Premium Economy that is the most profitable. However, as it may not be possible to fill the aircraft with high-paying passengers those in

Economy make an essential contribution, especially by allowing more frequent flights that business people appreciate.

MRO (Maintenance, Repair, and Overhaul)

This is a key aspect of safe and efficient airline operations, and one where IT and data mining are playing an increasing role. (See **SMS**.)

Multifunction Display (MFD)

A multifunction display (MFD) is a screen (LCD) surrounded by multiple "soft keys" that can be used to display information to the user in numerous configurable ways, thus allowing the same screen to be used for multiple purposes.

~ N ~

N1 and N2 (Engine Performance)

Turbofan jet engines consist of two essential moving parts:

1. A relatively slow-moving fan at the front.
2. Fast-moving turbines at the rear.

In a new engine developed by Pratt & Whitney, called the "geared turbofan," they are linked, allowing a lower fan speed, which is theoretically more efficient.

In normal circumstances—that is, when the blades are not damaged, say, by a bird strike—the rotational speeds, N1 and N2 (and notably N1, producing most of the thrust), are an excellent indication of the amount of thrust.

Unlike the rev counters in an automobile, N1 and N2 express the rate of rotation as a percentage of their rated or maximum normal revs. To maximize engine life, pilots try to keep significantly below 100 percent. However, in an emergency they might even go above 100 percent and risk sacrificing the engine!

For the new geared turbofan engine, with gears linking the two, the ratio between N1 and N2 would, of course, be constant.

See **EPR**, the other way engine performance is measured.

Nacelles

In jet aircraft, the nacelles are the engine inlets and housing, including the thrust reverser cowling and outlet. Although seemingly unremarkable, nacelles are a critical factor in how the engines perform and how the aircraft flies.

Nanodrones

A rather confusing term used by NASA and the military for microsatellites. These are somewhat like footballs—say 20 cm in diameter.

See http://theuavdigest.com with all sorts of info about drones and their myriad uses.

Nanosatellite (Microsatellite)

Similar rather confusing term used by NASA and the military for microsatellites. They are also somewhat like footballs—say 20 cm in diameter—and weigh some 5 kilos and can be deployed from spacecraft to make inspections with cameras.

Narrow-Body

Traditionally, airliners had a single aisle, with two or three seats on either side. This could be achieved with a fuselage (tube) diameter of up to a maximum of about 4 meters (12 feet). The thinking was that a larger-diameter tube would increase drag and hence fuel consumption.

However, with airlines wanting to carry more passengers per aircraft and more space, the commercial and other advantages of a wide-body design became apparent for some applications, but by no means all, when the Boeing 747, the DC-10 and TriStar trijets, and Airbus A300 twinjet appeared on the scene in the 1970s. One problem with single-aisle airliners is that one passenger blocking the aisle can hold up everyone.

See **Wide-Body**.

NASA (National Air and Space Administration)

Photos and videos of astronauts landing on the moon and Space Shuttle launches, not to mention other publicity, give the impression that NASA is primarily concerned with astronautics. In fact, the contribution NASA is making in the aviation domain is considerable.

National Air and Space Museum (NASM)

Located in Washington DC, it is the most visited museum in the US, with many younger visitors. It has an extensive collection of exhibits, ranging from the Wright Brothers' craft to actual spaceflight capsules.

The annex to the museum, situated near Dulles International Airport, has a great collection of the more sizeable aircraft. Coaches ply between the main museum and the annex.

Entrance is free.

National Airborne Operations Center (NAOC)

The "Doomsday Plane." See entry for **Air Force One**.

Navaid

Radio navigation aid.

For years, radio beacons have been the key to navigation, with air traffic controllers sending pilots from beacon to beacon or to points where the radials from two beacons cross. The new **NextGen** air traffic control system, which relies largely on GPS, will mean aircraft can fly without depending on the beacons, though these will surely be kept as a backup.

NDB (Nondirectional Beacon)

Navigational beacons broadcasting on medium frequencies, whose bearing can be detected using the aircraft's **ADF** (automatic direction finder). The use of medium radio frequencies instead of VHF makes them visible over the horizon and therefore valuable for low-flying aircraft.

"Near Miss"

A near miss is defined as aircraft coming within a certain distance. It very often merely signifies the aircraft were closer than they should have been but not really in danger of colliding. Britain's Ministry of Defence defines it as cases where pilots *think* they were dangerously close.

Neutron Bomb

Low-yield nuclear bomb whose lethality derives from radiation damage caused by the neutrons it emits. Also called an enhanced-radiation weapon (ERW).

Next Generation Air Dominance Program (NGAD)

This is something of a US buzzword referring to possibly a super sixth-generation fighter jet in a decade's time or a series of manned and platforms working together and involving much computer technology and artificial intelligence.

The US Navy said in late 2000 it could not wait many years to develop a sixth-generation aircraft as was the case for the F-35, taking it to the mid-thirties. On the other hand, the US Air Force's Ultra-Sophisticated Munitions would likely play a part with a standoff stealth manned or unmanned aircraft launching a weapon which on nearing the target would launch its own weapons.

NextGen (Next Generation ATC System)

FAA's program to revolutionize every aspect of the United States' ATC system, which is in many ways quite antiquated and unsuited to the increasing demands likely to be placed on it. The FAA must continue to finance and upgrade present equipment while investing in the system for the future.

At the heart of NextGen is ADS-B, where aircraft continuously broadcast their position based on GPS to air traffic control and aircraft in the vicinity. (See **ADS-B**.)

Ninja Bomb—Can Slice Through Housing or Motor Vehicles

A special US precision warhead for **Hellfire** missiles carried on drones. Fitted with six sword-like blades that allow it to slice through flimsy accommodation or motor vehicles. Development started as far back as 2011 under President Barack Obama, but it wasn't until 2019, that some officials started referring to a nonexplosive munition, calling it the R9X, and describing it as the "Flying Ginsu" or "Ninja Bomb."

One of the great difficulties in using missiles launched from drones to take out "unsavory" individuals is that they often purposely surround themselves with innocent people and notably children. Even when the target is in relatively open areas, nearby people are often killed or injured, hence the desirability of a nonexplosive weapon. The ninja bomb was not often used except in 2020 in Syria where avoiding collateral damage with Russian victims was of primary concern and perhaps even more recently in Afghanistan.

nm (Nautical Mile)

1 nm = 1.16 statute miles or 1.853 km.

One minute of arc of latitude on the Earth's surface. (See **Coordinates**.)

Used for navigation in the sky as well as on sea. Airspeeds are always in knots (nm/hour).

No Fault Go-around

When in the slightest doubt about the advisability of proceeding with a landing, the best choice is usually to go around and make another attempt or divert. To avoid pilots feeling under pressure, many airlines have a "no fault go-around" policy whereby pilots do not have to explain to the airline the reason or even mention it. They may have to tell air traffic control the reason, but ATC normally keeps the information to themselves.

No Fly List

According to the ACLU:

> *If you are denied boarding on a flight, you can submit a standard form to the Department of Homeland Security Traveler Redress Inquiry Program (DHS TRIP). They normally transmit complaints and any supporting information you*

provide to the Terrorist Screening Center (TSC), which determines whether any action should be taken.

The response will differ depending on whether you are a US citizen or permanent resident so see their website. ACLU also mentions the fact that US citizens have the right to return to the US and the relevant officials at embassies abroad must facilitate this.

Again, it is complicated, so see the following ACLU link for full details: https://www.aclu.org/know-your-rights/what-do-if-you-think-youre-no-fly-list

Also see **"SSSS" (Secondary Security Screening Selection) [US]**.

"Normal Accident"

In a famous book entitled *Normal Accidents (Living with High-Risk Technologies*, organizational analyst, and sociologist, Prof. Charles Perrow coined the apparently contradictory term "normal accident." He exhaustively studied many super-accidents, such as Bhopal (chemicals), Three Mile Island (nuclear), Chernobyl (nuclear), and a ferry accident in the North Sea (Maritime).

Perrow concluded that accidents are inevitable in complex systems, and that sometimes it is the very systems installed to promote safety that become the cause, with linkage between different subsystems leading to cascading linked events resulting in disaster.

On the other hand, he has argued that despite these negative aspects, on balance the impact of these new technologies, such as nuclear power, is positive, since achieving the same goal by other means would be even more risky.

Normal Law

See **"Alternate Law"** as opposed to **"Normal Law" [Airbus]**.

NORTHROP GRUMMAN

The world's fifth-largest defense contractor in terms of arms sales, with 78 percent of its revenue deriving from defense.

Interestingly, the way it uses the word "systems" to define its wide range of activities in the aerospace domain is a lesson in new terminology. Rather than mentioning the name of the actual products, such as fighters, bombers, and so on, it classifies them, logically enough, as systems or platforms.

It has strategic systems, theater systems, and so on.

Areas of activity are:

1. Information and services.
2. Electronics.
3. Aerospace—including the B-2 Spirit stealth bomber in the 1990s, and now the B-21 Raider long-range strike bomber.
4. Shipbuilding.

Expertise in one area is applicable in the other areas, and one cannot help but see some parallels as regards terminology strategies with France's Thales, except that Thales itself does not build actual aircraft of any size.

Nose Wheel

The nose wheel prevents the aircraft from tipping forward onto its nose and is used to steer the aircraft on the ground. Unlike the main undercarriage bogies which can sustain a considerable shock and bear much weight, it can do neither. This means that if the pilots are **Floating** down the runway they cannot "get the aircraft down" by pushing the nose down so much that the nose wheel hits first. Doing so could shatter it completely.

It is a rule that pilots are never penalized for a go-around, so they do not normally risk landing with the nose wheel touching down first.

Except for a few Boeing 737s, nose wheels do not have brakes, and even there they were only to be used in emergencies and soon abandoned because the lack of weight on them meant they were not very effective.

NOTAM (Notice to Airmen) Has Become (To Air Missions)

As it will always be referred to as NOTAM, no one will really care that the FAA is now in accordance with the desire for **Gender Inclusive Language** and has changed "Airmen" to "Air Missions."

NOTAMS are official warnings and updates regarding situations and, notably, potential hazards at airports and en route. Typical information would be runways/taxiways closed for repairs, faulty runway lighting, presence of birds, and temporarily restricted airspace due to flights by heads of state, air shows, and parachute drops, airspace closed for military activity, and inoperable navigation aids.

Unbelievably, in January 2023, US airspace was shut down for a couple of hours in the early morning because the FAA's NOTAM and backup computer system failed. Even when flights resumed some flights had to be cancelled because aircraft were out of position and some pilots had exceeded their allowed hours. Some allege the computer system was out of date because the FAA must live hand-to-mouth with handouts from Congress and finds it difficult to budget long term.

There are moves to make NOTAMs that can go up to 200 pages for long-haul international routes easier for pilots to follow. They can even miss something simple, such as "left runway out of use," as did pilots of an Air Canada flight landing at San Francisco with the result they almost landed on a taxiway on which several aircraft were taxiing to take off—one of them turned on its lights to warn them of their presence and their mistake.

Novichok, Nerve Agent

See **Atropine.**

NTSB (National Transportation Safety Board)

Although the NTSB figures prominently in air crash investigations, one should not forget its remit covers all modes of transport. Notably, its work recently included issuing warnings about the possible use of cell phone texting prior to a railroad crash and cases of teenagers texting before automobile crashes.

In the aviation sphere, the NTSB is viewed as the noble guy, while the FAA must steer a difficult course between various interests. The NTSB can "recommend," but it is the FAA that must implement those recommendations or otherwise.

In the aviation field the NTSB is respected worldwide for its accident investigation expertise and integrity. Sometimes it unfairly receives a lot of flak when doing its duty questioning the actions of "heroes."

Somewhat along the lines of the most-wanted felons list, the National Transport Safety Board has a wish list of the things it thinks would most contribute to airline safety.

The three most notable items on "MOST WANTED" list some years ago were:

1. Stop runway incursions/ground collisions of aircraft by giving warnings of probable collisions or incursions directly to flight crews in the cockpit.

2. Improve the present audio and data recorders and seek mandatory installation of video recorders, requiring that cockpit voice recorders (CVRs) retain at least two hours of audio;

   a. requiring backup power sources so cockpit voice recorders collect an extra ten minutes of data when an aircraft's main power fails;

   b. inspecting and maintaining data recorders yearly to make sure they operate properly;

   c. and finally installing **Cockpit Video Recorders** to give investigators more information to solve complex accidents.

3. Reduce accidents and incidents caused by human fatigue by setting working hour limits for flight crews and aviation mechanics based on fatigue research, circadian rhythms, and sleep and rest requirements.

The NTSB updates the list from time to time and is now contemplating a three-year interval.

On April 6, 2021, the Board approved the following 2021–2023 "Most Wanted List" of Transportation Safety Improvements. See https://www.ntsb.gov/Advocacy/mwl/Pages/mwl-21-22/mwl-making.aspx.

NTSB Defines "Fatal Injury" as Causing Death Within 30 Days of Crash

A fatal injury was once defined as one resulting from injuries occurring within three days of the crash. From October 2016, it was changed to 30 days. This is rather arbitrary as someone suffering from burns could expire due to them after somewhat longer than that.

NUCLEAR WEAPONS

In the early days of nuclear weapons, the US vastly overestimated the number required, say, to obliterate Moscow, because they only considered the effect of the blast generated. In fact, it is the heat generated rather than the blast that is devastating, and over a very much wider area.

~ MAD (Mutually Assured Destruction)

Not long ago, at age 97, the late Henri Kissinger said:

> *In my day the threat of a devastating nuclear war was daunting enough, but with weapons having become many times more powerful it was now considerably more fearsome.*

Not only that. Nowadays there are more players, not speaking of the possibility that terrorist groups could avail themselves of one or more nuclear weapons. To further complicate matters, there is the introduction of artificial intelligence to consider.

In Kissinger's day, as now, people lived with the concept of **MAD (Mutually Assured Destruction)** whereby neither side could defeat their opponent without risking themselves being obliterated (see **Nuclear Triad** below). There were some close shaves including one where a Russian operator prevented a strike on the US based on false data that the US had launched missiles. We have perhaps become too complacent for the risk of even accidental nuclear war is increasing with more countries with nuclear weapons or aspiring to have them.

~ Nuclear Triad [US]

The US bases its *strategic* deterrence on three components called the nuclear triad that hopefully will never ever have to be used, for if they are, it will mean MAD (Mutually Assured Destruction) has failed.

The three components are:

1. Ground-based strategic deterrent (**GBSD**).

2. Submarine-launched ballistic missiles.

3. Bombers.

The great cost is ongoing, for each component must be upgraded and replaced from time to time. The latter poses a problem, for by "embroidering" the deficiencies to convince Congress of the need to upgrade, the military could tempt an enemy to consider a first strike.

Snapshot of the US's nuclear weapons making up the nuclear triad and those in storage in 2022:

1. Land-based leg: 400 Minute Man III intercontinental ballistic missiles dispersed in underground silos around the US.

2. Sea-based leg: 14 Ohio Class nuclear submarines each armed with 24 Trident II intercontinental missiles, each of which can have up to 14 independent nuclear warheads.

3. Air-based leg: 46 nuclear-capable B-52H Stratofortresses and 20 B-2A aircraft.

A total of 1,744 deployed nuclear weapons for the three legs of the triad:

However, at least one critic thinks the **Nuclear Triad** creed is "hogwash" and only advanced so all arms of the US military can benefit handsomely. He maintained the sea-based nuclear submarine leg alone was largely sufficient to assure peace.

~ Dead Hand, Called "Perimeter" by Soviets

In the 1980s the Soviets feared that United States submarine-launched missiles were becoming so accurate the US might be tempted to perform a first strike against their missile silos with little warning as the submarines could be close. They therefore developed a Doomsday system they called "Perimeter," and the West called "Dead Hand" for them to automatically make a second strike even should the military command structure and leadership be taken out by the US first strike.

It worked on the principle that should sensors such as seismometers and radiation detectors determine there had been a nuclear strike with the chain of command broken, the system would automatically launch all the country's missiles. This would include having a missile overflying the country setting off the missiles wherever they might be.

Strategic Missile Forces Gen. Sergey Karakaev confirmed its existence to a Russian newspaper in 2011, saying the US could be destroyed in 30 minutes.

It appears the system can be switched on and off according to the level of threat.

~ "Madman Theory" in International Relations

The theory that a great tactical advantage could be gained by a leader, such as the president of the United States, pretending to be unhinged enough as to launch a nuclear strike.

A case of it being applied might be the Nixon administration forcing the North Vietnamese to the negotiating table in the Vietnam war. In October 1969 the US sent nuclear-armed bombers on sorties for three days close to the Soviet Union—

without telling the American people—but letting the Russians know Nixon was something of a madman, and there was no knowing what he might do.

President Trump seemingly, intentionally, or by nature, effectively exploits a form of madman strategy in his dealings not only with potential enemies but also allies. President Putin's saber rattling could fall into the same category.

Experts think the theory does not really work because the target often misinterprets the madman's intentions with it raising the risk of an actual nuclear exchange followed by counterstrikes or worse.

Incidentally, there was an instance where Nixon was not resorting to madman strategy but was completely drunk, and after an EC-121 reconnaissance aircraft had been shot down, ordered a tactical nuclear strike. Henry Kissinger, the National Security Advisor, induced the Joint Chiefs who were about to finalize a precise target to put a hold on that order until Nixon woke up sober the next morning.

~ McMahon Act

Having contributed so much towards the realization of the Manhattan Project to produce the first atomic bomb, and not least encouraging the Americans to undertake the task in the first place, the British were shocked to find the US Congress's passage of the McMahon Act at the war's end meant none of the knowledge would be shared with them after the war, despite prior agreements to the contrary. The act forbade the passing of any of the technology to other countries, including the British.

The British, under the postwar Labor government with Clement Attlee as prime minister, consequently had decided in secret to produce a British bomb. Interestingly, the physicist, Klaus Fuchs, later arrested and imprisoned for revealing some secrets of the Manhattan Project to the Russians, was back in England working for Britain. No official mention has been made of the fact he was probably able to explain to his British employers—as well as the Russians—how the plutonium bomb's imploding mechanism worked!

~ "Nuclear Football"

Wherever the US president goes he is accompanied by a military officer carrying the so-called nuclear football—a briefcase with the codes for launching a nuclear strike. It is not so widely known that the vice-president has one too.

~ Thermonuclear Weapons

Thermonuclear weapons are second-generation nuclear weapons where a nuclear *fusion* reaction (rather than a *fission* reaction) produces a very much more powerful explosion. However, as things now stand an old-fashioned fission nuclear trigger is required to generate the heat necessary for fusion to start—hence the name "thermonuclear." Sometimes referred to as the hydrogen bomb.

It is scary to think that although nuclear weapons have not been used in combat since 1945, not only missiles but many military aircraft and even drones are classed as thermonuclear-capable as well as conventional-ordinance-capable.

~ Detonation of First Russian, British, and Chinese Devices

The Russians detonated their first nuclear bomb on August 29, 1949, the British, a plutonium implosion device on 3 October 1952, and the Chinese on October 16, 1964.

~ "Broken Arrow": Accident Involving Nuclear Weapons

"Broken arrow" is a US term for an incident or accident involving nuclear armaments or their components *not risking nuclear war*.

There is another term, "Nucflash," for an incident involving nuclear armaments or their components, only differing from broken arrow in that *it might trigger a war*.

Brokken Arrows have mostly involved aircraft that have accidentally released nuclear weapons or crashed with one or more on board. The loss of a submarine with nuclear weapons on board, as has happened, would be a broken arrow incident.

We can mention four major broken arrow incidents:

1. This involved an incident involving an enormous nuclear device in 1957 where a B-36 transporting it was caught up in turbulence over New Mexico, USA, causing a crewmember to grab the manual bomb release lever to steady himself, resulting in the release of the weapon. The conventional explosive trigger exploded on hitting the ground to create a 12 feet deep crater. Fortunately, a terrible disaster due to the magnitude of the bomb was averted because the fissionable plutonium component was stored separately aboard the aircraft.

2. Perhaps the closest to a major disaster in the US was in 1961 when a B-52 bomber—kept routinely airborne so as not be destroyed in the event of a Soviet first strike—carrying two Mark 39 nuclear weapons crashed near Goldsboro, North Carolina.
   The aircraft began leaking fuel and descended with the pilots losing control. Before the aircraft crashed, the two bombs dropped with one plunging more than 20 feet into the muddy ground without even the conventional explosive material for the trigger detonating.
   Much more dangerously, the parachute on the other bomb did deploy meaning initiation of the arming sequence. No one is sure how many stages of the arming sequence were gone through, though some think all but one last switch operated. In the end the parachute got caught in the branches of a tree leaving the bomb itself virtually intact—and without the bomb detonating and destroying a large area of the eastern United States—for the bomb had 250 times the power of the one dropped on Hiroshima!

3. A major and very expensive broken arrow incident occurred at Palomares in Spain (1966) during the midair refueling of a B-52 bomber armed with four Mark 28 hydrogen bombs.
   While flying at an altitude of 31,000 feet, the B-52 bomber closed too fast with the KC-135 tanker resulting in it impacting the tanker's fueling boom, and an

explosion caused the tanker loaded with fuel to ignite, killing all four of its crew. The B-52 in turn began to break apart with four of its crew members ejecting and parachuting safely to the ground, while the remaining three were unable to save themselves.
Thankfully, the four 1.5 megaton thermonuclear bombs it was carrying were unarmed and could not go off as such, but the conventional explosives in them could, and that is what happened in the case of the two whose parachute did not deploy, resulting in the scattering of their radioactive material.
The two other bombs whose parachutes did deploy floated downwards with one coming down gently in a field and the other floating under its parachute and ending up under water in the Mediterranean.
Within a day, the US had secured and recovered the three bombs that had come down on land but could not locate the one several miles out to sea. It was of great concern and took almost three months to find and recover in an operation involving two submarines and various ships. Finally, a fisherman told the Americans he knew where the bomb had fallen, and following his guidance the bomb was eventually found and recovered.
The US had been worried that the Soviet Union might find it first!
In 2000 there was even a movie about it called *Men of Honor*, starring Robert De Niro.
The clean-up that followed was to prove exceedingly difficult and costly. Topsoil contaminated with plutonium was shipped to the US, and compensation had to be paid to locals claiming to be suffering problems due to the incident. American servicemen involved in the cleanup were later to say their health had been impacted. Furthermore, the US had to pay Spain for years of monitoring. Overall, the local people were not properly warned of the dangers.
The enormous cost of these flights and this incident led to the winding down of the Chrome Dome operation

whereby 12 B-52s, each with four thermonuclear bombs, were kept permanently aloft making runs on various circuits taking them close to the Soviet Union with the one over Spain. There was some resistance to this on the part of the Strategic Air Command and the joint chiefs with the number finally reduced to four flights at a time.

4. A couple of years afterwards a B-52 similarly carrying four thermonuclear bombs caught fire and crashed at Thule in Greenland with the conventional explosive on all four exploding and spreading radioactive contamination. In places the heat of burning fuel and debris melted the ice causing items and contamination to sink to the sea floor, making recovery of everything impossible.

Incidentally, the Thule circuit had been secretly maintained to keep an eye on the early warning station there in case the Russians knocked it out before launching a missile strike.

~ O ~

Octas

The amount of cloud cover is expressed in octas (eighths).

"Odd-One-Out" Logic Not Always Valid

Many key systems are in triplicate, and the computers are usually programmed to ignore the odd-one-out on the natural assumption it is faulty.

However, there are rare cases where this is inherently not true, say, when water penetrates the mechanism of two angle-of-attack sensors, and when the aircraft climbs, they freeze at the same time and remain locked at the same angle. This happened in the case of the *Not All Pilots Can Conduct Air Tests* crash of an Airbus 320 off the coast in the South of France*.

Oleo Pneumatic Shock Absorber

It may be partly thanks to this shock absorber incorporated in the landing gear that landings can feel so gentle when, in fact, the very heavy aircraft has come down quite hard.

The shock absorber consists of a sturdy tube in which a piston bearing the weight of the aircraft pushes hydraulic oil through an aperture to a reservoir of nitrogen which is also progressively compressed to attenuate the remaining shock.

"Open Descent" and "Open Climb"

This is an Airbus term for descending or climbing to a given height without conditions other than say changing the airspeed. Increasing the airspeed increases the rate of descent. If climbing the opposite is true—reducing airspeed increases the rate of climb.

Boeing and other companies call this "Flight Level Change."

Open-Jaw (Ticket)

A ticket where there is a break in the series of stopping-off points. In simplest form, it could mean flying, say, from London to Washington, DC, and flying back to London from New York, with the leg between the two cities accomplished by another means, such as train or even another carrier.

Open Skies Agreement

As mentioned under the **Freedoms** entry, the negotiation of flying rights between governments can be a cutthroat business. At one time a country's airline was often its flag carrier and government owned, with that government doing everything possible to protect it.

With the worldwide move toward more freedom, many restrictions have been reduced, with the US pushing its own version of Open Skies, which to some does not appear so open in that it restricts cabotage in the US. In addition, the Fly America Act, whereby federal employees, and subcontractors, are obliged to use US airlines, is deemed restrictive.

OPERATION PAPERCLIP

At the end of World War II, the Americans had a program, later known as Operation Paperclip, to spirit away top German scientists, aeronautical designers, and rocket engineers. The name arose because paperclips were attached to the files of those thought to be of value to the US.

The highest profile one was Wernher von Braun who besides developing the **V-2** missile was to play a leading role in the US space program, taking Americans to the moon. With blinding ambition, he went from fraternizing with top Nazis to consorting with US presidents. He was once arrested by the Gestapo on the pretext he had said he had developed rockets with the aim for them to go into space, not as weapons.

So valuable were these people deemed by the US, that orders were given to neither question their Nazi pasts nor their exploitation of slave labor involved in producing the weapons they created.

Other similar operations included one called LUSTY (**LU**ftwaffe **S**ecret **T**echnolog**Y**) where the US Air Force brought aircraft and aeronautical engineers from Germany to the States, even rustling some from the British and French occupied zones. In the US's favor was the fact those working on the **V-1** and **V-2** thought it preferable to surrender to the Americans as they were likely to treat them better than the British, who would remember their creations falling on London and killing many civilians.

See **WORLD WAR II**.

Operational Flight Data Monitoring (OFDM)

While **ACARS** is designed to report technical faults and not what the pilots are doing (wrong), there is a valuable system called Operational Flight Data Monitoring (OFDM) that cheaply, and confidentially, flags up flight deviations from the norm to an outside organization after the aircraft has landed. This can include a landing with the aircraft coming in too fast, or way above or way below the glideslope, and not going around. Ryanair uses this system, which partly explains why it is very safe. OFDM can not only flag up places where pilots need advice and extra training but pick up on worrying trends.

OSG (Open Systems Gateway), U-2, F-22, and F-35

The venerable U-2 equipped with OSG (Open Systems Gateway) has been tested as an intermediary enabling the F-22 with its Intra-Flight Data Link (IFDL) to communicate with the F-35 with its Multifunctional Advanced Data Link (MADL), thus permitting the two to fly and work together in battle.

Open Skies Agreement

As mentioned under the **Freedoms** entry, the negotiation of flying rights between governments can be a cutthroat business.

Otto Aviation's Celera 500L's Remarkable 22:1 Glide Ratio

Outer Marker

When approaching a runway to land, the aircraft passes over radio beacons called markers that emit a vertical signal with an identifier showing which one it is (outer, middle, or inner). A light in the cockpit, accompanied by a sound, tells the pilots where they are.

The outer marker is four to seven nautical miles from the runway threshold and roughly marks the point at which an aircraft enters the glide slope under normal circumstances and represents the beginning of the final part of the landing approach.

In the US the outer marker is often combined with a locator nondirectional beacon, which aircraft can home in on, but with the increasing use of GPS these are becoming unnecessary.

Overrun

An overrun really means being unable to stop before reaching the end of the paved runway either when landing or after aborting a takeoff. It is sometimes referred to as "overshooting."

As technology has improved and reduced the incidence of other types of accident, such as controlled flight into terrain (CFIT), runway overruns, especially on landing in bad weather, are coming to the forefront as the cause of major accidents. There have been several potentially disastrous cases, such as the Qantas 100 mph overrun at Bangkok in 1999, and the Air France overrun at Toronto in 2005, which could have resulted in great loss of life. In addition, there have been others where there was indeed loss of life.

There are recommendations regarding overrun areas beyond the end of runways, but with increasing urbanization many airports do not have enough space, especially after a runway has been lengthened. One solution is to have **EMAS (Engineered Material Arresting System)** at the end of the runway. This is an area of special material into when the wheels sink and slow the aircraft. These have prevented disasters but only work if the aircraft comes straight off at the end, which is often not the case, as the aircraft has skidded off to one side.

Although it was not actually EMAS that saved the day, the rain-sodden soil acted similarly when a Qantas's 747 exited the end of the Bangkok runway at 100 mph in 1999*.

Overspeed

Each aircraft type has airspeed limits for various configurations, ranging from clean configuration to maximum flap and undercarriage lowered. Should the airspeed exceed these limits, the instruments give an *overspeed* warning. Exceeding the limits by a significant amount or for a lengthy period could result in major damage to the aircraft or worse.

Over-Wing Exit

To comply with the regulation that the aircraft can be evacuated in 90 seconds with only half the exits usable, airliners have over-wing exits. Some are combined with slides as at the doors. However, many are not. With the wings often being high up off the ground, having the flaps fully deployed can make it easier, and more importantly safer, for passengers, as they can slide down on them and have less height to drop down.

Oxygen Masks for Aircrew

The oxygen masks for aircrew are fed from a compressed oxygen supply and unlike those for passengers can be set to prevent intake of surrounding air which could be contaminated by smoke or lethal gases.

Most are of the quick donning type able to be put on with one hand in five seconds.

Prior to departure pilots must check the oxygen supply and at times whether the microphones work.

See **Hypoxia.**

Oxygen Masks for Passengers

The oxygen masks above the passengers' heads drop down if for some reason the cabin depressurizes or even fails to pressurize on the climb out after takeoff. Passengers must remember to pull the mask (and the attached tube) hard downwards to break the seal in the oxygen generator so the chemical reaction generating the oxygen can start.

The oxygen is generated by a chemical reaction between sodium chlorate, barium peroxide, and potassium perchlorate. Once the chemical reaction starts, it cannot be stopped giving one between 10 and 12 minutes of oxygen, which is quite enough time for the pilots to take the plane down at up to say 7,000 feet per minute to below 10,000 or 12,000 feet, where one can then breathe unaided.

To allow for a child on someone's lap, there is always an extra one in a row of seats together, with, say, four masks for three seats. This means there should not be two children on laps together in a row. Those looking after children should set up their own masks first and once their own mask is working properly attend to them.

Cabin crew have oxygen masks they can use while walking about—and notably telling passengers what to do.

~ P ~

Padeye or Pad Eye

Metal ring standing on a flat plate welded or bolted to the deck of say of an aircraft carrier. Used to tie down aircraft, etc., and must be very strong to keep them in place in a storm.

Pairing of Aircrew

Just as airlines avoid rostering two pilots who have little experience on a given aircraft type together, they also try to avoid having two extremely old pilots flying together. See **Age of Pilot** and **Hierarchical Gradient between Pilots**.

"Pan-pan!"

Like Mayday, Pan-pan, also repeated three times, indicates to air traffic controllers that the aircraft is in trouble, but that it is not a crisis requiring immediate priority action by air traffic control and emergency services. However, controllers should have the aircraft having trouble in clear airspace and not pester the pilots while they try to sort out the problem and decide on a course of action.

See **"Mayday."**

PAPI (Precision Approach Path Indicator)

Array of four directional lights near the aiming touchdown point on the runway that appear to change color according to whether the pilot is below, on, or above the glide path. If there are two reds and two white, they are precisely on the glide slope. All red means far too high, all white means far too low, with three red slightly too high, and three white slightly too low.

VASIS is similar but less precise and sometimes installed instead because it is less expensive to make and maintain.

Parachutes for Light Aircraft

A few light aircraft and ultralights have parachutes attached to their fuselages that can be deployed to bring the aircraft down relatively gently in a crisis to save the lives of the occupants.

Passenger Etiquette

Mostly it is a matter of common decency, but a key point is not to recline one's seat without consulting the person sitting behind you and ensuring they are not having a meal or working on their laptop.

People who have paid extra to book a certain seat often complain about being made to feel guilty or even being publicly shamed with rude remarks on refusing to give up their seat in exchange for an inferior one for someone with children, or couples who want to sit together who were unwilling or did not think to reserve. One should not expect to exploit the good nature of others who might have a special reason, medical or otherwise, for choosing that seat.

See **Air Rage**.

Passports (Biometric)

With passports being stolen, forged, and altered on the one hand, and the fear of terrorist action rising following 9/11, there have been general moves to introduce biometric passports incorporating electronic chips giving physical and other information.

PCU (Power Control Unit—Controlling Rudder)

In the longest NTSB investigation ever, it was eventually found that a freak malfunction of the power control unit (PCU) controlling the Boeing 737 rudder had caused two crashes, and almost a third, with Boeing initially blaming the pilots*.

The NTSB had all along suspected the PCU was responsible but repeated tests of the unit failed to show any problems.

Three and a half years elapsed between the crashes, with the first having few fatalities, and so many 737s were flying safely worldwide and the NTSB were in a quandary for recommending a different rudder system would be very expensive, and anyway, the lengthy approval system meant that it would take time to come into effect, especially if it were ordered without convincing proof of its necessity.

Finally, after a third case where the rudder behaved oddly but where the pilots were able to recover, a "thermal shock test" involving cooling the device was carried out and showed that in some certain circumstances the 737 PCU could not only jam but reverse so that if the pilot forced down the left rudder pedal the aircraft would turn right. This meant once the PCU malfunctioned any attempt to recover would make the situation worse.

Pentagon

Term used to denote either decisions made at the top of the US military or the vast five-sided five-story building complex housing the Department of Defense in Virginia State. Though situated adjacent to Washington, DC, it has special Washington, DC, zip codes for its mail.

With the possibility of war looming in 1941 there was a pressing need to bring together the various branches of the military scattered around the capital. Workers constructed the building in only 16 months following a groundbreaking ceremony that September.

It owes its shape to the fact that the US Congress originally granted approval for a design configured for a plot nearby bounded by five roads. Following objections that placing the building on that plot would block the view of the Capitol that one had from the tomb of Pierre L'Enfant (the French architect-engineer who planned Washington) in Arlington National Cemetery, a second site on nearby land with the same designation was chosen. Keeping the original overall design avoided lengthy Congressional approval.

Freed from the geographical constraints of the original roads meant it was possible to construct a regular pentagon.

People Ejected Due to Decompression

There have been incidents, for instance where a bomb has made a hole in the fuselage or a part of an engine has flown off and struck a window, where people have or could have been ejected due to the air rushing out.

Whether someone is sucked out very much depends on the altitude and how near an unbuckled person is to the gap through which air is rushing.

See **Decompression (Loss of Cabin Pressure)**.

Fortunately, a number of those decompression incidents in the US occurred at a relatively low altitude (around 10,000 feet) with the consequences much less grave than they would have been at 35,000 feet.

Perrow, Prof. Charles B.

See **Normal Accident**.

Petrov, Stanislav: The Man Who Saved the World (1983)

Lieutenant Colonel Petrov was duty officer at command center for the Soviet nuclear early warning system in 1983 when his screens indicated five incoming missiles from the United States. Believing it to be a false alert and fearful of the implications, he did not pass the information on to higher ups according to his orders, thus averting the risk of nuclear war.

Years later he was invited to the US for the filming of a documentary called *The Man Who Saved the World.*

It is claimed it was several triggers away from the launch of a nuclear, but the situation so tense and the Soviet Union so fearful of an attack one cannot be sure of what might have happened had Petrov obeyed orders thinking like a military man rather than a civilian.

There was another case where one man possibly prevented a thermonuclear exchange during the 1962 Cuban Missile Crisis where a Soviet submarine that had lost contact with its command structure was being hounded by US ships and was on the point of launching a nuclear-tipped torpedo in the belief that war had already started. However, it needed the approval of three people, the captain, the political officer, and the executive officer, Vasily Arkhipov, who was very high-ranking, being flotilla chief of staff.

The captain and political officer wanted to launch but Arkhipov refused to give his consent thus preventing what might have escalated to a thermonuclear exchange.

Unlike Petrov, he went on to have an illustrious military career.

PF (Pilot Flying)

The pilot handling the aircraft, as opposed to the Pilot Not Flying (PNF), sometimes called the pilot monitoring. To avoid confusion, a pilot taking control will always say, "I have control," and expect confirmation that the other pilot has understood.

Usually, pilots take turns—on short flights, one pilot might fly the outward leg and the other the return leg. In an emergency or extremely difficult weather conditions the captain would normally take over, provided enough time remains for them to get the feel of the aircraft. The PNF (pilot not flying, also called the pilot monitoring) is usually busy too, handling radio communications with ATC and so on.

PFD (Primary Flight Display)

When working, as is usually the case, the Primary Flight Display enables the pilots to see in one place information that used to be shown on several separate instruments. A further advantage is that airspeed and altitude are shown on tapes on either side of the artificial horizon and can have bugs (markings) on them showing speeds at which the aircraft could stall, or go too fast for the flap setting, or V_{REF}, the speed they should aim for on landing.

PRIMARY FLIGHT DISPLAY

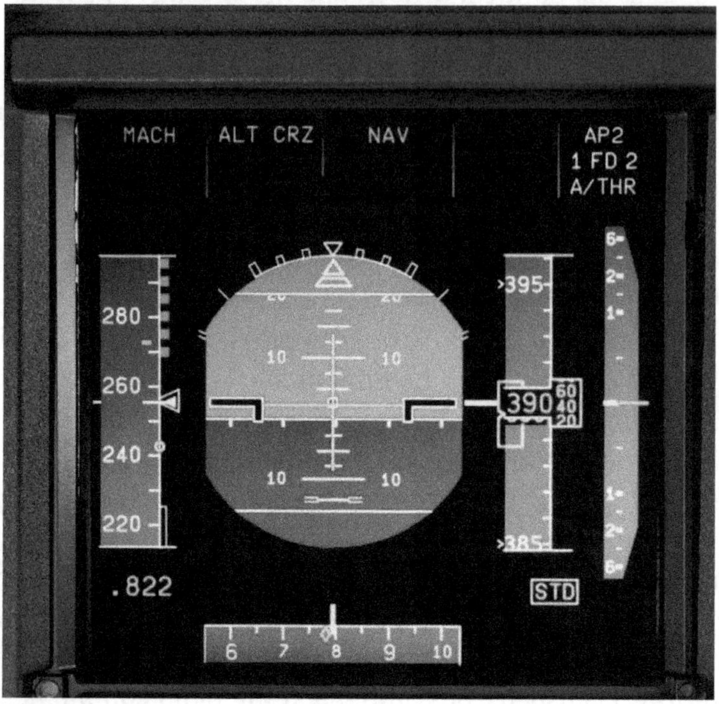

[Canva]

The above is a relatively simple version. The “tape” on the extreme left shows the airspeed in knots, and the one on the right shows the altitude in feet, with the large figures representing the **Flight Level**. The vertical speed, up and down, is shown on the extreme right, and is centered as the aircraft is flying level. The course (aircraft’s direction) is shown right at the bottom.

There are good explanations on YouTube, and anyone with a **Flight Simulator** would likely have it.

See **Glass Cockpit**.

Phonetic Alphabet

The present-day globally used NATO version is as follows:

Alfa, Bravo, Charlie, Delta, Echo, Foxtrot, Golf, Hotel, India, Juliet, Kilo, Lima, Mike, November, Oscar, Papa, Quebec,

Romeo, Sierra, Tango, Uniform, Victor, Whisky, X-Ray, Yankee, and Zulu.

According to the FAA's *Aeronautical Information Manual* (AIM), numbers should be enunciated as:

Wun, Too, Tree, Fow-er, Fife, Six, Sev-en, Ait, Niner, and Zero.

Apparently, *nine* enunciated by US speakers can be confused with *five*, hence the use of *niner*.

The use of whole-word call sign prefixes, such as *United* for UA or *Speedbird* for British Airways, and suffixes such as *heavy* to denote a jumbo-sized aircraft, and *super* for the A380 superjumbo makes exchanges with air traffic control not only longer but more easily understood, since the listener has time to lock in mentally.

Phugoid

The natural inherent porpoise-like motion of an aircraft manifested in the absence of functioning elevators to halt it. The nose pitches up, and then, as the speed consequently decreases, the nose pitches down. Then the same happens in reverse, with that pitching down of the nose making the speed increase and the nose pitch up, and so on, and so on. In large aircraft the oscillatory period can be several minutes.

There have been attempts, some successful, to control this by engine power alone, but this is difficult because when the aircraft speeds up and slows the pilot is liable to reduce or add power just when doing so will only accentuate the phugoid motion.

This has featured in crashes such as the worst single-aircraft crash ever, Japan Airlines Flight 123*, where lack of hydraulics caused by the rear bulkhead failing made the aircraft uncontrollable.

Phugoid Mode [Airbus]

A little-know Airbus programmed mode that prevented Sully maximally flaring in the "Miracle on The Hudson" incident.

See **Famous Feats of Airmanship**.

Piddle Packs, Bladder Relief Devices

With fighter pilots flying longer sorties, they must have a way to pee. Reducing fluid intake to avoid the need to urinate on missions has resulted in pilot dehydration and reduced mental performance.

The solution has been what is called the "piddle pack," a bag containing crystals that turn to gel on contact with urine into which the pilots relieve themselves. However, there have been accidents brought about by pilots interrupted by some urgent task—combat—just when contorting themselves to use piddle packs.

Also, with more female pilots in the military a practical device is seen as a necessity. This has led clever designers to try to come up with something better using their technical expertise.

One called the AMXD (Advanced Mission Extender Device) starts with a cup or pad that goes under the flight suit to which a hose with a pump is attached. When sensors in the cup or pad detect urine, it is pumped into a bag alongside the pilot.

It has received mixed reviews from pilots, with some thinking it too sophisticated.

Interestingly, the B-2 bomber that flies extremely long missions not has only a toilet but also a microwave so a pilot resting can heat up a meal.

Piggybacking

This is where manufacturers avail themselves of approvals and certifications obtained for earlier models or similar models of an aircraft to get approval for derivatives, thus avoiding the cost of retesting or new trials. The Boeing 737 is an example. Like **Grandfathering**.

Pilot Salaries

Especially in the US, the power of the unions means pilots' salaries within an airline depend on seniority rather than ability/performance.

Internationally, the following airlines are said to pay the most:

1. Emirates
2. Qatar Airways
3. Singapore Airlines
4. Cathay Pacific
5. Delta

Emirates and Qatar Airways, for example, not only offer the top salaries but favorable tax regimes to residents. Of course, all those airlines including Delta offer additional benefits including health insurance and free upgrades when traveling on the airline.

Pilot Training (Becoming a Pilot)

Traditionally, in the US and in other countries, a pilot not trained by the military would have gone to flying school, become an instructor, and then after gaining a certain amount of experience have got a job at an airline, often starting with a regional.

While the FAA favors this system, several aviation training experts at flying schools, and US universities say, "Such training prepares you for flying an aircraft by yourself, while at an airline you don't really fly anything by yourself."

Another expert argues that resources should not be spent on students who do not have what it takes, saying, "Now the only qualification to train as a pilot (in the US) is a Mastercard," adding that the free market is the wrong selection process.

In view of the shortage of pilots in the US—partly because of demands from China and India—changes will come about. Other countries have various schemes.

In the US, one route for someone without the necessary funds to become an airline pilot is to join the military, but can mean signing on for ten years and starting at the bottom of the ladder at an airline.

Pinger

The locator beacon on the "**Black Boxes**" (**Cockpit Voice Recorders and Flight Data Recorders**) that emits a ping for 30 days after a crash. There are plans to make the period longer as there are some cases such as **MH370*** where they (or the wreckage) have never been found under deep water.

Pipeline Flying to Build Up Flight Hours

Though it would seem drones would be ideal for checking pipelines, the distances involved and the sophisticated cameras required mean that very expensive ones would be required. As a result, recourse in the US and Canada is made to young pilots just out of flying school keen to build up their mandatory flying hours to be accepted by a small business jet company if not a major airline.

Often, they will be flying a Cessna 172 Skyhawk on which they were very likely trained at flying school.

Although pipeline flying nominally gives the pilots the hours needed to seek proper jobs, it is pretty useless because it is nothing like flying at an airline where people skills are important. Also, it is under VFR rules in good weather with little reference to instruments. Some say the skills learnt at flying school atrophy during pipeline flying.

PIREP (Pilot's Reports)

Reports by pilots regarding weather conditions encountered en route and even on landing (braking), retransmitted by land stations in standard format for the benefit of other aircraft. They normally have the prefix UA, but this is changed to UUA when hazardous conditions are involved.

Pitot Tube (Pitot Probe)

Based on an idea of Henri Pitot (a French scientist and hydraulic engineer born in 1695) for measuring the speed of liquids, the pitot tube in its simplest form is a forwards-facing hollow probe with a hole at its tip into which the oncoming air (referred to as ram air) rushes. By having a sensor compare that pressure with the static pressure from an inlet facing sideways, the airspeed can be determined.

In the case of the modern airliner, that setup is more complicated, with the static ports not necessarily on the pitot probe and there always being three sets, so that the one that fails can be identified as the odd man out. This strategy would not work should two of them fail, say, due to the buildup of ice crystals even though the nozzles are fitted with heaters to prevent them icing up, or even in one case where an aircraft cleaner forgot to remove masking tape he had used to protect the inlets while cleaning.

Planetary Gears/Cyclic Gears

Extremely rugged and compact gearing system, having a "sun" gear in the middle meshing with often three or five "planet" gears around it, which in turn mesh with a ring (annulus) embracing them all. Used in many applications but most notably, as regards aviation, in the Pratt & Whitney geared turbofan engine, which is expected to be more fuel efficient, since the gearing will enable the fan and turbines to operate at optimum speeds.

Plus-Size Passengers

A journal article advocating the rights of plus-size passengers said 41 percent of Americans and 25 percent of all UK adults are considered medically obese. Besides that, it raised the question as to whether the individual is just "overweight," "obese," or finally "morbidly obese" with the latter not only needing extra space to be able to sit in the aircraft but classed as handicapped and able to benefit under current laws to protect those with disabilities.

Indeed, the FAA has recently issued an order that toilets in new narrow-body aircraft and toilets newly installed therein be spacious enough and fitted with handrails for the handicapped.

Incidentally, **Sam Chui** has a short video showing an aircraft toilet design that takes up the space of a standard toilet but has sliding parts at the front so it can be "stretched" to provide space for a wheelchair-borne person and someone to assist them in privacy.

The question of passenger size and weight is a minefield, with flight attendants even claiming discrimination if deemed too heavy, and a dangerous topic to broach in a book like this.

US airlines pay lip service to fairness for the average passenger, claiming any passenger so large that they would encroach on the space of the person in the next seat will normally purchase two seats, or upgrade to Economy Plus or Business. At least one US major surreptitiously tells the very obese to purchase two seats on the understanding they will be reimbursed for the extra one afterwards.

Simon Calder, Travel Correspondent for the UK's *The Independent*, has pointed out that Air France has come up with what is perhaps the most logical and fairest solution saying that the plus size should purchase two seats on the understanding the second will be refunded if there is a spare seat, meaning the aircraft is not full.

However, in practice, two enormous people traveling together simply avoid sitting next to each other and often book an aisle seat and a window seat in the hope or expectation that the middle seat will be left empty or later vacated. If the unfortunate in the middle seat cannot move because the aircraft is full, their journey can be hell, even to the extent that their physical well-being or mental health may be compromised.

Quite a few years back, a woman traveling on Virgin Atlantic from London to Florida was unremittingly crushed by an obese woman—whose husband had chosen to sit in the row behind—with the result that her ribcage was displaced, and she had to be treated by a doctor on arrival.

The American doctor recommended she be upgraded on the return flight to help her recover, but the airline would not countenance it. Angry, she sued them and received very significant damages.

One solution would be to have an FAA regulation that a passenger has the right to "sole occupancy" of the seat they purchased and be able to claim a ticket refund should that not be the case. It could even be partial, depending on the degree of encroachment.

Those fighting for the rights of the plus-size passenger believe they are being discriminated against on the ground and are adamant they should not be required to pay for two seats or for extra-large ones in the air.

Safety

Air crashes are so few and far between in the US and elsewhere that little attention is paid to the safety implications of having more than a few ultra-obese or handicapped passengers abord.

Again, it is a sensitive topic, as exemplified by the fuss made some years ago when a Ryanair captain declared there were too many mobility-impaired passengers for safety aboard his plane and some had to be left behind.

Mobility-impaired passengers would not only find making their way to an exit in an emergency evacuation difficult if not impossible but could well block other passengers with them all losing their lives.

There are no easy answers except to say it may be safer in countries such as Japan with smaller and disciplined people—when an Airbus A350 caught fire on hitting a light aircraft on the runway on landing at Tokyo's Haneda Airport in 2024, everyone evacuated safely with, unusually, no one collecting their carryon luggage!

See **Weighing Passengers**.

PNF (Pilot Not Flying/Pilot Monitoring)

See **PF (Pilot Flying)**.

Pod Strike

Like a Tail Strike, a pod strike simply means an engine pod striking the ground with varying amounts of damage.

Pods under the Wings

These fairings are to cover the powerful mechanisms that move the slats and flaps.

Polaris Award [IFALPA]

Award recognizing “Acts of exceptional airmanship, heroic action, or a combination of these two attributes” given by the International Federation of Air Line Pilots’ Associations.

Pork Barrel

Derogatory term in the US for defense, agricultural, and infrastructure projects and the like paid for by all taxpayers but benefiting the electors of the Congressional member supporting the legislation allocating the funds.

Airbus counters Boeing’s objections to the governmental loans it receives by saying that Boeing benefits from pork barrel military contracts and funds received for research, not to mention the skewing of military procurement decisions, such as the unfortunate choice of the Boeing 767 as the next midair refueling tanker.

This adds to the cost of military procurement as work on a project must be spread over many electoral districts.

PPL (Private Pilot’s License)

License (certificate) allowing an individual to pilot certain aircraft not for hire. It requires a minimum of 40 or 45 hours’ flight (training) experience and involves written and practical tests.

PPRuNe (Professional Pilots Rumour Network)

Interesting UK pilots’ blog about topical aviation matters. See http://www.pprune.org/.

Pratt & Whitney

One of the big three aeroengine manufacturers, the others being General Electric and Rolls-Royce. All three have subsidiaries, sometimes in alliance with each other.

PRD (Pilot Records Database)

The Colgan Air crash* in February 2009 was the most notable example of crashes that could have been avoided in the US had airlines hiring pilots been able to learn their true history—failed tests and other problems. Congress finally ordered that the FAA maintain a pilot records database so that the person responsible, such as the chief pilot, at the airline could consult it on hiring a new pilot.

However, its implementation was dragged out and although that should have been done, a cargo flight, Atlas Air Flight 3591 flying from Miami to Houston on February 23, 2019, crashed largely due to the hiring of a copilot with a deceitful résumé and a history of failures, including a tendency to push buttons at random in a crisis without thinking, to show he was doing something.

See https://www.youtube.com/watch?v=keQSpUq6Vis.

Preighter

A new term to describe passenger aircraft used as freighters with the seats in the cabins removed and replaced with freight. Differs from a "combi" where part of the passenger cabin is permanently blocked off for freight.

With the higher prices for freight even some of the latest ultra-efficient aircraft such as the A350 were so adapted at the height of the Covid-19 crisis.

Prevailing Wind

The compass point *from* which the wind is blowing is defined as its direction. Therefore, a north wind is one that is blowing from the north towards the south and easy to remember because it is thought (in the Northern hemisphere) as being very cold.

Winds are caused by a combination of factors, the most important being that the sun heats up the terrain and air most at the Equator leaving the poles cold. Air rises at the Equator and descends at the poles and this effect combined with the rotation of the Earth produces winds. See **Coriolis Effect**.

In the case of eastwards and westwards routes there can be a difference of more than an hour between the average duration of the outbounds and inbound legs.

See **Headwind/Tailwind**.

Project Sunrise (Qantas Ultra-Long-Haul Flights)

After a delay due to Covid-19, Qantas is going ahead with its planned Airbus A350 nonstop flights to London and New York from Sydney and Melbourne on Australia's East Coast. It is called "Project Sunrise," apparently because two sunrises might be seen.

A significant number of parents with young children apparently liked the Perth–London nonstop service as it avoided the stress of taking little ones who had finally fallen asleep out to the terminal on the refueling stop—children weighing little could be very profitable passengers on ultra-long-haul flights!

It is said the nonstop flights from Australia via Perth to London use 30 percent more fuel than those with an intermediate stop, but considering the costs of stopping at Singapore the overall cost is slightly less. At a guess, Project Sunrise flights further away from Australia's East Coast might use 40 percent more fuel than one with a stopover. Even so, they should be even more profitable with the A350 having more premium seats and with flights starting from Sydney, Melbourne, and Brisbane likely to provide more premium passengers. However, much will depend on there not being a great rise in the cost of fuel.

Propulsion-Controlled Aircraft (PCA—NASA)

There have been two notable and incredible cases where pilots have succeeded in landing aircraft without hydraulics by juggling engine power alone. One was the DC-10 that landed at Sioux City Airport after the engine in the tail disintegrated cutting all three hydraulic circuits and the other was the DHL Airbus freighter that had been struck by a shoulder-launched surface-to-air missile on taking off from Baghdad. Both feats are described in *Air Crashes and Miracle Landings*.

NASA subsequently developed software that could perform this almost impossible task for the pilots, whereby they told the flight management computer what they wanted the aircraft to do (without hydraulics), and the computer would adjust the engine power accordingly. It was called PCA—propulsion-controlled aircraft. NASA even installed it on a DC-11 and found it could make a pretty good landing without using hydraulics. However, the FAA decided that instances of total loss of hydraulics were so rare that it was not worth requiring its installation.

Psychometric Evaluation of Pilots

From February 14, 2021 EASA's regulations come into force regarding psychological assessment of pilots:

In compliance with EASA directive EDD 2018//012R (relating to CAT.GEN.MPA.175 Endangering Safety)

1. *AOCs must ensure that pilots have undergone a psychological assessment before they commence line flying in order to:*
    a. *Identify psychological attributes and suitability of the flight crew in respect to the work environment; and*
    b. *Reduce the likelihood of negative interference with the safe operation of the aircraft.*
2. *This should be undertaken at least within the past 24 months before commencing line flying unless it is deemed that a previous assessment is still adequate for the risk mitigation required by ORO.GEN.200(a)(3).*
3. *A psychological assessment performed by one AOC may subsequently be accepted by a different AOC, provided that the latter is satisfied that the assessment has been*

performed in accordance with AMC1 CAT.GEN.MPA.175(b).

4. *The regulation requires a psychological assessment process overseen by a qualified psychologist or someone under their direction.*

While the above shows how important the EASA thinks a psychological assessment of pilots is, much of it has been practiced for years by the best flying schools not wanting to have anyone who could damage their good name with the airlines that take on their students after qualification. Of course, the US military are a great fan of these evaluations for their pilots and have highly paid psychologists.

Assessment

We are not really dealing with what is separate, namely the *clinical* judgment to determine whether someone suffers from a mental illness such as schizophrenia, bipolar disorder, anxiety disorder, paranoia, or depression, though the latter can be treated with medication with the pilot allowed to return to duty subject to certain FAA conditions.

While it is one thing to determine someone's current state of mind it is quite another to predict what it might be in the future in certain circumstances. It has been suggested that AI coupled with close monitoring of pilots' actions and behavior could give warnings of problems. However, acting on them in the face of anti-discriminatory legislation and the need for proof would be difficult.

See "Pilot mental health—Can AI-powered psychological assessment help?" *AEROSPACE* Vol. 51 No. 6 June 2024, https://aerosociety.com (Royal Aeronautical Society).

~ Q ~

Q Codes (QNE, QFE, and QNH)

Remnants of codes left over from wireless telegraphy days, where a three-letter code would be used to represent a frequently used phrase or sentence. These used to be transmitted in Morse code, but because they are short and avoid misunderstanding, they are used in speech by air traffic controllers.

The three most used Q codes for altimeter settings are:

1. QNE, the regional pressure setting so the altimeter will indicate the altitude of the aircraft, namely the height of the aircraft above mean sea level in that region.
   Those are the heights shown on aviation charts for terrain, mountains, and high buildings and are important for avoiding obstacles not only on landing but also on a go-around.
   A mnemonic for it is "Nautical Elevation."

2. QFE (setting to give the height above a given datum, normally the airport or airfield). When landing, the pilot wants to know their height above the runway. When the aircraft is on it, the altimeter reading would be zero. However, it is not often required as the radio altimeter can tell the pilots that even more accurately.
   A mnemonic for it is "Field Elevation."

3. QNH is the standard reference pressure for setting the altimeter for flight levels and is always 1013.2 hPa. In the US and Canada this applies above the transition level at 18,000 feet. In the UK the transition level is 3,000 feet.
   A mnemonic for it is "Nautical Height."

See **Barometers (Barometric)**.

QACVR (Quick Access Cockpit Voice Recorder)

As early as 1970, a detachable cockpit voice recorder that can simply be read by connecting to a desktop has been developed but not much used on privacy grounds. Modern versions can hold 3,000 hours of cockpit voice and sounds with access possible via 4G, 5G, and Wi-Fi. While traditional CVRs run on a maximum two-hour loop and can even be erased by the pilots, these cannot be deleted.

QACVRs must now be fitted to all airliners in China which is hardly surprising as surveillance is the norm there. However, unlike the traditional CVRs these are not designed to survive high-G or fiery crashes or long periods under deep water. However, the ability to look back and see how a pilot in an accident performed in the past could provide some valuable information.

Qantas

This odd-sounding name for Australia's national carrier actually stood for Queensland and Northern Territory Aerial Services founded in 1921.

See **Operation Sunrise** for planned non-stop 20+ hour flights between Sydney on Australia's East Coast and London and New York.

QUANTUM PHYSICS, THE BASICS

"Quantum" is increasingly mentioned in the context of cutting edge domains ranging from the design of aircraft engines using quantum computers to quantum radar that tell whether the signal returned supposedly from the target has been interfered with.

We attempt to give some idea of what it is all about knowing that is hardly possible. Some things we mention without fully explaining them, so the reader can at least try looking them up on the Internet.

Quantum physics concerns the most elementary particles, such as photons and electrons which it sees both as particles and waves showing possibilities. These can have what to us are almost mystical properties, including the following phenomena:

1. Quantum Entanglement
   Instantaneous interaction between two particles even over vast distances. Since this is faster than the speed of light, it in theory raises interesting possibilities for lag-free communications.
2. Schrödinger's Cat
   Paradoxical situation of a cat being both dead and alive at the same time.
3. Heisenberg's Uncertainty
   Impossibility of precisely determining a particle's position and momentum simultaneously.

Key words in quantum physics applications are qubits, superposition, entanglement, and decoherance.

See https://www.technologyreview.com/2019/01/29/66141/what-is-quantum-computing/.

~ Quantum Computers

Classical computers, and even so-called supercomputers, use "bits" that either 1 or a 0.

Quantum computers use "qubits" which are subatomic particles such as photons or electrons. Not only are these extremely small, but they can also have almost mystical properties and by a process called "superposition" rather than simply being 1 or 0, can with "entanglement" be simultaneously in multiple states thus permitting exponentially greater computing power.

Methods of achieving this are in their infancy and include the use of supercooled circuits and constraining electromagnetic fields.

Before we mention the great advantages, we should mention the one great disadvantage of quantum computers is that "decoherance" means that they are more error-prone than classical computers and will not replace them. Decoherence is where vibration or a change of temperature, and the like, cause superposition to break down. Adding more qubits and having extra algorithms can compensate but make the quantum computer less efficient.

Companies in the aviation industry are claiming quantum computers are helping them in many areas including the development of new materials, design of parts and wings, and simulation of flight characteristics. Aeroengine maker Rolls-Royce claims it will help greatly with the design of its new engines, especially regarding the materials used and their longevity.

~ **Quantum Communication and Cryptology**

Quantum physics is used in two ways in encrypted communications:

1. To encode the encryption key with a system called Quantum Key Distribution (QKD).
2. To encode the entire message. This is also called quantum teleportation.

The great problem with quantum communication is that the photons are very sensitive and even with fiber-optic cables repeaters are needed perhaps every 30 or 40 miles, and these must de-encrypt and re-encrypt making them complex and a possible target for hackers.

Also, the possibility that quantum computers will increasingly be able to break the keys of material encoded traditionally (historically) poses problems for governments that have been keeping files electronically rather than on paper.

Since the setting up of communications networks is more obvious it is possible to see that the Chinese are very advanced in this field as in other quantum physics applications.

Although it is something of a longshot, there are suggestion that quantum phenomena could be used to communicate instantly—faster than the speed of light—at great distances, using quantum entanglement mentioned above under **QUANTUM PHYSICS, THE BASICS**.

~ Quantum Inertial Reference Systems

Inertial navigation systems for aircraft started with devices using mechanical "spinning top" gyroscopes. These have been superseded by more robust and accurate ring-laser gyroscopes. However, even those can sometimes drift by as much as 10 miles or 16 kilometers over a very long transoceanic flight, with the result that very accurate GPS has taken precedence and is even used to reset IRS systems.

However, GPS can be disabled, say, in time of war or spoofed, so that IRS which is independent of anything outside the aircraft would be better were it possible to make it much more accurate over long distances.

It seems that this may be becoming possible using quantum physics, and already a quantum IRS test device is being tested and giving almost unbelievable accuracy. The test rig took up three shelves of space making it impracticable for use in aircraft today, but as with so many modern devices it should be possible to miniaturize it, given time.

See **Inertial Reference System** and **ADIRU (Air Data Inertial Reference Unit)**.

~ Quantum Radar

The basic principle is to modulate the probing signal applying quantum physics and compare the return signal applying the same techniques to see if it is the same. The great advantage is that the quantum matching does not depend on the strength of either signal. Thus, in theory, the probe should not be strong enough to be detected and stealth aircraft are likely to show up regardless of how small their signature is.

See **Digital Radio Frequency Memory (DRFM) Systems [Electronic Warfare]**.

Quick Access Data Recorder (QADR)

A data recorder that can be accessed quickly and easily by maintenance staff to obtain data of value for immediate servicing, even revealing such details as tire pressures. Unlike the flight data recorder, it is not designed to withstand extreme heat, shock, and great depths of water, and has no locator transmitter. Mentioned in connection with the 100 mph Qantas overrun at Bangkok in 1999 where the Thai authorities were irked the airline had removed it.

QRH (Quick Reference Handbook)

This is a handbook (also in electronic form) that is always kept in the cockpit that can be accessed quickly and easily telling the pilots the procedure to follow in an emergency or an abnormal situation.

~ R ~

RAAS (Runway Awareness and Advisory System)

Radar (Radio Detection and Ranging)

A similar system to that used by bats, but using radio waves instead of sound waves, to detect objects and determine their distance.

The British did much of the initial development work on radar, referring to it as *radio detection and ranging.* The term "radar" was proposed in 1940 by Commander S. M. Tucker of the US Navy and soon adopted in the US. It was not until 1943 that it officially replaced the term "radio detection and ranging" in the UK.

Many very capable scientists worked on radar projects during World War II, with them conceiving devices able to produce high power at very short wavelengths featuring prominently in the latter part of the war. These were even able to detect a submarine periscope in the near distance.

Aviation radar developed during World War II depended on the detection of radio waves bounced back from the surface of an aircraft. When used after the war for civilian air traffic control, controllers would have to order an aircraft to make a turn to make a positive identification. Another problem was that the weakened signal bounced back from a single aircraft could easily get lost among the clutter also on the monitor.

It was then that another idea developed during the war, namely the IFF (identification of friend or foe) device, came into play. The IFF device was simply a transponder that on receiving the radar pulse replied that the aircraft was friendly. Adapting the principle for use in air traffic control, the transponder now not only indicates the identity (flight number) but also the height of the aircraft and whether the situation is normal or an emergency. Furthermore, the signal "squawked back" is many times stronger than one merely bounced off the surface of the aircraft.

Thus, ATC has two types of radar at its disposition:

1. *Primary radar* based on the old military principle of the aircraft's body reflecting the signal transmitted. This is the radar the military employs to detect enemy aircraft.

2. *Secondary surveillance radar* depending on a transponder on the aircraft to return an enhanced "echo" with valuable additional data such as identity, altitude, and status code—7700 for an emergency, 7500 for a hijacking, and 7600 for a radio failure.

It should be remembered that the secondary radar used by ATC relies on the data supplied by the aircraft's transponder, in turn fed to it by the aircraft's altimeter, so it is no good a pilot asking ATC to confirm their height, other than, say, in the event of their being blinded by smoke and unable to see their instruments.

In one tragic case involving a Peruvian airliner in 1996*, an aircraft polisher had left protective tape on the pitot and static port inlets (used for measuring airspeed and altitude).

Over the sea at night, with nothing visible and their instruments giving crazy readings, and myriad warnings sounding because the computers were receiving absurd data, the pilots asked ATC to confirm their altitude and were told they were at 9,700 feet as shown by their altimeter, when they were, in fact, almost at sea level and moments later flew into it. Their *radio* altimeter would have been giving the correct height, but with no mountains conceivably there and being inundated with spurious warnings, the pilots were not watching that.

Radar has one great military drawback, namely that switching it on makes the interrogator visible to the enemy. In peacetime it is generally switched off, and countries trying to detect the location of a potential enemy's radar stations sometimes send aircraft near them to provoke them to switch it on, or on the odd occasion persuade pilots of civilian airliners to stray into their vicinity—an extremely dangerous practice as the airliner could be shot down.

Although we are all familiar with the rotating radar antennas at airports and on ships, these are mechanically vulnerable and slow. Now, the great progress in electronics and software is allowing their place to be taken where necessary by phased-array systems, consisting of multiple fixed antennas triggered and scanned electronically to simulate rotating antennas.

Radial

VOR (Very High-Frequency Omnidirectional Range) radio beacons emit, among other things, two signals that when combined indicate the magnetic bearing in degrees of the aircraft from them, with each bearing quite sensibly being called a radial.

Radials are like the spokes of a bicycle wheel. An indicator in the cockpit shows whether the pilots are flying toward or from the beacon.

VOR beacons often indicate airways, and air traffic control can tell pilots to fly toward them after reaching a given radial (spoke).

Radiation (Ionizing Radiation)

This is a subject too complex to cover here and the best place for a comprehensive view is perhaps Wikipedia.

In the military–aviation context, the word "radiation" can be misunderstood in that an anti-radiation missile is not one designed to combat nuclear radiation but one that detects the radiation from a radar site and destroys it.

See **HARM (High-Speed Anti-Radiation Missile)**.

Radio Altitude as Opposed to Barometric Altitude

Altitude given by radio altimeter, which is especially important when landing since certain automatic features and alerts depend upon it.

See **Altimeters, Radio Altimeter and Barometric Altimeter**.

Radio Communications with Air Traffic Control

Pilots normally use FM to converse with ATC which has one disadvantage apart from only working within line of sight—that is not obstructed, say, by a mountain—namely that when two people talk together there was either a heterodyne making neither understandable as happened in the case of the worst aviation disaster ever at Tenerife or more often only one being heard. While nowadays there would not be a heterodyne, someone keeping their call link to ATC open could block other pilots.

This applies to other methods of communication including digital, with old-fashioned AM being the exception. In the case of AM, the sound waves mix just as if one were talking to someone in a room.

Rail Gun

The US Navy spent a considerable sum experimenting with a rail gun using electromagnetic forces to propel a projectile at such high speed that its kinetic energy alone would be great enough to destroy the target without an explosive charge.

Although it could be made to work, the tremendous amount of electrical power, either stored expensively or generated, was deemed to make it impractical.

Directed Energy Weapons seem more promising.

RAM (Radar Absorbing Material), for Stealth Aircraft

Polymer-based material that can be applied in the form of tiles or paint to the surfaces of stealth aircraft to reduce their radar cross-section. They are extremely expensive and must be maintained carefully. One problem is that their effectiveness varies according to the frequency of the radar beam.

Ram Jet

See **Hypersonic Missiles, Ramjet, Scramjet, Rocket**.

Ramp

Area where aircraft are parked, serviced, refueled etc. at an airport. Includes the loading and unloading of passengers and cargo.

Term derived from maritime world whare boats were hauled out of the water onto remps.

Ramp Freeze, Airport Brought to Standstill for VVIP

In the US, and especially since **9/11**, a “Ramp Freeze” can apply at an airport in the presence of a VVIP (Very Very Important Person), such as the president, in which case no foot, vehicle, or aircraft traffic will be allowed on the main terminal ramp, in or around the concourses or on the **FBO (Fixed Base Operator)** ramps.

SIDA doors, roll-up doors, vehicle gates, and GA hangar doors must remain closed during the freeze. Construction activities in the Airport Operations area must cease; equipment remains stationary, and contractors may remain in their enclosure, and indoors if possible.

During the ramp freeze window, no traffic is allowed to move except for the “One” aircraft. That would be mostly Air Force One, but also Japan One, Germany One, Shepherd One (the Pope), France One, etc. This might last 15 annoying minutes.

Ranging

The term used to describe equipment that determines *distance*.

For example, there is the VOR (very high frequency omnidirectional *range* beacon).

“Rapid Dragon” Air-Launch Cruise Missile Deployment [US]

Cruise missiles with pre-set target data and coordinates are loaded into containers and dropped from a huge transport aircraft such as the **Lockheed C-130** or **Boeing C-17**. On being jettisoned, parachutes on the containers deploy and hold them steady, whereupon the missiles are released vertically, and fly under their own power to their targets.

As a C-130 can carry up to five of these containers, with each having six missiles, a grand total of 30 missiles can be launched in a single salvo. Furthermore, this may be done relatively safely from outside the enemy’s air defense zone.

This is a cheap option using ordinary platforms.

RAT (Ram Air Turbine)

An electric generator with windmill vanes attached deployed from the fuselage to generate emergency electrical power by windmilling should the engines stop functioning. This simple device can be a lifesaver but originally only provided enough power for basic instruments as in the case of the Boeing 767 Gimli Glider (see **Accoladed Pilots ~ Captain Pearson**) incident.

Nowadays, large airliners have more sophisticated RATs as they must provide enough hydraulic pressure for the hydraulically controlled rudder, ailerons, and elevators. However, they do not provide enough power to raise the undercarriage immediately after takeoff as the terrible Air India Boeing 787 crash in June 2025 perhaps showed.

RDD (Radiological Dispersal Devices)

"Dirty bombs," which can be quite simple and small enough to be placed even in luggage.

Rear Bulkhead

The rear bulkhead on an airliner is like a champagne cork preventing the pressurized air in the cabins from flowing out of the tail. However, unlike corks inserted in the small-diameter neck of a bottle, rear bulkheads in airliners are usually just behind the toilets, where the diameter of the fuselage, including the cargo space under the floor, is still considerable. To minimize weight for something of such large diameter, these bulkheads are traditionally made of thin aluminum panels cut into complex shapes before being joined together to produce a dome that is concave when viewed from the cabin side.

Failure of the rear bulkhead can be disastrous as happened in the case of the worst-ever crash involving a single aircraft, which was Japan Airlines Flight 123 with 520 fatalities in 1985.

Rear-Mounted Engines

It was once the fashion to have engines mounted at the rear of the aircraft, partly to reduce cabin noise and partly because that would theoretically give a cleaner, more efficient configuration. Now the fashion is to have the engines in pods under the wings, which makes servicing easier and facilitates the installation of the new extremely powerful, large-diameter engines. One disadvantage of high-tail/rear-mounted engine configurations is that the tail drops in a stall, making it difficult, and sometimes impossible, to recover.

The rear-engine configuration was once much favored by Russian designers.

Reason, Prof. James

James Reason, Professor of Psychology at the University of Manchester, in the UK, has done much to change the way safety is viewed not only in aviation but also in many other industrial settings. He developed his model initially for the nuclear power industry. (See **Swiss Cheese Model**.)

Red-Eye

Colloquial US term for a flight departing between 9:00 p.m. and 5:00 a.m. Reporting of the O. J. Simpson trial made the term known outside the US.

Redispatch

See **Fuel Reserves**.

Because regulations mean a long-haul flight would have to carry an excessive amount of fuel, an airport short of its intended destination may be cited in the flight plan. Then on nearing it, usually with more than adequate fuel, it is "redispatched" to the intended final destination.

Reduced Vertical Separation Minimum (RVSM)

Although barometric **Altimeters** were very accurate at low altitudes, they were less so at higher ones, which meant the vertical separation had to be kept at 2,000 feet for safety. However, with increasing traffic and aircraft often funneled along certain routes, it often could not be handled leading to delays. An aircraft taking off, say, from London, and missing its slot might have to wait ages to be allowed takeoff as renewed permission has to be granted for it to overfly an area on its route thousands of miles away.

Fortunately, advances in technology have made it possible to accurately determine altitude even at high level and has meant that suitably equipped aircraft can fly with a vertical separation of 1,000 feet according to Reduced Vertical Separation Minimum (RVSM) rules. This applies between FL290 (29,000 feet) and FL410 (41,000 feet). The aircraft must have the right equipment and crew qualified for it.

Redundancy: FAA Definition, EASA Definition

Key systems on airliners in particular have redundancy incorporated so that if one fails there is another as backup, and often a third if that fails as well.

The FAA are happy if both or all three are the same, whereas Europe's EASA in some cases insists that they be from different makers, with the FAA having to concede.

Reference Design for New Aircraft, Airframe Optimization

When a new aircraft is designed there is a reference design with proportions optimized for a given wing length, engine size, etc. making it the most efficient on a per seat basis. To meet clients' needs the airframer will shrink or stretch the fuselage accordingly producing different variants.

When the fuselage is stretched the impact on efficiency is not great as when it is shrunk, for it still must carry those heavy engines, wings, and so on. Thus, very often, the smallest variant often fails to achieve significant sales.

Regional Turboprops

This domain is dominated by **ATR**. However, the Brazilian maker Embraer is consider introducing a competitive 70–90 seat turboprop with the claim there will be no middle seats.

Rejected Takeoff

Pilots may abort a takeoff for any number of reasons but are trained to only do so before reaching V_1, the takeoff decision speed. After attaining that speed, the aircraft should be able to take off and climb away even with only one engine of a twin-engine aircraft working properly. Taking off after V_1 with too little runway left could lead to a disaster.

Even aborting the takeoff on nearing V_1, powerful braking, generating much heat in the brakes, may be required to arrest the aircraft, which in turn means a considerable time must be allowed for the brakes to cool before a new takeoff attempt is made. There is also considerable stress on the undercarriage and braking mechanism.

Up until an airspeed of 80 knots, pilots may reject the takeoff even for relatively minor reason—a call out on reaching that speed is made—but for the reasons mentioned above a takeoff will only be rejected thereafter for something serious.

Relight

A jet engine that has flamed out needs to be relit to supply power again. Sensors detect the extinction, and ignitors provide a powerful electric spark to immediately relight it.

In the air, an engine can only be relit at certain airspeeds. On the ground, an engine is often started using bleed air from the APU (auxiliary power unit) to make it revolve, after which fuel is injected and ignited using the ignitors.

Repairs to Fuselage, Especially After a Tailstrike

The fuselage of an airliner expands and contracts every time it goes up and comes down in what is called a cycle. This results in the metal failing due to fatigue at places where bending is concentrated or where there are initially minute cracks.

Also, metal can behave strangely when continuously bent; anyone can see this by taking a paperclip, unwinding it, and repeatedly bending and unbending it. At first it can be done with little resistance yet remaining intact. Finally, it suddenly hardens at the flexing point and snaps.

The age of an aircraft with its fuselage made of traditional aluminum alloy is calculated in terms of the number of cycles it has undergone, and the more it has, the greater the attention that must be paid to finding hidden defects. Ironically, the "blessing" of no longer having smokers on board has meant cracks are more difficult to find as the yellow-brown coloration on the fuselage around one after many years of smoke-filled cabins could, and sometimes should, have revealed them.

Finding developing cracks under panels that have been placed over places that have been improperly repaired after a tailstrike is difficult and time-consuming, especially when a plate has been stuck over them. Ultrasound checks can help but are not easy to perform correctly.

One of the worst disasters due to covered up improper fuselage repairs was China Airlines (Taiwan) Flight 611, a Boeing 747-209B, on May 25, 2002*, killing all 225 people on board. The aircraft, delivered to the airline in 1979, had 64,800 hours of flight time at the time of the accident.

Some of the grazes from the tailstrike had been deep enough for the whole section to be cut out in accordance with Boeing's instructions. However, the area was just sanded down and a plate placed on top, without the operation being properly written up and giving anyone reviewing the repairs the impression it had been done properly in accordance with Boeing's recommendations.

Therefore, no one realized the potential danger, though following the crash an investigator reviewing earlier photos of the area did see telltale brown tobacco smoke marks even though smoking on airliners had been banned several years earlier.

The other accident, the worst single-aircraft crash ever, after improper repairs after a tailstrike was Japan Airlines Flight JL123*, again a Boeing 747. There it was not the fuselage but the aft bulkhead, a kind of plug made up of sheets of aluminum alloy joined together keeping the air in from the tail, that failed. It had been improperly repaired by technicians from Boeing following a tailstrike seven years earlier.

Interestingly, when Boeing and Japan Airlines were trading accusations as to who was responsible, one of Boeing's arguments was that Japan Airlines should have picked up on the brown tobacco smoke deposits where the bulkhead was about to fail. JAL countered that it was in a rather inaccessible place and there were other stains including those of coffee from the gallery. In the end they decided the slanging match was doing neither any good, and with JAL being a major client seeking peace was better. What JAL might have picked up, and prevented the worst ever single-aircraft disaster, was the whistling sound heard at the rear of the aircraft for several weeks prior to the disaster.

Replicator: Pentagon Plan for Thousands of Cheap Drones

A Pentagon program unveiled in 2023 claimed to be based on "Thousands of attritable autonomous platforms that will be characterized by being small, smart, cheap, and many."

The objective is to counter the presumed threat from China with more ships (at a time the US must deploy its assets in numerous theaters whereas China can concentrate on, say, one), more missiles, and more people.

Deputy Secretary of Defense Kathleen H. Hicks identified the particular challenge posed by China's rapidly diversifying **Anti-Access/Area-Denial Capabilities (A2/AD).**

Resilience

It is obvious that as a pilot one needs resilience to be able to deal with crises. In the broad sense it can cover the ability to deal with all sorts of situations.

Resilience is too broad a subject to cover properly in a book like this. However, the topic of resilience both in aviation and other domains including business is perhaps best covered by Richard de Crespigny, the captain of QF32 cited in the entry >>> ! **Famous Feats of Airmanship.** He has written a book called *Fly—The Elements of Resilience* covering resilience in multiple situations with the QF32 incident as starting point.

It has input insights from remarkable personages who have shown resilience. These include Sully Sullenberger of *Miracle on The Hudson* fame, astronaut Neil Armstrong, and NASA's Gene Kranz.

His book is available in Australia on https://www.amazon.com.au/Fly-Life-Lessons-Cockpit-QF32/dp/0670078735 but not readily outside. However, there is a YouTube promotional video for the audiobook https://www.youtube.com/watch?v=IGSpiXWNUwU that anyone can watch and really gives the feel of what happened, and is worth watching even for those mainly interested in QF32.

For seeing how to obtain Richard de Crespigny's book outside Australia see https://flythebookcom.wordpress.com/buy-fly/.

Resonance Phenomena

Unforeseen disasters particularly for bridges have occurred due to applied oscillations harmonizing with the natural frequency of the structure. Troops break step on marching across bridges.

These phenomena are better understood nowadays, but before they were, a very promising airliner, the Lockheed L-188 Electra—the only turboprop airliner made in the US (1957–1961)—suffered early accidents that affected sales prospects. The two most dramatic ones were due to vibratory harmonics (flutter) at high speed that no one could have predicted at the time, and which were easily dealt with once recognized. However, it should be noted that while these caused the wing to fall off, wings fell off in two other cases in storms—Braniff International Airways Flight 352 in 1968* over Texas and Braniff International Airways Flight 352 in 1971 where a teenage girl survived after falling onto forest treetops, spinning like a mulberry leaf in her seat. (See **Survivors Die Waiting for Help**.)

This can even apply to engines with the trigger outside. There were three instances of a fan blade failure in the low-pressure compressor of Pratt & Whitney engines on Airbus A220s. Tests identified a mechanically coupled mode excitation between the stage-one and stage-three rotors of the low-pressure compressor, driven by an "acoustic coincidence" with the 2.5 bleed-valve duct cavity. At high engine speeds in specific operating conditions, stated the inquiry, the low-pressure compressor rotor blade tips could cause turbulent airflow that generated an acoustic tone as it passed over the bleed air cavity situated immediately behind the compressor. This tone excited a stage-three rotor blade bending mode which was then mechanically transferred, through the low-pressure compressor module, to the stage-one rotor—generating a bending mode in that rotor which exceeded the stress limits on its blades.

The aircraft were able to keep on flying with those engines by having software modification including a thrust limitation at high altitude. These failures occurred following a change in the vane schedule, and engine-control software was also introduced to restore the original vane schedule.

Furthermore, Pratt & Whitney is also modifying the geometry of the 2.5 bleed-valve duct to increase frequency margins and eradicate the problematic resonant response. This shows how complicated these phenomena can be.

"Retard" [Airbus Auto-Callout Say at 20 Feet on Landing]

When Airbus airliners are in the final stage of a landing the synthetic voice calls out the height above ground in the case, say, of the A330, always adds the word "retard" at 20 feet radio height, regardless of the position of the levers. This is not an order but a reminder that the thrust levers must be pulled back to idle at some point.

Reverse Engineering

Where a country's or a company's technology lags, they can attempt to take, say, a competitor's aircraft to pieces and try to copy it. China has done this quite successfully with Russian aircraft. It has been suggested China even managed to get plans for the F-35 from Lockheed Martin's computers; however, there is much more than the physical aircraft to the F-35.

Russia did the same when Rolls-Royce supplied it with engines for MiGs in the Korean War, which so surprised the Americans.

Reverse engineering modern aircraft engines is very difficult as getting them to work properly depends on analytical techniques and experienced engineers. However, China is quickly catching up.

RIMCAS (Runway Incursion Monitoring and Collision Avoidance System)

System in the control tower that automatically warns the controller of possible collisions on the runway even in the case of an incoming aircraft risking hitting one still there or entering.

RFID (Radio Frequency Identification)

Tags, which are in effect mini-radio transponders, are increasingly being integrated into items—even the family pet—so that they can be identified by a reader in their proximity. Their use is being tried in aviation for identifying anything from aircraft parts to passengers' luggage.

Ring Laser Gyroscope (RLG)

In traditional gimbal-mounted gyroscopes, the spinning wheel remains at the angle it was to start with. This can be used as a reference for an inertial navigation system. However, it is delicate and liable to "drift" over time. Fortunately, it was found possible to dispense with the mechanical system in favor of one using lasers based on a phenomenon called the Sagnac effect.

In the ring laser gyroscope two laser beams travel in opposite directions over a ring-shaped path defined by three mirrors. The gyro measures the frequency difference between the two beams to define angular velocity, which helps determine the direction that the aircraft or other platform is headed.

Honeywell is a top player, having produced half a million of its GG1320AN ring laser gyroscopes. It claims that advanced digital electronics have improved the gyro's accuracy to point where it typically "drifts" only 0.0035 degrees per hour, which equates to about a mile of drift per hour of flight.

Instead of having mirrors one can have optical fiber in coils with a total length as much as several kilometers. These are called fiber optic gyroscopes (FOG) and do not require precision machining to make and are cheaper. These are used for drones, missiles, and many other applications.

It is a much too complex and difficult subject to fully understand and to explain here, albeit fascinating.

RNAV (Area Navigation)

This was originally called Random Navigation, hence the "R."

Traditionally, airliners used to fly on airways between navaids (radio beacons). This worked well but meant they had to fly somewhat circuitous routes wasting time and fuel.

Now, with the advent of GPS or more officially GNSS (Global Navigation Satellite System) of navigation under IFR (Instrument Flight Rules) an aircraft can proceed on any path within the approved area with aircraft using RNAV.

"Roger"

In the early days of flying, radio communications were of poor quality and a simple word was needed to confirm a message was received and "Roger" was chosen as it was the phonetic code at the time for the letter "R" meaning received. It has been kept to indicate a message has been properly received, even though the modern NATO phonetic code for R is "Romeo."

Rollover (Helicopter)

Situation where one wheel or skid of a helicopter is snagged on an object, or stuck to the ground, causing it to tip and start revolving around that point. A stage is reached at which point no amount of control input can correct this, and the craft keels over disastrously. The only solution is for the pilot to push the craft down onto the ground—something they might be loath to do in a military confrontation, or when trying to pluck someone from a place where they have been involuntarily held.

Rollovers can also occur when the helicopter pilot is spatially disorientated in a brownout, as can occur when the downwash from the rotors blows up a cloud of sand, to produce almost zero visibility and no points of visual reference. Rollovers due to brownouts have been responsible for several helicopter losses in desert conditions, though with the helicopters being near the ground, they were not necessarily fatal for the occupants.

Rolls-Royce

UK engine-maker that almost went bankrupt in developing the RB-11 engine for Lockheed's **Tristar**.

Recently had problems over the possible engines for the Airbus A-350 1000 for Emirates where the hot sandy conditions might limit the time on wing. In the "dispute" Rolls Royce's decision to ensure contracts were profitable may have played a part. However, it seems Rolls Royce is succeeding in upgrading those engines to perform well in the harsh conditions Emirates aircraft are subjected to—high temperatures requiring maximum thrust settings, wind-blown sand, and short turn-arounds.

ROT (Rate of Turn)

GA aircraft flying relatively slowly usually have a turn indicator indicating a standard rate turn, which is 360 degrees in 2 minutes, and which is equivalent to 3 degrees per second.

Airliners do not really use standard rate turns as such but use the maximum bank angle setting, with normally a limit of 30 degrees for manual flight and 25 degrees for the flight director. Thus, the rate of turn would depend on the bank angle setting and, of course, airspeed.

ROWOPS (Runway Overrun Warning Overrun Protection System) [Airbus]

An automatic system that warns pilots as they approach the runway if it thinks the runway is too short by giving the visual and aural warning, "Runway Too Short!" in which case the pilots would normally perform a go-around. If the aircraft has touched down it would say, "Max Brake … Max Reverse!"

The calculation is done with a 15 percent safety factor, which might lead some pilots to ignore it.

However, a greater problem is that the runway lengths in its database are not updated for temporary situations such as shortened runways for maintenance or resurfacing, with this especially true for diversion airports. Of course, such information is in the **NOTAMS**.

RPAAS Remotely Piloted and Autonomous Aircraft Systems

This subset of UAVs (Unmanned Aerial Vehicles) is a domain of increasing importance with the need for legislation to ensure the safety of manned craft is not threatened.

The terms "manned" and "unmanned" had come to be viewed as politically incorrect with "pilotless" suggested, but "unmanned" seems to have regained favor.

See **Gender Inclusive Language, UAV = Uncrewed Aerial Vehicle**.

RPK (Revenue Passengers Kilometers)

Paying passengers multiplied by kilometers flown by them. FTK (freight tons kilometer) is the equivalent term for freight.

RTX (Raytheon)

On April 3, 1920, Raytheon Technologies and United Technologies merged to form RTX, making an even grander conglomerate, with the three major units, Collins Aerospace, Pratt & Whitney, and Raytheon.

Raytheon

Raytheon was founded in 1922 in Cambridge, Massachusetts, as the American Appliance Company initially concentrating on refrigeration before moving on to electronics. It developed a gaseous rectifier with an electron tube the company called Raytheon after "rays from God." In 1925 the name of the company was changed to the more catchy Raytheon Manufacturing Company.

By the 1930s it had become one of the world's largest manufacturers of vacuum tubes and in World War II became the key manufacturer of magnetrons benefitting from know-how gained from the UK. Interestingly, a member of staff, Percy Spencer, accidentally found microwaves used for radar could quickly heat up food and invented the microwave oven.

After the war, the company expanded into defense work acquiring the defense interests of various companies on the way, while divesting itself of most non-defense business. Following its great success with the Patriot missile, it became the beneficiary of grandiose radar related contracts.

United Technologies

In 1934, after the US government had decreed that aircraft and aircraft engine manufacturers should not also have holdings in airlines, the United Aircraft Corporation was formed from the breakup of United Aircraft and Transport Corporation, with the transport division becoming United Airlines.

It included Pratt & Whitney, Chance Vought, and Sikorsky. Manufacturing interests west of the Mississippi were hived off to become integrated with Boeing Aircraft. In the lead-up to the war there was much demand for items the corporation could supply, and far greater demand during the war with Pratt & Whitney especially benefitting on the engine front.

In the 1950s, Pratt & Whitney became very profitable thanks to its powerful J57 jet engine which for a while dominated the market.

In the 1960s, Pratt & Whitney supplied the engine for the new Boeing 747 but encountered problems partly due to the large diameter of the engine meaning it flexed and the fact the aircraft was somewhat underpowered. Eventually, the problems were ironed out with the engine proving successful.

In the 1970s, the United Aircraft Corporation became United Technologies Corporation that in 2020 merged with Raytheon Technologies, which ultimately became RTX.

Rudder

The rudder is the wide expanse of airfoil behind the vertical stabilizer (tailfin) that makes the aircraft yaw to the left or right as its name implies.

Surprisingly, it is little used in turning the aircraft for that is primarily done by using the ailerons and spoilers to dip the wing in the direction one wishes to turn.

Though not much used in normal operations, it is relatively large in area so that at low airspeeds such as when landing and taking off it can correct the yaw due to asymmetric thrust should an engine fail. It is also used when landing in a crosswind to partially compensate for the deviation. In cruise it is trimmed so the aircraft flies straight.

However, overuse of the large expanse of rudder at high airspeeds can be disastrous, say when encountering wake turbulence, and can even cause the rudder to break off as happened to an Airbus taking off from New York's JFK*.

In trying to correct the jostle due to **Wake Turbulence**, the pilot will very likely make the degree of bank worse for their corrective action will come too late coinciding with the new direction of tilt thus accentuating it.

Some are surprised that rudders are so large these days. The reason is that with twin-engine aircraft as opposed to four-engine ones, failure of a single engine at takeoff can result in a very strong pull to one side which needs a large area of rudder to counteract it at relatively low airspeed.

~ Rudder and Nosewheel Steering on Runway

On landing. pilots have to ensure they get the nosewheel properly down on the runway well before the rudder becomes ineffective due to loss of airspeed. Should an engine fail with the nose up in the air and not enough speed for the rudder to be effective, the aircraft could veer off the runway.

Ruddervators/V-Tail

Empennage (tail assembly) where the traditional rudder and elevators are combined in a V-shape. The control system is complex and therefore rarely used. The main advantage is a military one in that the radar echo is reflected at an angle. Used in the F-117 Nighthawk stealth fighter and the MQ-1 Predator UAV (inverted V tail).

Runways and Their Markings

The length of runway an airliner needs to take off depends on not only its weight but also on the temperature and elevation—the higher and the hotter the location, the greater the length of runway required.

Dallas/Fort Worth, with seven runways and more planned, has some extremely long ones for this reason, with four being 13,400 feet (4,085 meters) long. JFK is famous among nonmilitary facilities for its 14,752-foot (4,441-meter) 13R-31L runway, which, being at sea level, is "worth" rather more than the equivalent length would be at DFW.

Major airports in Europe, such as Paris's Charles de Gaulle, and Frankfurt, have runways approximately 13,829 feet (4,215 meters) long. London's Heathrow's longest runway is merely 12,802 feet (3,902 meters) long, despite it being the world's busiest international airport.

RUNWAY MARKINGS 1

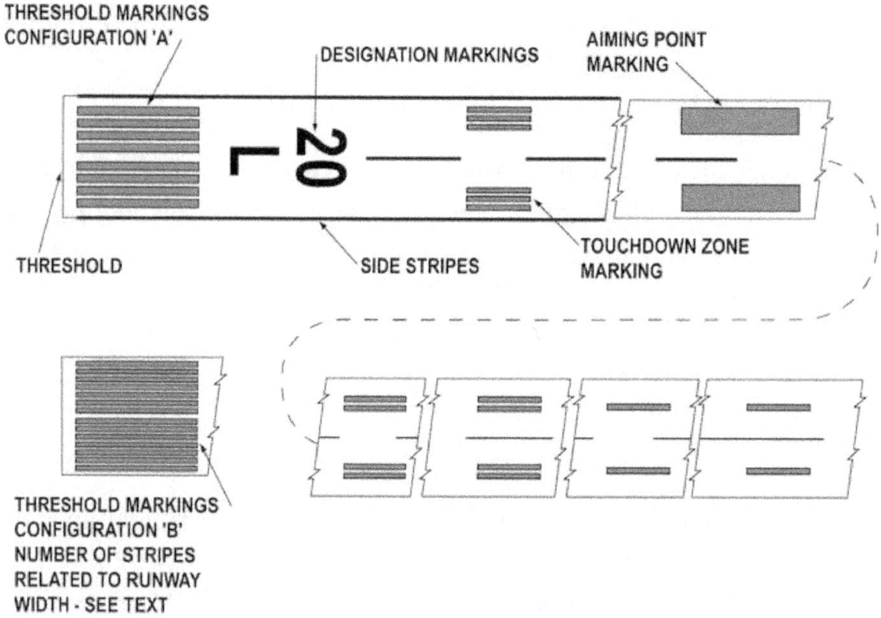

[FAA]

The most important markings are:

1. Designator
   At the beginning of every runway is a two-digit number indicating in ten-degree units, its magnetic compass orientation in degrees. Thus, a runway pointing due east

at 090 degrees is called *Runway 09*, except in the United States where it would be called 9 with the initial 0 removed. The other end of that runway, pointing due west would have the designator 27. Some airports have two or even three parallel runways, with the left one designated, say, 27L, the center one 27C, and the right one 27R. On the actual runway the letters (L, C, R) are placed underneath the numbers, not alongside.

There are some runways in places like the north of Canada where the magnetic bearing to the North Pole shifts considerably and it would be very inconvenient to have to modify the markings and charts every few years. The designation markings for these add the letter T to the three digit bearing to show the True North bearing. So Runway 32 might be 314° T.

2. Centerline
   A continuous dashed white line with the dashes often 120 feet long and the interval between them 80 feet.
   CAT II and II (see **Category**) runways usually also have embedded runway centerline lights for very low visibility conditions.

3. Threshold markings
   These long stripes in line with the runway mark somewhat like a barcode not only indicate the beginning of the runway but also its width.
   4 bars 60 feet (18 m)
   6 bars 75 feet (23 m)
   8 bars 100 feet (30 m)
   12 bars 150 feet (45 m) e.g., London Heathrow
   16 bars 200 feet (60 m) e.g., Darwin**
   ** Darwin's extra-wide runway was an emergency landing option for the US space shuttle.

4. Aiming points
   150 feet long prominent solid-white rectangles some way from the threshold designating the place to land.

On the grass or terrain off to the side of them will be two key pieces of ground equipment, namely the **VASI** or **PAPI** lights and the antenna for the glide slope.

5. Touchdown zone markings
   These are thin stripes before and beyond the aiming point. They are on either side of the center line indicating the aircraft can touch down there as well as at the aiming point. Sometimes there are three either side of the center line just before the aiming point, and a series of two or one after it, with the pair with one indicting touchdown zone is nearing its end. If the runway is long this touchdown zone can extend for the first third of the runway—airlines normally insist pilots go around if unable to touch down in the first third of the runway.

6. Other markings include ones for:

   a. A displaced threshold where the first part of the paved runway say cannot support the impact of repeated landings but can support an aircraft taking off with no impact;

   b. Green chevrons indicating a blast pad area or overrun area that is normally weak and can only be traversed in an emergency;

   c. Red chevrons marking an **EMAS** area consisting of special energy-absorbing material into which the aircraft wheels sink and crush it causing the aircraft to slow quickly in the case of an overrun that could be disastrous.
      The aircraft would not be able to exit from there under its own power, and the airline would be responsible for the expense of the replacement of the material.

Besides the runway markings there are those on the taxiways leading to and from the runway including the most important one indicating where aircraft should hold before getting permission to enter the runway.

Runway Excursion

A runway excursion is when the aircraft on landing or taking off departs from the end or off to the side of the paved part of the runway designated for use. There was a case in June 2020 when a FedEx DC-11 freighter landing at Mumbai in rain with a slight tailwind stopped on the paved part at the end of the runway, yet just beyond where it should have stopped. There were no casualties, no damage to the aircraft, and it was able to turn round and taxi. The incident was nevertheless classed as a "runway excursion."

While technology has meant incidents such as midair collisions or controlled flight into terrain have been virtually eliminated, this is not true for runway excursions. This is because there are so many factors involved, some of which are difficult to predict.

Runway Incursions

The improper or incorrect entry or presence of an aircraft, vehicle or person onto areas designated for the takeoff and landing of aircraft—possibly resulting in a major disaster because of the likely high speeds involved.

Runway Length, Longest Runways in US ###

How good a runway is for taking off depends not only on its length but also on the temperature and elevation (height above sea level).

Thus, the longest commercial runway in the US, the 16,000-foot runway at Denver International Airport, is so long because its elevation is almost 5,500 feet above sea level.

The Bay runway at New York's JFK is 14,572 feet long and at sea level meaning it is perhaps the best. It is so long because it has so many long-haul flights, not to mention having one so long is an asset with the airport subject to difficult weather conditions with a high volume of traffic.

Another busy commercial airport with almost as long a runway is Harry Reid International Airport in Las Vegas, with a 14,515-foot runway. This is needed because of the 2,181 foot elevation and high desert temperatures.

The longest in the world is at Shigatse Peace Airport (RKZ) in Tibet with a length of 16,404 feet, being at an incredible altitude of 12,467 feet. It is a dual civil/military airport with only 14 civil departures a week of which 7 are to Chengdu Shuangliu International Airport (CTU) in China.

Runway Safety Area (RSA)

According to the FAA, commercial airport runways should have an area 1,000 feet long and 500 feet wide beyond them to allow for overruns. However, many existing airports cannot meet this requirement because of encroaching suburbia and other reasons. (See **EMAS** and **Overruns**.)

Runway Visual Range (RVR)

Distance that can be seen looking down the runway, and a key element in determining whether a given aircraft can land. Constantly reported to ATC and pilots, nowadays mostly by means of automatic equipment.

Roster

All pilots should be legally qualified to fly the aircraft—and able to do so on their own if necessary. However, pilots can be confronted with emergencies or difficult situations, and their combination of qualities, combined with **CRM (Crew Resource Management)**, is critical for safety. And that combination depends on how they are rostered.

For example, a pilot with relatively little experience on an aircraft type should be rostered with someone with considerable experience. Surprising as it may seem, having two extremely senior management pilots rostered together, as often happens when a VIP is to be a passenger, is dangerous. In some cases, they might think they know it all and there might be a poor command structure, with both acting like prima donnas.

Another aspect of rostering is the scheduling. Although low-cost carriers in, say, the UK and Ireland have schedules that work the pilots hard within the regulations, it is reassuring that their route structure is such that most pilots manage to sleep properly in their own beds at night, which is not necessarily true in the United States, with its much greater distances, where some pilots must find crash pads or slumber in crew lounges.

RUSSIA AND EASTERN EUROPE including Ukraine

Russia had some very good aircraft designers and rocket engineers coming up to the end of **World War II**, but Stalin had treated them abysmally with some sent to camps in Siberia or made to work under strict political supervision. Towards the end of the war, they were, nevertheless, developing some very advanced aircraft which did not come to fruition because Stalin could see the war could be won without them.

Here are some particularly interesting Russian and Eastern Europe (including Ukraine) aircraft:

~ Antonov An-124 Ruslan 1986/55

One of the world's largest cargo aircraft, designed when Ukraine (where the Antonov design bureau was situated) was part of the Soviet Union. It is like the Lockheed C-5 Galaxy but able to carry a 17 percent greater payload. Able to "kneel" like the Galaxy for easier loading.

~ Antonov An-225 Mriya 1988/1

Six-engine giant cargo plane developed from the An-124 that turned heads wherever it went. This, the largest cargo aircraft in the world, was used on special assignments for transporting massive items, such as 150-ton generators, or relief supplies in emergencies.

Although only one has been completed there is talk of a second partially completed one that was possibly going to be finished for

China. However, parts from it could be used to repair the one wrecked in 2022 during the "war" in Ukraine. There are doubts as to whether that would be financially viable.

~ MiG-15 1949/13,130 Plus 4,000 or More under license

In 1946, Stalin, ordered the Mikoyan design bureau (that he had shortly before established with Mikoyan in charge) to create a fighter that could fly high enough and fast enough to intercept US bombers carrying nuclear weapons attacking.

Mikoyan's earlier copies of the German's Messerschmitt Me 262 and reverse-engineered German jet engines had proved disappointing.

The Russians like the Americans (see **Operation Paperclip**) had German experts and even their latest aircraft. While Mikoyan admits his bureau looked at many in designing what was to become the MiG-15, it is possible that he benefitted most from a design that the Americans did not have that never flew but had been tested in Germany's advanced supersonic wind tunnels. That design had had sweptback wings as did the great design he finalized. He still needed a powerful and reliable engine.

Fortunately, Great Britain, impoverished by the war, was willing to consider allowing Rolls-Royce to sell such an engine to them. With Stalin doubtful, saying, "Who in their right mind would give away their technology," the Mikoyan people visited Rolls-Royce and their executives' homes, playing billiards, with the winner of the next shot deciding whether the deal would go ahead.

Mikoyan won.

Even before the Russians received an engine, they knew the alloys used because a member of the Mikoyan staff visiting the Rolls-Royce machining shops had special shoes with sticky soles to pick up debris from the floor, which they duly analyzed back in Russia.

With the Rolls-Royce engine which they reverse engineered and produced in great numbers and never paid for, they ended up with a superb all-round fighter albeit one initially destined to be bomber interceptor. It even had a pressurized cockpit, and ejection seat, a first for Russia.

The Americans were staggered and angry when, soon after the outbreak of the Korean War in the early 1950s, they found themselves up against superb MiG-15s with what appeared to be Rolls-Royce Nene or Derwent engines able to climb to 50,000 feet, far outshining their B-29 bombers, and even their fighters until they came up with the Sabre, which was just about equal, but not quite, to the MiG-15.

~ Su-34

But the Su-34 is a different breed, often tasked with missions that can last for many hours, and sometimes even beyond a day.

The toilet, or "sanitary container" as it's referred to, is a basic but crucial amenity. It's a simple setup, but it addresses a fundamental human need, especially during prolonged sorties where pilots are required to stay in their cockpit for extended periods. The galley, albeit rudimentary, usually consists of a heating unit to warm up food, ensuring that the pilots can maintain their energy and alertness levels without the need to return to base for sustenance

~ YARS ICBM

Though not as powerful as the new **Sarmat** below, Yars missiles use solid fuel making them easy to transport and be quickly launched *in situ*. While they can be launched from an underground silo, they can also be fired from a mobile launcher able to negotiate roads and forest tracks.

This makes them a formidable weapon for not knowing where most are would make it virtually impossible for an opponent to take them out preemptively.

They have a range of at least 6,800 miles (11,000 km) and can carry three to ten **MIRVs (Independently Targetable Reentry Vehicle)** nuclear warheads) *each* with a power between 500 to 150 kilotons.

Note that the bomb dropped on Hiroshima had a yield of about 15 kilotons and that on Nagasaki about 20 kilotons of TNT equivalent, so just one, say, of the ten MIRVs would have ten times the power of that.

~ Sarmat RS-28 (Satan 2) ICBM (Silo-Based ICBM)

Russia wanted to fast track the Sarmat, their latest heavy ICBM, because their previous one, the R-36 Voevoda (NATO reporting name: Satan 1) that was first brought into service in the 1970s was made in the Ukraine and despite upgrades was expected to be taken out of service in 2022. However, the Sarmat development program has been plagued by problems with it only put into service in limited numbers and not thoroughly tested.

It has a mass of 208 metric tons (of which fuel may represent 178 tons), a diameter of 3 m, and a length of 35 m. It can use the same silos as the Voevoda (Satan 1), which are well protected and can stand up to nuclear explosions.

Like the method used for launching missiles from submarines, the Sarmat is lifted out of the silo by a gunpowder charge to a height of some 65 feet above ground where it stays suspended to allow the four first-stage engines to start up and send the missile on its way.

For guidance it has inertial, GLONASS (Russian GPS), and **Astro-inertial**.

The second stage accelerates the missile with it controllable in a flat trajectory, giving it greater range and making it difficult to detect.

The range is said to be 11,000 miles (17,700 km), with it able to hit the US from any direction including from the South Pole unlike from the North Pole where it would encounter traditional antimissile systems. Only an impossibly costly missile defense system all round US coasts and borders could possibly counter it.

Sarmat was expected to carry 10 warheads with an explosive power of 550 kilotons each. Other sources cited 16 warheads on the Sarmat missile.

Having a payload exceeding 10 tons, it can have various combinations of warheads ranging from 10 large ones to 16 smaller ones, 3 hypersonic boost-glide vehicles which can maneuver at low altitudes, not to mention countermeasures and decoys. With the warheads independently targeted, a large area can be subjected to the effects of a nuclear attack.

See **Missile Defense**.

~ Avangard Hypersonic Glide Vehicle

Can be mounted on an **ICBM (Intercontinental Ballistic Missile)** and can attain speeds up to Mach 20 and can change direction mid-flight making it almost impossible to intercept.

~ Ilyushin Il-80 Russia's Doomsday Aircraft

With no cabin windows and shielding to resist nuclear blasts and **EMP**s, it allegedly can stay airborne for up to one week with refueling. One engine is kept on standby, with it ready whisky off top officials in a crisis.

Sometimes referred to as the "Flying Kremlin," it not only protects top officials such as President Putin but serves as a mobile command post, very much like the US's **Doomsday Plane**.

Russian Airspace Not Available for "Western" Carriers

The unavailability of shortcuts through Russian airspace has meant, especially for European carriers, having to take a complicated southern route to airports in the Far East. These southern routes can be made even more circuitous due to having to avoid trouble spots.

The extra range required has increased the popularity of versions of the Airbus A350, able to fly somewhat further than the Boeing 787 that before could cover most routes.

Particularly hard hit for flights to the Far East is Finnair, based right up in the north having to fly southwards instead of eastwards. Even British Airways says the extra distance makes it unprofitable to fly from London to Beijing, though the real reason may be that it is competing with Chinese airlines able to fly through Russian airspace making their flights not only cheaper but shorter.

RVR (Runway Visual Range)

Visibility measured along the runway, used as an indication as to whether it is possible to land or take off. Now measured automatically.

RVSM (Reduced Vertical Separation Minima)

Aircraft with suitably accurate controls (as regards altitude and lateral position) and reliable systems to warn them of any deviation are allowed to fly with only a vertical separation of 1,000 feet, above usually 28,000 feet in the US.

Mentour Pilot (see **How to Find Information about Aviation**) has an illuminating video (*THROWN to the SEA like a PAPER AIRPLANE!*) explaining this and how a Bombardier business jet was turned over by the wake turbulence generated by an Emirates A380 superjumbo flying towards it and just 1,000 feet above. The A380, fully laden with passengers and fuel for a long-haul flight to Australia, was producing much wake turbulence even in cruise.

~ S ~

Sabotage (First Proven Case)

The first proven case of sabotage of an airliner was on October 10, 1933, when an explosive device (highly likely in the luggage compartment) brought down a Boeing 247 operated by United Air Lines as the airline was then called. It blew off the tail, which was later found virtually intact a mile away from the main debris. The explosive seemed to have been nitroglycerine.

The aircraft was on the Cleveland-to-Chicago leg of a transcontinental flight originating in Newark in New Jersey and bound for Oakland in California. The actual crash was near Chesterton, Indiana. The four passengers, two pilots, and one flight attendant were killed.

Investigators never even found a suspect, leaving the possibility that someone just happened to be transporting the unstable explosive.

SAF (Sustainable Aviation Fuel)

This is fuel made from material that has recently removed carbon from the atmosphere unlike fossil fuels that removed it millions of years ago. Feedstock can be cooking oils and fats, municipal waste, and non-food crops.

SAF is more expensive than fuel from fossils and that is one reason why it is not used so widely. It is normally mixed 50/50 with traditional aviation fuel and it is claimed it could reduce CO_2 emissions by up to 80 percent.

Safest Seat

Crashes are so few and far between, and the circumstances so different, that statistically calculating the safest seat is not so meaningful. Also, with the risk being minimal, driving to the airport or falling down the steps when boarding or deboarding is 1,000 or more times more likely to be associated with harm.

However, if you want to increase your chance of survival in an unlikely crash, statistical analysis has shown that an aisle seat and more importantly having one within five rows of an exit (emergency or otherwise) is safest. A study of some accidents such as those where there is a fire, keeping one's wits and having the emotional *will* to survive are factors making survival more likely—some people freeze and in those few seconds miss their chance to escape.

Backwards-facing seats would be safer but do not really appeal to passengers, but, more importantly neither to airlines, for the floor attachments would have to be very much stronger meaning considerable extra weight. One argument against them is that flying objects might well hit people in the face.

Sitting in the rear third of the aircraft is slightly safer as that part tends to be strong and the concertinaing front may absorb some of the impact. Wearing "sensible" shoes such as trainers makes evacuation easier and protects one on the ground where there can be sharp objects often hidden in foam underfoot. See **Evacuation**.

Remember, the floor on the aircraft can be very dirty.

Also, it is advisable to wear comfortable, ideally cotton, clothes that cover bare skin. Professor Muir at UK's Cranfield University used to spice up her erudite talks on passenger survivability in crashes—a subject she had assiduously researched—by saying women should avoid wearing nylon panties (knickers). These are liable to melt due to the heat generated by friction on going down the slide in an emergency evacuation causing deep difficult-to-treat burns on their bottoms.

Rumor has it that the **Brace Position** is designed to kill passengers outright so the airline pays less compensation than they would for someone merely injured and needing lifelong care. This is not true and learning the brace position can be a lifesaver for passengers for it can mean the person suffers less harm and is more likely to be able to save themselves.

Since most accidents occur when landing or taking off, booking a nonstop direct flight is more meaningful safety-wise than booking what might be the "safest seat."

In recent years there have been quite a few emergency evacuations due to fire with occupants barely getting out in time with many passengers shamefully collecting their luggage. When an Air France flight overran the runway in Toronto and ended up on fire in a gully, a passenger blocked an aisle as he unpacked his suitcase despite the remonstrations of a flight attendant, forcing her to take the people she was guiding another way. In the end the last person got out with only seconds to spare*.

It has been suggested that overhead bins should lock automatically to prevent that, but this would mean extra weight and complexity.

At the beginning of the section "OTHER LINKS" on our website (https://flyingknowitall.com) there is a link to "BE FIT TO EVACUTATE" with interesting and valuable information about the safest seat and what to do in the event of a crash. It shows the potentially life-saving brace position even for seats so close that you must rest your head against the seat in front.

Danger of Infectious Diseases

With the chances of a crash being so remote, it might be worth considering the risk of infection, bearing in mind selecting a window seat reduces the risk of someone coughing over you on their way to the toilet. Sitting in an aisle seat at the front of the cabin (if the toilets are at the back) could be a wise choice, giving the best of both worlds.

If someone near you is coughing, try to move seats—it is said having an infectious person within two rows increases your risk to 80 percent. If your seat has an individual air vent above it, set it so it blows on your face and other people's germs away. Effective disinfecting wipes for tabletops and arm rests can help, as can using paper towels to avoid directly touching toilet latches, etc. Of course, traveling Business Class with greater separation is a plus in this respect.

Safety, "Accidentology"

The study of accidents and their avoidance is a fascinating subject demanding expertise in domains ranging from psychology to IT and management. The French have the term "accidentologie" for it. Theories and models for accidents and ways to prevent them often evolved in the context of the nuclear power industry, where an accident could have terrible consequences, and have been later applied to aviation.

Many valuable lessons have been learnt, and aircraft are now not only unlikely to crash but incorporate features that increase the likelihood that people will survive any crashes that do happen.

Surprisingly, some parts are made weaker for safety. Very often, photos of an aircraft that has crashed or touched down heavily—say, before the runway—with no one seriously hurt show the landing gear (undercarriage) lying close by. This is because it is designed to break off and not pierce the wings, containing vast amounts of fuel, which very likely would start a terrible fire. Even the engines are designed to break off should an exceedingly strong force push them backwards, as might happen if the aircraft ditches in water at high speed; otherwise, the aircraft might cartwheel.

Safety Briefing

Passengers should pay attention to the briefing explaining the brace position and so on. The airline should explain why the lights may be dimmed at takeoff and landing so people's eyes can get accustomed to the dark and that the blind be raised so it is possible to see the situation outside and notably not open a door where there is a fire, or an engine running that might suck people in.

Safety of Particular Aircraft

Comparing the current safety of airliners is difficult because there may have been accidents unrelated to the aircraft itself such as the terrible one at Tenerife where two Boeing 747s collided on the runway in fog*. Then there are cases such as the Boeing 737 MAX where the aircraft itself was intrinsically at fault but where it should be quite safe now that the fault has been highlighted and rectified.

This means that historical statistics are often a poor guide.

Interestingly, the public's perception may not be a good guide.

The original DC-10 indeed had its faults, but the crash that truly ruined its reputation in the eyes of the public was the later one, where an engine fell off when taking off. In doing so it damaged hydraulic systems in the wing causing that wing's flaps to retract. Then that wing stalled when the pilots reduced airspeed according to standard operating procedure, with the aircraft flipping over and crashing disastrously. The engine fell off because a pylon fixture was damaged due to maintenance not carried out according to the manufacturer's instructions. In fact, the DC-10 ended up a very safe aircraft. See **"Faster-May-Be-Safer."**

Safety of Particular Airlines

Comparing the safety of airlines is tricky, too. Even airlines that once had an accident-prone culture and a run of accidents can turn around and become very safe. Take the case of Korean Airlines, which changed its culture under the tutelage of Delta.

See **Hierarchical Gradient between Pilots** and **CRM**.

Safety Pilot

This is a pilot able to act as pilot in command (PIC) in the event of the pilot flying (PF) becoming incapacitated. They must have all the qualifications, including medical, to fly the aircraft in question and that aircraft must have dual controls.

Sometimes elderly citizens unable to pass the FAA medical test will opt to fly with a safety pilot. While the person being covered does not have to reveal full medical details, they must inform the safety pilot of the likely nature of any incapacitation.

Salary Pilots, etc.

As mentioned elsewhere, salaries at airlines depend on seniority rather than ability.

It has been suggested that the extremely high salaries granted to senior pilots at the major airlines in the US may lead to airlines seeking ways to get rid of them, first having just one pilot with backup pilots on the ground handling multiple aircraft.

As the advantages of having two people in the cockpit are obvious for crosschecking and restraining suicidal pilots, one could have perhaps less highly qualified people such as cabin attendants with extra training doing the simple tasks.

Sam Chui

Probably the most successful flight-reviewing aviation YouTuber, and so successful he is often not only invited on inaugural flights but also on private jets including the new Dassault Falcon 6X. However, this does not prevent him exploring less glamorous aspects of aviation such as what happens in North Korea.

He combines a friendly easy-going laid-back style with great knowledge and experience, not to mention love of everything aviation. He has 3 million subscribers.

Sarin

One of the original nerve agents that the US started producing as early as 1953 and removed from its stockpile in 2023 having signed the Chemical Weapons Convention in 1993.

See **Novichok, Nerve Agent**.

SARPS (Standards and Recommended Practices)

SARPS are technical specifications adopted by the Council of ICAO in accordance with Article 37 of the Convention on International Civil Aviation.

Satellites

1972 was the year that satellites became good enough to make dangerous intelligence-gathering overflights of countries by, say, the U-2, less unnecessary.

See **Drones** and limitations of satellites.

“Scope” Agreement Limiting Seats Regional Airlines Allowed

These are agreements between the major US airlines and their unions aimed to protect their pilots from “undue” competition from so-called regional airlines with lower salaries and less strict working conditions. Ironically, these regional airlines are sometimes owned directly or indirectly by the majors and anyway mimic them as regards livery so the passenger may well believe they are the same.

The restrictive conditions include the number of seats, maximum takeoff weight (MTOW), the routes flown and their length, and notably their number. This means that airframers such as Bombardier, Embraer, and Mitsubishi must be careful not to exceed the number of seats, often 70 or 76, sometimes taking a large aircraft and stripping out some seats, though they risk problems over the MTOW.

This can lead to waste and inconsistencies as in the case of the Embraer E175-E2 designed for the heavier Pratt & Whitney GTF engine, which is more economical, quieter, and less polluting than the GE CF34 on the venerable old E1s. The extra weight of the Pratt & Whitney put it over the MTOW Scope limit.

It has been argued that the current air traffic control problems in the US are because this Scope agreement means there are too many aircraft flying because larger more efficient ones are not allowed.

Screening of Passengers, Luggage, and Freight

One advantage of the much-hated screening of luggage and persons is that passengers cannot so easily bring dangerous items, such as Primus stoves, on board and even light them up, as has happened on some pilgrimages to Mecca.

Nowadays, the greatest danger regarding what passengers bring on board relates to items brought on board either to blow up the aircraft or to hijack it, with the possible intention of using the aircraft itself as a flying bomb. With aircraft and systems generally becoming safer and safer, the fate of today’s passengers may depend mostly on security measures, and consequently passenger screening has become a key facet of every flight.

The screening may begin even before the passenger reaches the airport, in the form of checking names against watch lists. In the case of US-bound flights, names of passengers and certain details must be sent to the US authorities before departure. If the passenger happens to have the same name as someone on a watch list, they may have to undergo extensive questioning or worse, leading to them missing their flight. This has resulted in absurd situations, with seven-year-old boys being treated as potential hijackers. In one case, a boy finally had to have an official insert in his passport certifying he was okay to prevent him being stopped every time.

Screening also takes place as passengers enter the airport, with their demeanor and even their route being a factor. Some of the techniques used are not new. Many years ago, customs at Paris's Orly Airport used psychology to arrange the paths through the customs checking zone so that one would be most likely to be chosen by someone with something to hide.

With "terrorists" able to devise all sorts of ways to defeat the physical examination of baggage and persons before boarding, the authorities may well find themselves fighting the last war rather than the present one. For instance, following the shoe-bomber incident in 2001, much attention has been paid to checking people's shoes, even though terrorists are unlikely to use the same method again.

A great tool was expected to be SPOT (Screening of Passengers by Observation Techniques), using a list of 94 signs of anxiety and fear, like lack of eye contact or sweating, conceived by Paul Ekman, a psychology professor at the University of California Medical School in San Francisco. However, when tested it only showed marginal success except where the signs were blatant. Were it integrated with cameras and instruments measuring temperature and pulse rate in combination with artificial intelligence, it might do better. Even then, there remains the problem that the cues vary according to culture—not a problem when used internally in China.

According to an article by Jessica Hullinger, "Another method of screening passengers is simply to talk to them."

A 2014 study found that asking open-ended questions—known as CCE (Controlled Cognitive Engagement)—is 20 times more effective than trying to monitor them based on behavior. For example, an agent might ask a passenger where they're traveling to and why before prodding them with a random question like where they went to college and what they majored in, then watch for signs of panic. "If you're a regular passenger, you're just chatting about the thing you know the best—yourself," says researcher Thomas Ormerod, PhD, head of the School of Psychology at the University of Sussex in England. "It shouldn't feel like an interrogation." In the study, officers using conversation-based screening caught 66 percent of deceptive passengers, compared to just 3 percent who used behavior-based screening.

Many believe terrorists will always be one step ahead, and so at airports and elsewhere great efforts are being made to develop software programs to identify automatically suspicious behavior using CCTV (closed-circuit television), heartbeat measurements, and so on. CCTV can identify suspicious behavior, such as *lurking with intent* in shopping malls, and would be useful in aircraft themselves. However, there is a long way to go before reliable systems can be developed, and there would be objections on human rights and privacy grounds.

One interesting technique is for security to expose passengers on the way to the gate to images and words of significance only to terrorists so that sensors in cameras focused on their faces can detect any consequential changes in pulse rate, body temperature, or blood pressure.

See **"SSSS" (Secondary Security Screening Selection) [US]** and **Behavior Detection Officers**.

Seaplane, Flying Boat, and Floatplane

A seaplane can either be a *flying boat*, where the hull (fuselage) sits on the water, or a *floatplane* where the aircraft is supported by floats with the fuselage out of the water.

Seat Belts

Clear Air Turbulence that could throw you to the cabin ceiling and dent it is always a possibility, however slight. Even if you are prepared to take that risk or allow your child to do so by not keeping your seat belt loosely buckled, there is another aspect to it. When you or your child falls back down, you or they could easily injure someone, even a child or baby.

Therefore, one should keep one's seat belt buckled, even loosely, whenever possible.

Seat Pitch

The seat pitch is the distance between a given point on a seat and the same point on the seat behind or in front. The amount of space the passenger has depends not only on the pitch but also on the thickness of the seatback. Some modern seats are made of high-tech materials and can be thinner, thus giving the occupants more space for the same pitch.

Seat Selection

People are often advised to avoid the seats right at the back of the aircraft where they are near the toilets with people standing nearby and also near the galley with a lot of activity there. They may be the last to be served with food with the likelihood what they want will no longer be available. Furthermore, the overhead bins may already be full of items belonging to the cabin crew. In addition, they will be more subject to being shaken around by turbulence, and lastly the last off the aircraft on landing.

That said, for those reasons, there may be more empty seats at the rear of the aircraft.

Middle Seat

Recently a captain addressing the passengers unusually explained in detail how they should conduct themselves vis-à-vis each other and obey the cabin crew in an emergency. One interesting point was his declaring the unfortunates in the middle seats had the right to *both* armrests. This seems a sound idea and perhaps should be made official. It could make their lives so much better at no cost to those on either side who would only have to lean the other way.

Often two people traveling together in Economy (Coach) will book an aisle and a window seat in the expectation that no one will want the middle seat and it will be empty, giving them extra comfort without paying. This does not work when the aircraft is full, in which case they have been known to give the unfortunate in the middle seat a hard time instead of graciously changing places.

See **Weight and Dimensions of Passengers.**

Seats in Airliners

Though there had been bunks and couchettes on airliners before now, and even in the golden age of the flying boat in the 1930s, the introduction of fully reclinable lie-flat seats was first introduced in first class by British Airways. Four years later they and then other airlines introduced them in business class.

Now some airlines have started equipping their business class with seats with partitions and (low) doors giving a great deal of privacy and making first class redundant. In fact, with sometimes extra charges for these, first class is virtually disappearing, apart from on airlines such as Emirates with a super-first class with shower and so on.

While passengers appreciate the doors, the airlines have mixed views with some, for now, firmly against.

SEC (Spoiler Elevator Computer)

Computers for instance on the Airbus A-320 that controls the **spoilers** and **elevators**. There are several so that if one should fail a second and even a third can take over.

Self-Defense Training for Cabin Crew

In June 2021, the FAA, concerned about the increasing number of cases of flight attendants subjected to aggression from passengers—with 3,000 incidents of unruly passengers reported by airlines since the beginning of that year—announced it was restarting voluntary self-defense courses for them. These are quite hard hitting with elbow jabs, eye pokes, and how to kick a passenger in the groin!

Self-Healing

Aircraft suffer minor damage from all sorts of things—hailstones, for example, and even from knocks from vehicles and equipment when parked. Now that composites are increasingly used in their construction, thought is being given to the idea of making aircraft self-healing, just like the human body.

One idea is to include hollow fibers with a resin that would seep out and repair the damage. It would include a dye visible in ultraviolet light to help maintenance check on it later. There could even be a resin circulatory system that could be topped up as needed, rather like a blood transfusion.

Sensors, Not Only the Eyes, and Ears …

Sensors play essential roles almost everywhere in aviation and aerial warfare, with them even embedded in the composite aircraft structure to sense when pieces might fail. (See **SHM**.)

Sensors can be anything that senses something and sends information, and in considering the use of **Artificial Intelligence (AI)** in warfare, drones are sometimes referred to as sensors.

Service Ceiling, 43,000 Feet (Airliner), 51,000 Feet (Private Jet)

The maximum height at which an aircraft is allowed, or is even able, to fly depends on many factors. See **Coffin Corner** and **Step Climb**.

In fixing the certified service ceiling at 43,000 feet for an airliner, and at 51,000 feet for a private jet, the need to be able to quickly descend to 14,000 feet in 4 minutes in the event of decompression is a major factor considered. A massive 747 cannot descend as quickly as a relatively small (expensive) private jet.

Sharing Expenses (GA Pilots Must Be Making Trip for Own Reason)

The FAA is concerned regulation provision §61.113 (c)/Part 135 allowing general aviation pilots to share expenses with passengers under certain conditions is being abused and to quote, "Putting the flying public in danger, diluting safety in the national airspace system, and undercutting the business of legitimate operators."

Simply put, a general aviation pilot may share expenses with passengers providing they are in the first place making the trip for their own reason—say visiting a sick relation—and they contribute a pro-rata share of the operating expenses. Furthermore, those expenses can only be for fuel, oil, airport expenditures, or possibly rental fees, with no element of profit involved.

SHM (Structural Health Monitoring)

SHM is an emerging technology whereby systems, such as sensor-actuator networks, might automatically assess the integrity of aircraft structures. According to *Flight International*, Airbus views SHM as a key to its "intelligent structure philosophy," and Boeing sees it as an emerging technology that may provide significant improvements in operational efficiencies.

Shoe Bomber (Richard Reid)

One unexpected advantage of the no-smoking rule in airliners was that it made someone trying to light some primitive explosive device more obvious.

This was so in the case of Richard Reid, the shoe bomber, who was found trying to light an explosive device in the heel of his very large shoe on American Airlines Flight AA63 from Paris to Miami on December 22, 2001.

Airport security had prevented the six-foot-four-inch (193 cm) Reid from boarding a flight the previous day because of his disheveled appearance and lack of check-in luggage. He had been handed over to the French National Police, who finally gave him the "all-clear" and the American Airlines ticket to Miami. He duly passed through the screening process, which at the time ignored people's footwear.

It was after the meal service that a flight attendant thought she smelled a match and went to see if someone was lighting up. A passenger indicated it was Reid, sitting alone, and she duly told him smoking was not permitted. He promised to comply, but a short while later she was annoyed to see him in a huddled position as if hiding something, which she thought might be a cigarette. He remained huddled and did not respond when she asked him what he was doing.

When she approached even closer and insisted, Reid tried to ward her off, incidentally revealing his lap, where he had a shoe with a fuse he was trying to light. She called for help, with Reid wrestling her to the cabin floor. With another female flight attendant joining the fray, she was able to slither away and get some water, which she brought back and threw in Reid's face. Passengers by then had come to the assistance of the second flight attendant, who had been bitten by Reid.

Reid was subdued, tied up, and given a sedative by a physician who happened to be on board. The aircraft was then diverted to Boston's Logan International Airport, where Reid was arrested.

Though the timely actions of the flight attendant are reported as having prevented all 198 people on board that flight from losing their lives, one could say it was Paris's airport security that saved the day. By not allowing Reid to fly the previous day, they forced him to wear his shoes overnight with the result that his sweat made the fuse too damp to light.

Reid was sentenced to a long term of imprisonment, but does his case teach us any lessons regarding what makes someone want to kill 198 people, including themselves?

He had a difficult childhood with his father in prison. He became involved in petty crime and was in and out of young offenders' facilities. He became a Muslim on the advice of his father, who said Muslims were treated better and got better food when incarcerated. The young Reid had a feeling of being discriminated against, attended a mosque where extremist views were prevalent, and began to express more and more extremist views himself. It was then that a scout probably identified him as suicide bomber material.

In an interview Reid said:

Of course, I would have been sad to have those people die, but I knew that my cause was just and righteous. It was the will of Allah that I did not succeed.

Showers (On Board)

Emirates installed a couple of showers in first class on some of its new Airbus 380s and had hoped to reuse the water for hand washing after filtering and treatment with ultraviolet light but could not get approval from the European authorities, despite tests showing there should be no problem. As a result, the showers are timed to last no longer than five minutes.

Aircraft toilets nowadays are very sophisticated and use vacuum flushing and little water. This has the advantage that they can use narrow-bore piping and be installed more flexibly, with less danger of leakages damaging the aircraft over time. Leakage of water, even from the galleys, can be dangerous since it can cause corrosion or cause avionics and electrical systems to malfunction.

SIDA (Security Identification Display Area)

These are sensitive areas airside, such as the flight ramps, at US commercial airports where only properly trained and approved—background checked—personnel may have access with identification showing them as such.

Sidestick, Sidestick Controller (SSC)

A feature of Airbus cockpits, where the traditional control column between the pilot's legs is replaced by a control stick rather like one for playing computer games situated at the left of the captain and at the right of the other pilot. A key point about the sidestick is that its position usually reflects what the pilot wants the aircraft to do rather than what the controls are doing.

See **Smoothest Landings**.

Sidesticks have the advantage of leaving a clearer view of the instruments and more space for additional controls but make it difficult for one pilot to see what the other is doing as is possible with the control column favored by Boeing. In the crash of **AF447*** where the pilot flying pulled back on his stick and stalled the aircraft (and kept pulling back, preventing it recovering), the nonflying pilot could not see exactly what he was doing. Linked sidesticks (sometimes called "active sidesticks") might have made a difference.

Linked sidesticks are being installed on some business jets such as Gulfstream G500 and G600 (linked electronically), while much more common in military aircraft. The Airbus sidesticks have a priority button, and a pilot pressing it for 40 seconds takes over with the other sidestick disconnected, though it can be restored with the priority button.

Though linking the sidesticks in airliners seems simple, it could introduce much more expensive complexity.

One point is that most of the maneuvers are done with the autopilot, not the control column or sidestick. In coming in to land, air traffic control will tell the pilots at some point to change course, in which case they just turn a knob to set the new course rather than maneuver with the sidestick.

In 2023 there was news that Airbus was considering using and retrofitting linked (active) sidesticks.

SIGINT (Signals Intelligence)

Intelligence gathered either through the capture and analysis of messages between people called COMINT (communications intelligence), or between machines called ELINT (electronic intelligence).

As the term SIGINT is sometimes used to refer to COMINT, there is room for confusion.

Traffic analysis, whereby much can be learnt just from the flow of signals even without breaking codes, can be surprisingly valuable as proved at Bletchley in World War II before the breaking of Enigma.

SIGMET (Significant Meteorological Information)

Warning issued by aviation meteorological service about suddenly deteriorating weather conditions. These can be mandatory with flying in certain areas (of the US) forbidden due to a dangerous storm and other conditions including volcanic ash.

Single-Pilot Operation

Fleeting single-pilot operation is already allowed, say, when one of the pilots needs to go for a toilet break. In exceptional cases even that poses problems, as in the Germanwings case where the copilot locked the captain outside the cockpit as he proceeded to intentionally fly the Airbus 320 into a mountain*.

Sixth-Generation Fighters, US, China, Russia, and Joint Projects UK, Etc.

While it is largely agreed that the current top-of-the-line fighters such as the US's F-22 and F-35, the Russian Su-57, and perhaps the Chinese J-20 are fifth generation, the classification over the years has differed somewhat according to the authority. Some fighters have been classed as 4.5-generation.

The sixth generation should represent a very great leap forward and though development programs have already been embarked on, significant entry into service is not expected until the 2030s. It is thought that air operations will change in nature with dogfights being passé since engagements between aircraft would be initiated with them far apart using missiles. There will be more emphasis on ground attack and reconnaissance. Aircraft will be designed to be able to operate manned or unmanned, and when manned the pilot will have an (expensive) helmet enabling them to see through the body of the aircraft in any direction.

Size of Pilot

Pilots come in all sizes, though perhaps not in extreme sizes.

Not only can the seats be moved backwards and forwards, and up and down for the best view, the unit at the pilot's feet with the rudder and brake pedals can be adjusted to be nearer or further down.

However, there have been cases where a very short pilot may be relaxing in cruise with their seat too far back to immediately seize the controls. This happened on an Aeroflot Airbus flight* between Moscow and Hong Kong where the captain allowed his teenage son to sit at the controls thinking the autopilot was looking after everything.

The strong boy exerted a very strong force on the sidestick causing the autopilot banking mode to disconnect as it was designed to do since it could be an emergency.

The boy continued playing around at the controls with the autopilot still managing the heading and pitch, and with neither the captain nor the short first officer realizing something was amiss.

The aircraft banked more and more with the boy who was obviously watching the instruments asking his father why the aircraft was turning, as is normal when an aircraft banks.

Finally, it was another pilot—the one who was operating camera from which we learnt what happened—who shouted a warning.

With the first officer too far from the controls to take over, various orders were shouted at the perplexed boy. Things went from bad to worse, and when the father finally managed to take the boy's place and regain proper control, they were dropping so fast it was too late. Had they been higher he could have saved them.

This is an example of a **"Cockpit Video Recorder"** showing why an accident happened.

Skidding and Sideslipping

If rudder is applied without any banking, the aircraft will skid outwards. Conversely, if the aircraft is banked without any rudder input, it will sideslip in the direction of the wing that is lower.

Usually, by coordinating the operation of rudder, ailerons, and elevators the aircraft can be made to turn without either happening and the passengers merely feeling slightly heavier.

Skunk Works

The term "Skunk Works" that started at Lockheed has now entered general usage to describe any innovative project using small teams of talented people working outside the normal constraints of the company hierarchy. Here we will limit our coverage to "Skunk Works" pertaining to Lockheed, now Lockheed Martin.

Towards the end of World War II, a small team of aircraft designers at Lockheed was secretly based in a circus tent next to a plastics factory on the outskirts of Lockheed's Burbank facility. They were working on a project so secret that staff were told not to reveal who they were when answering the phone. Unable to think of anything original they would say "Skunk Works" on account of the horrible smell wafting over from the plastics factory. Despite reservations on the part of the team leader, the name stuck, with it creating some of the most outstanding aircraft.

Though special work began with the P-38 Lightning in 1939, according to Kelly Johnson **Skunk** projects started with the P-80 Shooting Star in 1943, and after the war the U-2, SR-71 Blackbird, and F-117 Nighthawk. The same engineers worked on the F-22 Raptor and F-35 Lightning II with the projects being so extensive they can hardly be called skunk-like.

The P-80 Shooting Star was the US's first jet fighter. Starting in 1943, a select team under Clarence Kelly Johnson produced the first US jet aircraft in just 143 days.

Though it did not play a part in World War II, the P-80 continued in service in various air forces for some 40 years. With the US lagging in the domain of jet engines, it had to get the first engine from the UK.

A point that will be of interest to many in management is the philosophy of Kelly Johnson who made the Skunk Works so successful: "Have the fewest people possible but ensure that they are the very best."

The term "KISS" is also attributed to him, with various interpretations of which one is the real one, with it possibly being, "Keep It Simple Stupid!"

See under **US Military AIRCRAFT** for details of the aircraft he and the Skunk Works created.

Skyborg Weapons Systems

According to *Aviation* Week:

By fiscal 2023, the (US) Air Force expects to deliver the first operational versions of a new unmanned aircraft system (UAS) called Skyborg—a provocative portmanteau blending the medium of flight with the contraction for a cybernetic organism. The Skyborg family of aircraft is expected to fill an emerging **"Attritable"** *category for combat aircraft that blurs the line between a reusable UAS and a single-use cruise missile.*

Leidos is managing an autonomy mission system.

As the aircraft are developed, Skyborg will also serve as the test case of a radical change in acquisition philosophy, with ecosystems of collaborative software coders and aircraft manufacturers replacing the traditional approach with a supply chain defined by a single prime contractor.

SLOP (Strategic Lateral Offset Procedure)

The primary purpose of this procedure whereby aircraft may be given permission to fly some distance to the left or right of the proscribed track is to avoid the risk of aircraft flying on the same airways colliding. This is now more likely because GPS means they fly so precisely and any error by the controller could put them on a collision course.

Second, now that RVSM (reduced vertical separation minima) allows vertical separation of aircraft to be 1,000 feet rather than 2,000 feet, allowing aircraft to be offset from the centerline reduces the danger from wake turbulence. This is not possible on all airways.

Slot

One meaning of slot is the allocated takeoff time from an airport and agreed with ATC. The late boarding of some passengers can result in the slot being missed and the need to request a new one. This can result in a long wait, not only because of congestion at the airport in question but also because permission may be required to overfly zones thousands of miles away.

Another meaning of slot is the right for an airline to fly from an airport within a certain time frame (and, in turn, at a certain time). At popular airports, such as London's Heathrow, the slots are highly prized due to their limited number, and an airline possessing many slots, such as British Airways with just over 50 percent, is in a very enviable position commercially, as is Air France at Paris Charles de Gaulle.

See **Open Skies Agreement**.

"Smart Seat"

Looking up the term "smart seat" will take one to numerous entries about hi-tech designs for passengers' seats, and even seats with sensors that can note what the occupant does, and even their health. One idea was that they could detect their mental state, and if "shifty" flag them up to the crew as representing a possible threat.

An interesting but little-known meaning of "smart seat" is the one occupied behind or near the pilots by someone savvy who can give advice in crises, and who in some notable cases has saved the situation. This is not because the captain and copilot flying the aircraft are less than proficient but because overwhelmed by the task of handling the aircraft, they may not be able to see the full picture and analyze.

In the worst-ever multi aircraft crash at Tenerife* where two jumbo 747s collided on the runway the flight engineer in the "smart seat" had his doubts but instead of saying straight out that the other 747 might well still be taxiing in the fog down the runway, he expressed them as a feeble-sounding question.

SMO (Since Major Overhaul)

The term used in general aviation where regular overhaul, say, of the engine is very expensive and a buyer would want to know when they will have to do the next one.

Smoking on Airliners

Finally, smoking in airliners is forbidden all over the world with the last holdout being Air China and other Chinese airlines in 1917.

It was partially allowed in the US until the year 2000, with a period prior to that where it was forbidden on domestic flights but allowed on international flights over six hours.

Turkish airlines was the first airline to ban smoking in the mid-1970s.

See **Ashtray in Toilets**.

Smoothest Landings, the First Officer or Captain?

Passengers greatly appreciate a smooth landing, not knowing that a relatively hard landing might have been safer in a strong crosswind.

Someone has controversially suggested that younger pilots (who are generally first officers) should in theory be more likely to be able to make a smooth landing than captains. The reasons being that dexterity decreases with age, and at least on Airbuses, the sidestick is on the left of the captain making him or her use their left hand. Of course, how good someone is at using their left hand varies from individual to individual.

An ex-fighter pilot used to control sticks roughly centrally placed on various military aircraft, who then moved to Boeing airliners with control columns, and finally on to Airbuses in both the captain's and first officer's seat, said he had no difficulty in any of the situations.

It seems some people may be truly ambidextrous, while others find they are only at ease using one hand. Could training and practice overcome this in most cases?

Some years back, Boeing released an artist's conception of a grand cockpit for a fly-by-wire airliner which never came to fruition partly because of a drop in fuel prices. Interestingly, it had sidesticks rather than control columns, with the captain's stick on the right. Airbus must have thought that having the sidestick on the captain's left had the advantage of making access to the other controls easier.

It has been argued that since people's sharpness and dexterity decrease from their early twenties, pilots are at their very best at age 27 once they have gained enough experience! Of course, safety depends on many other factors where age and experience can be an asset.

SMS (Safety Management Systems)

A Safety Management System is a set of management practices and programs to proactively improve safety in fields like maintenance, airport management, and so on.

Accidents cited as ones that might have been avoided had SMS been employed include the Linate Airport accident in 2001*, where a straying Cessna wandered into the path of an airliner taking off, and the Taipei accident in 2000*, where a Singapore Airlines aircraft mistakenly took off from a disused runway and collided with construction equipment parked halfway down.

SOCOM (Special Operations Command) [US]

Founded following the disastrous attempt to free hostages held at the US embassy in Iran, where confusion between the different commands proved problematic. It oversees the Special Operations components of the US Army, Navy, Air Force and Marine Corps.

Somatogravic Illusion, Felt by Passengers and Pilots

This term describes the sensation that—without a visual reference—one is going up when accelerating and going down when decelerating. It is derived from *somato* meaning body and *gravic* meaning phenomenon related to gravity.

Passengers frequently experience the illusion on feeling the aircraft is dropping when the pilots abruptly reduce engine thrust after takeoff for noise abatement reasons, or later, on reaching cruising height. Their sense of inertia working from their inner ear makes the deceleration feel as if the aircraft is dropping even if it is flying level, if not still climbing.

The illusion made some passengers on Qantas Flight QF32 believe their A380 superjumbo was about to crash when Captain

Richard de Crespigny sharply reduced power when one of the engines disintegrated*.

Somatogravic illusion experienced by passengers does not put the aircraft at risk, but it is a different story when it comes to the pilots. A pilot flying at night or in cloud with no external visual reference and not watching their artificial horizon can panic when the aircraft accelerates, and they sense **wrongly** it is going sharply nose up and making them think it going to stall, and erroneously push the nose sharply down in the belief they are saving the aircraft. With the aircraft having really been flying horizontally or nearly so, it can easily enter a steep dive from which it might be impossible to recover before hitting the ground.

Several crashes have happened due to somatogravic illusion, notably when near the ground just after takeoff.

Sonic Boom

Apart from the cost implications of flying supersonically, the restrictions on flying supersonically over land greatly harmed the prospects of supersonic air travel. However, some believe there is a "sweet spot" where an airliner can fly supersonically with the sonic boom not reaching the ground because the sound wave bounces off the lower warmer air. However, this depends on the aircraft flying at the right height and speed, which varies according to the conditions.

The airspeeds we are talking about might be around Mach 1.2, which while not enormous could give the aircraft an edge when over land.

SOP (Standard Operating Procedure)

Step-by-step instructions mandating how procedures are to be carried out. Though it can apply to anything including maintenance and cabin crew procedures, it is particularly relevant in the case of pilots. For instance, there is the SOP for landings and takeoffs.

Speed Tape

With the development of liquid-crystal and CRT displays, information that used to be on different gauges can be presented graphically on the flight display all in one place, though the traditional gauges are retained, usually between the pilots, as backups.

Airspeed and altitude are shown on the left and right of the artificial horizon by means of tapes with numbers on them. Besides being easier to comprehend, these tapes can also have markers (bugs) on them showing a desired speed such as the reference speed for landing, maximum speed, stalling speed, desired height, and so on.

Spitfire (Legendary World War II Fighter)

Are no beliefs sacred? People in the United Kingdom who had experienced World War II and the Battle of Britain were brought up to believe that the Spitfire fighter (that saved England from invasion) had been uniquely the creation of the resolute aircraft designer R. J. Mitchell—sacrificing himself working on it prior to the war while dying from cancer.

Paul Beaver has written *Spitfire People*, based partly on records from the Supermarine company that employed Mitchell. The book claims that contrary to popular belief and the 1942 propaganda movie *The First of the Few*, starring David Niven, Mitchell, as the chief designer, was in effect more of a project manager overseeing a team of talented designers, who had more aerodynamic expertise than he had. According to Beaver, it was a Canadian, Beverley Shenstone, who designed the unique elliptical wing, and Alf Faddy who put the difficult-to-manufacture wing into production.

Spoilerons Instead of Ailerons

Ailerons can produce a considerable bending moment at the wing roots even though on airliners only the inner ones are used unless they fly at low airspeeds.

On the B-52 bomber, for example, it was found that extensive use of the ailerons was shortening the life of the wings and their roots. Consequently, the ailerons were dispensed with—they were welded down—and in their place "spoilerons" were inset on top of the wings. Smaller than spoilers but similar, they would nudge the wing down and more importantly reduce the wing's lift, and be made to operate like ailerons, hence the name "spoileron."

Spoilers

Flat panels inlaid on top of the wings and hinged at the front that flick up to "spoil" the flow of air over the wing. This not only produces an air-braking effect but also increases the sink rate, enabling the aircraft to lose height without pointing the nose downwards and risk making the passengers uneasy.

On landing they are deployed automatically as soon as the main wheels touch the runway. Besides slowing the aircraft when it is moving quickly, they push it down not only reducing the chance of it bouncing but also pushing the tires harder down on the runway to improve their grip while braking.

Once the aircraft slows their effect drops off dramatically and if the runway is slippery, or worse still if the aircraft is aquaplaning, only reverse thrust will really help slow it.

See **Flaps** for photo.

Spot Height

Elevation (height above mean sea level) of high points such as mountain peaks, buildings, and towers shown on aviation charts and, in the case of towers and buildings, with the height AGL in brackets underneath.

Spy Planes—Reconnaissance Aircraft

One might think that now we have spy satellites, there now would be no need for aircraft such as the **Lockheed U-2**. However, satellites must be reprogrammed, and this takes time, and in some cases can take as much as a day. Furthermore, people on the ground can work out when the satellite is due to pass overhead and take measures to hide what they are doing when it does.

Squirt

Term used for the burst of more comprehensive data/info given out by devices like ADS-B transponders.

Squawk Codes from Aircraft Transponder

In addition to the aircraft's identity, height, and even other data, the transponder "squawks" a four-digit code starting with a 7.

This can simply one assigned to it confirming its identity or an alerting code such as the three below which will show up in red on the air traffic controller's screen:

1. 7500 (Hijack). [I]

2. 7600 (Radio Failure) [C]

3. 7700 (Emergency) [E]
   One way to remember their order is the mnemonic "ICE," where "I" stands for interference with the aircraft, "C" for communications failure, and "E" for emergency.

If, say, a general aviation aircraft flying under visual flight rules and not being tracked by ATC is in trouble, the controller will usually tell the pilot to switch to 7700 to enable them to identify the craft which will show up in red on their screen. [In North America, a flight—often a general aviation one—flying under VFR (visual flight rules) will have the code 1200.]

The term "squawk" comes from World War II, when the transponder was first installed as an IFF (identification of friend or foe) device and given the code name Parrot. Turning it on (to indicate one was a friend) was called "squawking." The name stuck. See **Squirt**.

SSR (Secondary Surveillance Radar)

Secondary Surveillance Radar used by ATC identifies aircraft from the "echo" emitted by their **Transponders** and not that reflected from the body of the aircraft.

SSR can detect direction and distance but relies on the aircraft transponder data to show altitude. As mentioned, a Peruvian pilot, whose barometric altimeter was not working because an aircraft cleaner had left masking tape over the pitot probe inlets, asked ATC to confirm his height, which, of course, showed the same as the altimeter in the cockpit. He then flew into the sea certain that he was at 9,700 feet *.

If the transponder is deactivated, as in the disappearance of Malaysian Airlines **MH370** and for the aircraft in **9/11**, the aircraft disappears from the controllers' monitors.

"SSSS" (Secondary Security Screening Selection) [US]

If SSSS figures on your boarding pass you may not even be able to print it or obtain one from a check-in machine at the airport and have to go to a counter and then be subjected to strict screening.

It means that for some reason or other the authorities think you should be subject to that extra screening. The most likely reason is that you are on a watch list or your name is the same as someone who is. In the latter case it may be possible to obtain a certificate saying you are "OK" to prevent you being screened and given extra searches every time you fly.

See **No Fly List**.

SST (Supersonic Transport) US Response to Concorde

See **Boeing 2707**.

Stabilized Approach

For safety reasons, both the military and airlines insist pilots have their aircraft stabilized, that is lined up with the runway in good time at the appropriate speed and with the correct flap setting. No last-minute hasty maneuvers. One should be **"Ahead of the Aircraft."** When flying under **IMC (Instrument Meteorological Conditions)** the aircraft must be fully stabilized (including having flaps set and undercarriage down and locked) by the time it reaches 1,000 feet on its approach.

Stall

Loss of ability to fly due to too low an airspeed or a change in flaps or slats setting, meaning the wing can no longer provide lift at the given airspeed.

As the onset of a stall can be very sudden, manufacturers install sensors to detect signs of abnormal airflow over the wings that warn the pilots of an imminent stall. The warning can take various forms, including sounds, verbal warnings, and shaking of the control column.

In traditional aircraft with the engine(s) mounted on the nose or the wings, and with a low tail assembly, the nose may drop of its own accord in a stall, thereby helping the aircraft regain enough speed to recover. However, in high-tail aircraft with engines at the rear, the opposite happens—the nose goes up, inducing an even greater decrease in airspeed, in turn resulting in a catastrophic stall from which recovery might only be possible from a very considerable height, if at all.

To avoid this, aircraft with tail-mounted engines, as well as other aircraft, are equipped with systems—stick shakers, warning lights, and audible alerts—to warn pilots of an imminent stall. Should those fail to make the pilot take corrective action, a stick pusher pushes the nose down before the actual stall. The pilots override this at their peril.

While stalls usually occur due to a loss of airspeed—itself caused by insufficient thrust from the engines, climbing with insufficient power, or, say, a headwind suddenly changing to a tailwind—they can happen at constant airspeeds in the event of, for instance, the flaps, droops, or slats being inadvertently retracted. This is called a change-of-configuration stall.

See **Compressor Stall**, which concerns the engines and is quite different.

Standoff

Stealth technology can make aircraft difficult to see on radar, with even the most perfected becoming more likely to be seen the closer they come to the target. This logically leads to the concept of standoff weapons launched from beyond where the aircraft would be visible.

Furthermore, with long-range missiles, B-52 or transport aircraft can safely launch them from standoff positions under no threat from the enemy.

Starlink, Elon Musk's Satellites, and the War in Ukraine

These satellites in low Earth orbit have been of considerable help to the Ukrainians. They are difficult to jam because the beam can be quite narrow.

Startle Effect

This term used by air crash investigators including those covering **AF447 ONLY A MINUTE TO SAVE THEM** describes a pilot's reaction on being taken by surprise. Though sounding rather harmless, it can be a euphemism for fear and panic with the pilot not able to think properly. Experts have said this should be regarded as a possible normal reaction, and rather than deny it the pilot should recognize it and therefore handle it better.

Static Pressure

Static pressure is the pressure of the air outside the aircraft when static. In other words, that of air not being forced into an inlet due to the forwards motion of the aircraft through the air. The barometric altimeter depends on this, as does the calculation of the airspeed.

See **Pitot Tube**.

Static Wicks/Static Discharge Wicks

So-called wicks attached to trailing edges of wings, ailerons, elevators rudder, and vertical and horizontal stabilizer tips to discharge static electricity that could, among other things, impede operation of onboard navigation and communications systems. While they dissipate electric charges, they neither increase nor decrease the probability of a lightning strike. However, lightning strikes can damage them.

STCA (Short-Term Conflict Alert) [ATC]

If the computer at air traffic control calculates aircraft are in danger of colliding their returns on the controller's screen turn to red. They will immediately give instructions to eliminate the risk, but very often the **TCAS (Traffic Collision Avoidance Systems)** will already have given resolution advisories perhaps telling one aircraft to climb and the other to descend to avert a collision.

Stealth—Making Military Aircraft Less Visible

Hostile aircraft can be detected in various ways. They can simply be seen visually or heard as in the case of the Observer Corps in England in World War II, who would visually measure their height, or, if in cloud, locate them using crude listening devices able to roughly determine the direction from which their sounds were coming.

Aircraft also emit heat which can be detected with infrared sensors. Even the disturbance of the air though which they pass can be detected by the Doppler technology like that used to detect wind shear at airports.

In general, radar has been the most effective way to detect aircraft. Not only can the direction of the aircraft be determined but also its distance by the time taken for the radar signal to return.

The radar signals bounced back from the aircraft are called its radar signature and ways were sought to minimize it and, if possible, make the aircraft virtually invisible. Even a destroyer like the US's Zumwalt-class destroyer can, at great expense, have a radar signature making it appear to be a fishing vessel.

For aircraft, one way was to have facets reflecting the signal in other directions rather than straight back. Another way was to have the aircraft coated with material that absorbs the signal—many stealth aircraft have a black coating. However, this coating is very delicate and deteriorates if the aircraft flies too fast causing it to heat up. Rather than having a coating, the latest stealth aircraft are being built of material able to perform a similar function and therefore able to fly faster and less liable to damage such as from scratching on the ground.

Americans have made the most use of stealth aircraft, but mostly not directly against peer opponents, losing only one. In that case, a Nighthawk was downed in Serbia by a Soviet missile under unusually unfavorable circumstances in that the path it was likely to take was known beforehand; it was almost the only aircraft in the sky; and was not protected by an electronic countermeasures warfare (ECM) aircraft able to confuse missile launchers on the ground. It is alleged the precise position of that Nighthawk was able to be seen when it opened its bomb doors. The pilot ejected, called a Mayday on the way down, and was nearly caught, but he was finally rescued in a daring operation. The US military did not order a parallel operation to destroy the wreckage, apparently because it considered it old technology.

As stealth technology advances so does detection technology.

Longer wave-length radar is better able to detect stealth aircraft but cannot position them precisely enough for their destruction, so a switch must be made to shorter frequencies at the last minute, complicating matters greatly. Radar systems with multiple receivers can detect echoes bounced sideways. Perhaps the latest technique is over-the-horizon radar where the radar beam is bounced off the ionosphere enabling not only detection at greater distances but also greater effectiveness in that the beam is coming from above rather than ahead of the aircraft.

One idea is to create an ionized-air plasma around the aircraft to absorb radar beams. See **Plasma**.

Producing stealth aircraft is difficult (and expensive) because the contours and shapes must be extremely precise. Should the coating on a panel be scratched the whole panel has to be recoated. Maintenance is expensive as well.

Previous Stealth Aircraft

F-117 Nighthawk (Lockheed)
SR-71 Blackbird (Lockheed)

Operational Stealth Aircraft

B-2 Spirit (Northrop Grumman)
F-22 Raptor (Lockheed Martin)
F-35 Lightning II (Lockheed Martin)
J-20 (Chengdu Aircraft Corporation, China)

Many stealth aircraft are at the project stage including the US B-21 Raider.

It is impossible to make an aircraft completely invisible, and long-term plans in the US, for example, see making some aircraft as stealthy as possible and on occasion using them in a standoff role where they would be invisible firing a hypersonic weapon which on nearing the target would fire independently targeted missiles—perhaps using artificial intelligence to identify targets.

On the other hand, the aircraft, stealth or otherwise, would be integrated with manned or unmanned aircraft and drones linked by a sophisticated network. This brings us back to the fact that the only stealth aircraft the US lost was one operating in isolation.

Stealth over Speed, F-35 Lightning II

For the next generation of fighter, the F-35 Lightning II, planners decided stealth and superior sensor systems took precedence over raw speed such as found on the F-22 Lightning, though it meant limiting its speed to Mach 1.6 to protect its radar absorbing materials and at the same time cram an integrated sensor suite into a smaller aircraft allowing pilots an all-round view using a very sophisticated—and expensive—helmet.

It can simultaneously track numerous targets simultaneously and target them with missiles while remaining invisible.

Step Climb

Long-haul airliners are often too laden with fuel at takeoff to be able to be able to climb to their most efficient cruising height. They therefore increase their altitude in steps as fuel is depleted until they are able to do so. Of course, they are subject to air traffic control and may not always be able to climb at the ideal moment.

This is called "**Step Climb**."

Sterile Cockpit Rule

To prevent pilots being distracted at busy times there is now, particularly in the US, the "Sterile Cockpit Rule" whereby they are not allowed to engage in nonessential activities or allow themselves to be disturbed say by cabin crew under 10,000 feet.

It was introduced after a series of accidents due to pilots being distracted and often gossiping.

After takeoff, flight attendants stay buckled in their seats by the doors until the aircraft reaches 10,000 feet, notified to them by a "ping" that everyone in the cabin can hear.

Stick Shaker

To warn pilots of an imminent stall, the control column may be made to shake artificially. Many other means of warning pilots of such a possibility usually come into play. Many airliners have automatic recovery programs.

See **Alpha**.

Straight-In (Landing)

A landing where the aircraft comes straight in without joining a rectangular pattern.

See **Base Leg** and **Downwind Leg**.

Strakes

Small blade-like devices often mounted on upper surfaces engines nacelles to direct airflow over certain control surfaces at high angles of attack. Enhances control authority and lessens the chance of loss of control at lower airspeeds.

Strategic Bombers

Following the US strikes against Iranian nuclear facilities using B-2 stealth bombers, the Chinese who had concentrated on hypersonic glide vehicles, missiles and drones have realized their value.

Strike-Fighter (F/A)

A dual role combat aircraft that can strike ground targets and engage in air-to-air combat. Air-to-air and air-to-ground, e.g., F/A-18 Hornet.

Structural Health Monitoring (SHM)

Use of sensors and so on to monitor structural state of the aircraft, especially where composite materials that cannot be checked in the traditional ways are involved.

Submarines

In World War II the US Navy spent only 2 percent of its funds on submarines which sank more than 50 percent of Japanese shipping (including freighters), despite them for the first two years being issued with substandard torpedoes, which generally either failed to detonate or went in circles. The US torpedo bombers at the battle of Midway also had duds that went in circles and didn't even explode on hitting a ship. However, those pilots did not die in vain for they brought Zero fighters down low to fight them giving the dive-bombers a free run from high up and sinking three aircraft carriers.

Had there been more German U-boats at the beginning of the war before measures and systems evolved to combat them, the UK could have been crippled—it was said to have been Churchill's greatest fear.

In World War II the aim was to build submarines relatively cheaply so many could be made. Nowadays, they are relatively few and far between but extremely expensive, and sometimes fabulously so, though that is little compared with the destruction they can inflict.

Nowadays, the submarine is at once fearsome and a protector according to the principle of **MAD (Mutually Assured Destruction)**.

There is the conundrum that it may be impossible for the target of torpedoes or missiles launched from a submarine to identify the country to which they belong. This could manifest itself in the South China Sea should the mainland attack Taiwan, and lead to a dangerous escalation due to mistaken assumptions.

Nowadays, nuclear-powered submarines can stay submerged for extended periods, the only problem being victuals.

One should not forget the smaller intelligence gathering and special ops submarines that can cut or tap undersea cables, place sensors, or put operatives ashore.

United States Submarines

The US Navy classifies submarines (hulls) as "SS" where the two letters stand for submersible ship, even though submarines are called "boats" and not ships! The last letter stands for the method of propulsion, which for large vessels in the case of the US is always nuclear, "N." Thus, we have "SSBN" for nuclear powered ballistic missile submarine and "SSN."

The next letter stands for the type of missile or armament that they have, with "B" standing for ballistic missiles, which normally would be nuclear, but not necessarily so, and "G" for cruise missiles.

To sum up we now have for the US:

1. Nuclear Attack Submarines (SSNs)

2. Nuclear Cruise Missile Submarines (SSGNs)

3. Nuclear Ballistic Missile Submarines (SSBN)

When it was decided not so many of the ballistic missile submarines (SSBNs) were needed considering the tremendous yield of their multi-warhead nuclear weapons, four were redeployed as cruise missile submarines, but, of course, would be indistinguishable. This has led to fears that conventional warhead cruise missiles coming from those SSBNs could make an enemy think they were nuclear armed and result in nuclear war by mistake.

China, Russia, Great Britain, and France have submarines with nuclear weapons, and many other countries have various submarines with North Korea said to have by far the most.

The US is working on a new class of SSBN. The US has SRBMs (Short Range Ballistic Missiles) and IRBMs (Intermediate Range Ballistic Missiles) in addition to ICBMs (InterContinental Ballistic Missiles). France and the UK have smaller numbers of the same thing. Israel almost surely has IRBMs and SRBMs, as do India, Pakistan, China, and perhaps others.

Though being on or rather in a submarine might seem the safest place in the world in a nuclear war, launching its missiles would reveal its position as the many satellites overhead would detect them as soon as they broke though the cloud cover and be able to work out from where they were launched and respond with a weapon powerful enough to destroy it in the distance it might travel in the meantime.

Submarine Conning Tower, Now Called Sail, Fin, Stabilizer

In World War II, submarines had to spend much time on the surface to recharge their batteries with what were then "conning towers" often visible on the enemy's radar. They were called conning towers because they had the essential controls for the submarine controlled from there.

Nowadays, most of these ever more sophisticated controls have been transposed to the hull since nuclear submarines hardly ever operate on the surface. What was once called the conning tower is now called the "sail," and sometimes the fin or stabilizer elsewhere as it helps keep the submersible steady underwater.

Suborbital and Orbital—Meaning

These terms used in connection with spaceflight and by the military referring to missiles are confusing with suborbital seeming to mean something flying lower when it means something else.

The difference is mainly a matter of speed, with an orbital missile being one flying fast enough to remain in orbit around a planet (ours for the military), and a suborbital missile being one flying not fast enough to remain in orbit.

To orbit 125 miles (200 km) the Earth a rocket must fly at a staggering 17,400 mph (28,000 km/h), but to reach the same height and then just drop down a vehicle only needs to reach a speed of 3,700 mph (6,000 km/h).

See https://space.com.

Subsonic Flight

When an aircraft is flying below the speed of sound, air moves aside even before encountering, say, a leading edge. At and beyond the speed of sound this is no longer the case, and the air slams against the leading edges, creating a shock wave. This is the so-called sound barrier, which was once the subject of movies in which pilots risked their lives to break it. US Air Force Capt. Charles E. "Chuck" Yeager finally became the first person to fly an aircraft faster than the speed of sound in 1947.

In practice, everything is not so cut and dried, since the contours of an aircraft—a thick wing, for example, with a considerable bulge on top—mean that the speed of the air differs from one place to another, and one part of the aircraft may be flying subsonically and another part supersonically. The term *transonic* is applied to the range of speeds between pure subsonic flight and pure supersonic flight—roughly Mach 0.8 to 0.85, to Mach 1.15 to 1.2.

Subtle Incapacitation of Pilot

If a pilot has an obvious problem such as a heart attack, it is easy for their colleague to recognize. Subtle incapacitation is insidious and can be easily missed, especially in high-work situations. Good cockpit resource management, where pilots do not get in the habit of skipping call outs, means lack of response or comment can be noticed and make creeping incapacitation more evident.

In dire situations, the captain in newspaper parlance may seem to be "fighting at the controls to save the aircraft," when, in fact, it may be that they are not in their right mind or suffering from a mini-seizure, something the first officer may not realize, mistaking it for great effort and therefore fatally not intercede.

Super Guppy 1965/5 [Beluga, Beluga XL, and Dreamlifter]

At the beginning of the US space program there was a need to ferry large empty items that were not particularly heavy across the United States. Some items from the West Coast had to be shipped via the Panama Canal taking two weeks or more.

A bloated cargo freighter called the "Pregnant Guppy" based on Boeing's first postwar airliner, the Boeing 377 Stratocruiser, first flew in 1962. The need was so great that five "Super Guppies" were designed, again based on the 377 but fitted with turboprop engines.

Two were later used by Airbus for ferrying parts, with a French company given the right to manufacture two more. Subsequently, Airbus designed its own outsize cargo carrier based on its A300, calling it the Beluga. Now Airbus has the Beluga XL based on the A330. The first (of six) entered service in January 2020.

Boeing's latest outsize cargo carrier is based on the 747 and called the *Dreamlifter*.

Supersonic Airliners

As early as the 1960s, the supersonic airliners seemed to be the future, with the Russians, the French and British planning a Mach 2 airliner, and the Americans at the urging of President Kenedy not wanting to be outdone announcing a "superior" Mach 3 version (see Supersonic **Transport (SST) [US]** below).

The Russian version flew just first, but being a "rush job," was not successful and was withdrawn from service after only a relatively few shortish domestic flights.

Partly because the US SST was too ambitious—flying at Mach 3 was beyond what aluminum alloys at the time could support due to heat generated at that speed—the project was abandoned after considerable expense.

The Anglo-French supersonic airliner (see below) project was pursued with only 12 aircraft produced and given to British

Airways and Air France. They flew for a number of years mostly for the benefit of rich passengers before being withdrawn from service in 2003.

Since the Concorde was designed there have been great technological advances and improvements in the materials available. However, the great problem for supersonic airliners remains the sonic boom with them generally, and in particular the US, not allowed to fly over land.

There is another factor reducing the demand for supersonic airliners in that with people able to communicate by video on their phones and computers, with Zoom video conferencing there may be less need for people to rush to meetings.

NASA, in collaboration with Lockheed Martin's Quest program, is experimenting with designs such as the aircraft below aircraft with very long pointed noses that could, they hope, reduce sonic boom to tolerable levels, but it is still early days for that.

[NASA]

Several small companies with plans for smallish supersonic airliners have recently fallen by the wayside, but there is one called Boom that thinks it has fund a "sweet spot" for a relatively small supersonic airliner. ~

~ Concorde (Anglo-French Supersonic Transport)

A beautiful supersonic airliner developed at great cost to UK and French taxpayers. The British on several occasions wanted to pull out, but a clause in the original agreement to produce it meant they would have to pay enormous compensation.

It was not the first supersonic airliner to fly for the heavier Russian Tupolev 144 had its maiden flight a couple of months earlier in 1969. However, the latter was less technologically advanced, had even less range, crashed at the Paris airshow, and was withdrawn from service after a few revenue flights.

Only 12 Concordes were made, and these were virtually given away to the two countries' flag carriers. The aircraft had failed to sell despite airlines initially showing a great deal of interest. The reasons for airlines' failure to purchase it were:

1. The round-the-world demonstration flight proved a flop (people on its route were not happy to suffer the sonic boom of an aircraft they could not afford to fly in; the Japanese did not like its smoky engines; and flights over Australia upset Aborigine communities).

2. Its entry into service coincided with a great jump in fuel prices.

3. Its limited range meant that it was only viable on relatively short oceanic routes such as between London and New York, and just about Paris and New York.

4. Noise problems and the sonic boom meant that countries would not let their airports be used for stopovers.

5. Perhaps partly because of sour grapes, its use in the US was very restricted. See **SST**.

6. Fuel consumption was too high.

British Airways and Air France operated Concorde for prestige after being given them for free, on the understanding the respective governments would share any profit. Indeed, BA did find it quite profitable when it doubled the ticket prices and made the London–New York flight the height of luxury. Though the cramped seating left much to be desired, the champagne and caviar and special Concorde lounges made up for that. Surprisingly, extra income came from charter flights for "ordinary people," many of whom had saved up for the treat of a lifetime.

With no ongoing production, improvements were not made, and the tires, for instance, were not rethought, and when a Concorde crashed in flames after taking off for New York from Paris's Charles de Gaulle Airport in 2000*, statistically it went from being one of the safest airliners to one of the most dangerous.

It was temporarily withdrawn from service, but when it was returned with improved tires and the fuel tanks lined with Kevlar so they would not burst if hit by, say, tire fragments, it coincided with the aftermath of 9/11 and a decline particularly in business travel to New York. The result was that it was flying virtually empty.

Ironically, once its final withdrawal from service was announced—not for safety reasons but because the cost of maintenance and spare parts for such a limited fleet was proving exorbitant—it flew virtually full for the last six months.

See **SST**.

Supersonic Flight (>Mach 1)

Drag and behavioral problems increase sharply as the aircraft approaches the speed of sound. Airliners cruise at around Mach 8.5, since marginal gains in speed would not be worth the extra cost.

Interestingly, there is a further barrier that is again financial rather than technical, namely that above Mach 2.2 the heat generated by the compression of the air molecules surrounding the aircraft becomes so great, despite the bitterly cold temperatures at the operating height, that special heat-resistant materials are required. That is why the Anglo-French supersonic Concorde had a cruising speed around Mach 2 and a maximum speed near Mach 2.2.

As no other aircraft would fly up to almost 60,000 feet, Concorde was able to perform a continuous climb rather than the usual **Step Climb** and fly at the optimum height in accordance with the remaining weight of fuel.

Supersonic Transport (SST) [US]

Faced with the Soviet Union's apparent lead in space exploration, President Kennedy initiated the Apollo program, which succeeded in putting Americans on the moon and, luckily, bringing them back. Faced with the danger that the Europeans would beat the US in the aviation sphere with their supersonic Concorde airliner, Kennedy, in 1963, also initiated a project to design an airliner that would beat the European craft hands-down by flying at three times the speed of sound.

Kennedy was incensed when he learnt hours before he was to announce supporting the US SST project with funds, that Juan Trippe of Pan Am—who prided himself at being first with the most advanced airliners—had taken an option on six Concordes. He thought it was like stabbing the US in the back and vowed to make things difficult for Pan Am.

Boeing was awarded the contract for the SST project but found it something of a poisoned chalice, since it took resources away from others, such as the 747 jumbo. In 1971 Congress voted to cancel the SST, which was opposed by many for diverse reasons, including noise and the alleged danger of cancer because of damage to the upper atmosphere.

See **SST**.

Surveilling Entire City, Gorgon Stare (Wide-Angle Motion Imagery)

"Gorgon Stare" is the name the US Air Force gives to a system of cameras fitted to a drone that unlike the usual ones focusing on a particular spot with extremely high resolution, covers a wide area such as a *whole* city with, say, 1.8 billion pixels but very low resolution. This enables it to track the movement of thousands of people and vehicles, and though a person or vehicle of interest (say, robbing a bank or placing an IED) may only be displayed as one or two pixels making it impossible to immediately identify them, that is sufficient for them to be tracked going to and coming from the scene of the event. Other means of identifying them, say, with a high-resolution camera, can then be used, making it possible to question people about places they have been.

Of course, how to analyze so much information, especially preemptively, is a problem that only AI might be able to solve. For instance, people about to commit a crime often drive around in a nonsensible (odd) way, and AI might be able to pick that up.

Survivability [US Navy Vessels]

The US Navy has three levels of survivability for its vessels:

1. Level 1—Low Survivability
   Replenishment ships, mine, and patrol craft.
   Also Level 1+ vessels that in theory can use speed to evade being hit, but with the increasing sophistication of missiles speed is unlikely to preserve them from a direct hit.

2. Level 2—Medium Survivability
   Frigates and amphibious ships.

3. Level 3—High Survivability
   Cruisers, destroyers, and aircraft carriers.

With all vessels subject to a direct hit from a hypersonic missile with a powerful warhead, this classification begins to lose its meaning.

Sustainable Aviation Fuel (SAF)

See **SAF (Sustainable Aviation Fuel).**

Sustaining Military Aircraft

Especially in the case of stealth aircraft, with their delicate, expensive, and painstaking-to-apply **RAM (Radar Absorbing Material)**, they can cost several times the initial cost.

Swarm

There is increasing interest in the potential of swarms of drones and even missiles to overcome defenses with it thought that all sides could resort to their use in the event of China engaging Taiwan. See **Replicator**.

However, the use of swarms applies to all theaters.

Swing Wing Aircraft—Weight of the Mechanism

Having a swing wing that could be extended straight out almost perpendicularly to the fuselage for takeoff and landing and low-speed loitering would seem a good solution. However, though Russians have had some success the very considerable weight of the mechanism has meant that although it worked, much more fuel was required and high-speed performance was poorer.

Boeing's original concept for US **SST (Supersonic Transport)** that President Kennedy wanted to be better than the Anglo-French) Concorde had a wing with variable sweep but the mechanism was found to be too heavy and too complex and was abandoned, as was the whole project later.

Swiss Cheese Accident Model

Concept conceived by Prof. James Reason, whose book *Human Error*, published in 1990, and other work revolutionized the way accidents are analyzed, with the causes being sought throughout the system, even where the acts of the operators (for instance, the pilots) are not appropriate, often due to bad training.

The Swiss cheese model seems quite simple in that the idea is that there are various layers in a system or organization:

1. Organizational influences (latent failures).
2. Unsafe supervision (latent failures).
3. Preconditions for unsafe acts (latent and/or active failures).
4. Unsafe acts (active failures), say, by pilots.

If the holes in each layer—in the cheese—coincide, allowing a knitting needle to pass through, the result will be an accident. This is an oversimplification, and there is much more to Prof. Reason's work than that.

Some say that while the Reason model is good for looking at accidents retroactively, it is not so good for finding the failures proactively. Thus, the ICAO recommends use of the SHEL model. See **"Normal Accident"** (Prof. Perrow) and **Judgment Errors** (Wiegmann and Shappell).

Synthetic Airspeed System/Synthetic Air Data

Aircraft depend on their **Pitot Tubes** to determine airspeed, and on their **Angle of Attack Sensors** to determine the angle at which they are pitched up or down.

These are usually accurate and are usually in threes so if one is faulty, it can be detected on the odd-man-out principle and ignored. On very rare occasions this does not work, say, when two freeze and lock at the same angle, leaving the odd-man-out one—the correct one—to be ignored, as happened in the crash of an Air New Zealand Airbus on a test flight*.

Synthetic airspeed systems and synthetic air data systems can independently complement the pitot tubes and angle of attack sensors and prevent crashes such as those for the 737 MAX..

Synthetic Aperture Radar (SAR)

Synthetic aperture radar is where an antenna on a moving aircraft or satellite can create images of the ground below as if it were a very much larger antenna.

The concept was conceived years ago but there is less information now in view of its military applications. Modern computer technology has made it a very powerful tool with it possible to identify small objects through cloud. Interpretation of images requires expertise as sometimes a railway line can only be identified by the train shown lying parallel to it.

See **AESA (Active Electronically Scanned Array) Radar**.

~ T ~

TACAN (Tactical Air Navigation)

US term for a VOR air navigation beacon for military use. More accurate than the civilian one but less valuable now we have GPS.

It did come up in a civilian context when a US military aircraft reported the location of the world's worst-ever single-aircraft crash, JL123 in Japan in 1985*. It gave the TACAN bearing and distance.

Tactical and Strategic Thermonuclear Weapons

The difference between tactical and strategic nuclear weapons lies in their range and above all in their yield (power).

However, there has been a great change since the 1960s when terribly destructive high yield bombs were the norm. This was because missiles and even bombers could only have an accuracy of 10 miles (16 km) if that, meaning that to be sure to destroy a target they would have be able to take out an area of 15 or 20 miles (25 to 32 km) or more. Nowadays, with weapons having accuracy of, say, 30 yards (50 m) such powerful nuclear weapons are not normally needed.

Before going into that, we may ask why the US Air Force is making the F-35 thermonuclear-capable when other air-based, land-based, sea-based, and undersea-based *missiles* already provide so many effective options. The point is that an aircraft can be recalled even at the last second, enabling it to be used as a negotiating tool, whereas this is not possible with a ballistic missile.

The yield of a *tactical* thermonuclear weapon ranges roughly from 0.3 kilotons (300 tons of TNT) to 50 kilotons (50 tons of TNT). That of *strategic* thermonuclear weapons is of a quite different order and can range from 100 kilotons (100 tons of TNT) to a megaton (1,000,000 tons of TNT) or more. The bomb dropped over Hiroshima had a yield of between 12 and 18 kilotons, while the plutonium bomb dropped over Nagasaki was between 18 and 23 kilotons.

One should remember that **MIRVs (Multiple Independently Targetable Reentry Vehicles)** can have ten or more independently targeted warheads, each packing a considerable strategic punch.

The yield required can be lessened, for instance by making the bomb able to penetrate deep into the ground to destroy underground bunkers and facilities, say, in Iran or North Korea.

As mentioned under **Thermonuclear Weapons**, were it possible to make a lightweight fusion weapon without the usual "dirty" fission trigger, tactical nuclear weapons would seem more acceptable since troops would be able to immediately enter the targeted area without fear of radiation. Also, there would be little environmental impact.

Tailplane

See **Horizontal Stabilizer**.

Tailstrike/Tail Strike

This is where the bottom of the fuselage at the tail accidentally strikes or scrapes the ground on takeoff or landing. Though not immediately fatal itself, faulty repairs have resulted in terrible disasters many years later due to metal fatigue.

This happened twice:

1. In what was the worst-ever single-aircraft crash, namely flight JL123 in Japan on August 12, 1985*. It was a packed 747 used on short domestic routes with therefore many takeoffs and landings causing repetitive stress on the inappropriately repaired rear bulkhead following a tailstrike seven years earlier. A total of 520 lives were lost. Four survivors were found the next day.

2. In the crash of China Airlines (Taiwan) Flight 611 on May 25, 2002*, where the rear section of the fuselage completely broke off with the front breaking up due to the stresses encountered as it fell.
   This resulted from bad repairs after a tailstrike 22 years

before. A total of 225 lives were lost.
See **Repairs to Fuselage**.

Tailstrikes can be anything from very minor to very serious. On takeoff pilots can sometimes find it impossible to hear one over the noise of the engines, and this is when a flight attendant sitting at the rear should break the **Sterile Cockpit Rule** and call the pilots to inform them.

The pilot can sometimes sense a tailstrike by a pause in the nose rising or they may realize they are on the ground and pitching up at far too steep an angle. Quite often it will be the control tower informing them they have seen dust or smoke at the underside below the tail. Fly-by-wire aircraft and others can have software to largely prevent them occurring.

There are many cases where pilots have continued making a long-haul flight while aware they have suffered a tailstrike, when the recommendation is that even on short haul, they do not even climb to where they pressurize the aircraft as this can lead to further damage that can be structural.

Tailstrikes on takeoff can be caused by entering a too low gross weight into the flight management computer with the result there is insufficient thrust to achieve the necessary airspeed. Another reason can be pulling back too early. A certain airspeed margin is set for the rotation speed (V_R) to allow for variations in the wind or its direction and make a tailstrike improbable.

Tailstrikes on landing tend to be more serious but rare.

Takeoff Distance

Takeoff distance is defined as the path required for a plane to accelerate from a stationary position and reach a height of 35 feet, at which point the pilot monitoring would be calling out "positive rate" should all be going well. Thus it is more than the distance run from stationary to lift off.

Taking Off (Critical V-Speeds)

At takeoff, as at other times, there are various critical speeds (velocities) including those mentioned above.

For a Boeing 747-400 taking off weighing 800,000 lb (363 metric tons) with flaps at an angle of 20 degrees, V_1 might be of the order of 153 knots (175 mph/283 km/h), V_R slightly higher at around 169 knots (194 mph/312 km/h), and V_2 might be 181 knots (208 mph/334 km/h).

See item 2 in list below for meaning of V-speeds.

Taking Off (Taxiing, Takeoff Roll, Liftoff) and Landing

Taking off and landing are the times when the aircraft is nearest the ground with little margin in terms of height should anything untoward happen. The various stages are as follows:

1. Taxiing from the gate or from the stand to the runway. In the past, there have been cases such as SQ0006* where the aircraft has taxied to a disused or the wrong runway. Nowadays, with GPS and improved radar surveillance from the control tower this is unlikely. However, pilots must take care in constrained places not to hit the wing of another aircraft.

2. Before the takeoff roll there are checklists to ensure that the controls such as the flaps have been set correctly. In addition, the pilots must have worked out the key V-speeds:

    a. V_1, corresponding to the **decision speed** after which the takeoff must be pursued even if an engine has failed say due to a bird strike.

    b. V_R, the usually (but not always) slightly higher **rotation speed** at which the pilots rotate the aircraft, so it rises.

    c. V_2, the **safe flying speed** which the aircraft must attain and maintain when it is 35 feet off the ground.

    Before initiating the takeoff roll, they will have discussed the procedure to follow should something go wrong after liftoff—airports have set flight paths and heights for such circumstances to ensure there is no risk of colliding with other aircraft or obstacles. Before initiating the takeoff roll the pilots must spool up the engines to the desired takeoff thrust. Passengers will have noticed this is done in two stages with the engines seemingly spun up to half thrust, with a pause before the power is increased to the high power required. This is done simply because if they are spun straight up to full thrust from idle, one engine may reach that before the other resulting in an imbalance that could cause the aircraft to swivel.

3. A rolling takeoff
   There are cases where the controller may want to squeeze in the aircraft in question before one approaching to land and give permission for it to enter the runway and take off without stopping. In that case the aircraft will enter the runway with the engines already well spooled up, and then fully spool up to the desired thrust, and take off without stopping.

4. The takeoff roll and the call outs
   Although getting quickly and cleanly off the ground is desirable, if there is plenty of runway and the aircraft is not operating close to its limit, the pilots will usually avoid using maximum takeoff thrust to reduce wear on the engines (see **Flex Takeoff**).
   Though this normally proceeds uneventfully, it is a tense moment for the pilots as they must be ready for the unexpected and much can go wrong.
   The pilot not flying (the pilot monitoring) will make key callouts, the first one being 80 knots for Boeing aircraft and possibly 100 knots for Airbus aircraft. The pilot flying (PF) checks that their instruments show the same speed.
   Up until then, the takeoff can be rejected (aborted) for even a minor reason since the aircraft can be brought to a

stop without undue stress and overheating of the brakes.
The next key V-speed speed is the takeoff decision speed V_1, beyond which the takeoff must be pursued no matter what, for there may not be room to stop.
It is not so frightening, for once this speed is attained the aircraft should be able to take off safely even with one engine not working.
The next call out is V_R, the speed at which the pilot flying pulls back on the control column or sidestick to rotate the aircraft upwards, so it lifts off.
On Airbus aircraft with all engines operating, the target pitch attitude on take-off is 15 degrees. If an engine fails, this is reduced to 12.5 degrees to ensure the aircraft gets safely gets airborne.

5. Once the aircraft is off the ground, the pilot not flying (PNF) will confirm "positive rate" (of climb) and ensure they are flying at least at V_2, the takeoff safety speed.

6. V_{REF} speed for final phase of a landing.
Reference (desired) speed for landing, calculated according to weight, flap settings, and so on. Too slow and the aircraft might stall; too fast and the aircraft might not be able to land safely or land with enough runway to stop safely. If actual airspeed deviates from V_{REF} by more than the company allows, say, 20 knots, the pilots should abort the landing and go around. If the wind is gusty or there is wind shear pilots might well add 20 knots as a precaution.

Task Saturation

Judgment is a key factor in safety and cannot be exercised when overwhelmed by tasks. Bearing this in mind, Ryanair stressed the need for the pilot not flying (monitoring) to call, "500 feet go around" where the landing parameters are not met.

TAWS (Terrain Avoidance and Warning System)

While **GPWS (Ground Proximity Warning Systems)** and their upgrades did so much to reduce the incidence of **CFIT (Controlled Flight into Terrain)** crashes, the most modern systems are even more sophisticated and better covered under TAWS (Terrain Avoidance and Warning Systems) as follows, which must:

1. Display terrain information.

2. Have the traditional enhanced EGPWS function looking not only straight down but ahead of the aircraft considering its lateral and vertical flight path and issuing alerts as necessary.

3. When landing at an airport, not depend on the ILS glideslope but rely on navigational information and the airport database to determine whether the aircraft is below the typical 3-degree glide slope and issue a Premature Descent Alert (PDA) as necessary.

4. It must also warn of situations with a risk of hitting the ground such as excessive rate of descent, excessive closure rate to terrain, negative climb rate or altitude loss after takeoff, flight into terrain other than when landing, notwithstanding alerts in Section 3 not based on ILS, warn of excessive downward deviation from ILS glideslope, or warn of deviation from the RNAV glidepath.

5. Finally, it will give the "Five Hundred" when the airplane descends to that height.

TCAS (Traffic Collision Avoidance System)

Technology on the aircraft also warns pilots of the presence of nearby traffic and as a last resort tells them what avoidance action to take, ensuring say one aircraft goes up and the other goes down, and that they do not both do the same as sometimes happens when people come face to face in the street and collide. This system is called **TCAS (Traffic Collision and Avoidance System).**

It relies on the transponders in the two aircraft talking to each other, and in the event of a possible collision issues two alerts.

1. The first, called a traffic advisory (TA,) warns there is traffic in the vicinity that could possibly represent a danger. It will also give a TA when an aircraft below is climbing (perhaps routinely to a safe altitude set by the air traffic controller) if it is ascending so fast that that the TCAS thinks it risks colliding. In that case the pilot below will receive a warning too and realize the reason and normally reduce their rate of ascent.
   On receiving a TA, the pilots take over manual control, keep a lookout for the other traffic (position shown on the TCAS display), and are ready to comply with any much more serious resolution advisory (RA) that may follow. They avoid sharp maneuvers that might confuse the TCAS.

2. The resolution advisory (RA) issued when there is a real danger of a collision and is designed to avoid the aircraft veering in the same direction as so often happens when two pedestrians come face to face in the street.
   On receiving a resolution advisory (RA), the pilots should comply smoothly—no need for a violent maneuver lifting passengers and flight attendants up in the air. However, should the RA change to *increase* (descent/ascent) that is another matter.

Following several near disasters and one terrible one where the pilots of an airliner full of schoolchildren flying over Lake Constance between Switzerland and Germany ignored the TCAS resolution advisory and collided with a freighter in the middle of the night*, the need to obey TCAS rather than the controller was mandated worldwide.

However, the recent introduction of **ADS-B**, also able to prevent collisions, means TCAS has to some extent been superseded.

It should be noted that below 1,000 feet, TCAS does not issue climb/descend aural advisories. This is because being so near the ground there would be the danger of the aircraft hitting it. This would have been the case in the 2005 collision between a helicopter and an airliner landing at Washington Reagan National Airport, as they were at 300 feet.

TCDS (Type Certificate Data Sheets) [FAA]

The Type Certificate Data Sheets (TCDS) database is a repository of make and model information. The TCDS is a formal description of the aircraft, engine, or propeller. It lists limitations and information required for type certification, including airspeed limits, weight limits, thrust limitations, and so on.

Technical Air Intelligence (TAI) [Akutan Zero, World War II]

Despite warnings from Chennault of Flying Tigers fame that the Japanese had some remarkable aircraft in World War II, the US military were amazed to find just how good they were in combat after Pearl Harbor. Having no good sources of information in Japan, they knew hardly anything.

To rectify this, they established so-called Technical Air Intelligence Units (TAI) whose task was to recover parts of downed Japanese aircraft and try to recreate one and be able to test fly it. While many crashed into the sea, and the others crashed in deep jungle, retrieving parts was difficult and even hazardous work. Furthermore, wrecks were usually found to have been stripped by US troops seeking souvenirs.

In 1942 they had a stroke of luck. A US flying boat sighted the wreckage of a Zero on an island (Akutan) in the Aleutians. The inexperienced pilot had crash-landed with the plane turning over, breaking his neck. Pilots of other Japanese aircraft had been ordered to destroy any downed aircraft but on seeing it still intact assumed the pilot was alive inside and left it as it was. It was recovered by the US Navy and sent to the States, where once completely repaired, it was test flown. Its weaknesses such as poor high-speed roll performance, spluttering of the carburetor in a bunt, and a propensity to peel off leftwards rather than rightwards were determined. This information was invaluable and saved many US pilots' lives. However, another Zero had been recovered by Chennault in China and was being shipped to the States with shipping taking longer from there. Thus, even without the Akutan Zero the US would have soon learnt the fighter's secrets. Subsequently, other types of Japanese aircraft were recovered and evaluated by TAI units.

The Mitsubishi Zero fighter outclassed US fighters at the beginning as regards dogfighting agility but subsequently suffered great losses after the US had learnt much from the one recovered, and thenceforth avoided close dogfights in favor of high-speed passes.

Towards the end of the war in Europe, TAI learnt from codebreakers that two Japanese submarines were leaving Germany with some of the latest technology including plans and parts of aircraft jet engines. One of the submarines was tracked and sunk, but the other reached Singapore with the secret material shipped on to Japan. The Japanese then succeeded in making a jet engine but one not quite so powerful as the one on the Messerschmitt 262, partly because they did not have access to the alloys required. However, because the airframe they fitted it to was lightweight like the Zero, its performance would have been comparable. It is said it first flew five days before the war's end. Had many of these been ready for an invasion of Japan several months later and able to shoot down B-29s, losses would have been even more horrific than anticipated. There was also a rocket-driven craft based on a German one.

TERCOM (Terrain Contour Matching)

With electronic warfare disrupting GPS signals an old but clever way to guide missiles, such as the US's Tomahawk, is "terrain matching" whereby the missile has maps and images showing the contours of the terrain over which it is going to fly and can adjust its track and altitude accordingly.

Terrain Matching

See **TERCOM** above.

TFR (Temporary Flight Rule)

The FAA issue a TFR, say, when the US president goes somewhere, even on vacation, on Air Force One. There is an inner and outer bubble inside which aircraft must be on a flight plan, have an active transponder, and cannot loiter. Within the inner bubble, only US government flights, and regularly scheduled passenger and cargo flights between secured major airports may transit. Violators risk being intercepted by US military aircraft and incurring serious penalties such as suspension of pilot's license and fines. See **Ramp Freeze**.

THAAD Missile System (Terminal High Altitude Area Defense)

US anti-ballistic missile defense system intercepting and destroying missiles in their terminal phase at heights as great as 200 kilometers, that are considerably greater than Dome and Patriot systems.

It can target short-, medium-, and intermediate-range ballistic missilese.

Lockheed is the lead contractor with the X-band active electronically scanned array radar provided by RTX as one of the subcontractors.

It is a kinetic energy weapon destroying its target by striking it with no explosive involved.

THALES

Thales is the rebranded French Thomson-CSF, following its acquisition of the UK's Racal Electronics. The name reflects its increasing international presence and aspirations. Indeed, it has become one of the world's top ten defense electronics companies.

More than half its business is defense related, with the company's reach even extending to shipbuilding. This includes working with Britain's BAE on the design of future aircraft carriers.

It stresses dual use, in the sense that expertise gained in the domain of military technology can be applied in the civilian context, even for something quite mundane like public transport ticketing.

Thales is doing a great deal of business with Airbus, and their relationship could have been much more formal had the Germans not, in late 2004, blocked the joint CEO of EADS Noël Forgeard's plan for Thales to integrate with EADS. The German side feared this would give the French too much leverage—a constant fear.

"The Jet Business" Showroom for Sales of Business Jets

Having considerable experience in buying and selling jets to order, Steven Varsano has opened **"The Jet Business"** in London's Park Lane as the "world's first and only street-level corporate aviation showroom for the marketing and acquisition of corporate jet aircraft."

Next to the Hilton Hotel it even has mockups of cabins enabling prospective buyers to see the possibilities, and benefit from the advice of its team of experts.

See **Airports**.

Thermals

Updrafts of air that sometimes cause aircraft to rise uncomfortably. Birds and glider pilots love them. What passengers call "air pockets," where the aircraft seems to be falling relentlessly, are not pockets at all but simply downdrafts, and sometimes passengers imagine they are dropping thousands of feet when it is only 100 or so.

Thermobaric Weapons, Aerosol Bombs (Vacuum Bomb)

The thermobaric weapon, also known as an aerosol bomb or fuel-air explosive, is a two-stage munition. It is relatively simple and has been in existence since the 1960s and called the vacuum bomb for the reason we shall explain.

It is made up of two parts. The first part is a charge distributing an aerosol cloud consisting of very fine material made from carbon-based fuel and miniscule metallic particles. The second part is a charge to ignite that cloud, creating a fireball sucking in all surrounding oxygen thereby creating a vacuum and a huge shock wave lasting considerably longer than for a conventional explosive.

The aerosol cloud can penetrate entrances to buildings, making it fearsome as anyone caught up in it becomes part of the fireball able to vaporize human bodies. Second, while the shockwave can do tremendous damage not only to buildings but even to people at some distance, the removal of oxygen from the air produces a "vacuum" damaging people's lungs and even interior organs.

Thermobaric weapons have been used both by Russia and the West. The US deployed them in rather vain attempts to eradicate al-Qaida in their mountainous hideouts in Afghanistan. That said, Russia has probably used them more extensively than the West with an arsenal ranging from relatively small to very large ones.

They are largely used to attack defensive positions, with very large air-launched versions able to take out caves and tunnel complexes.

"Threat Level" Definition Enabling ATC to Grasp Danger

The **ICAO (International Civil Aviation Organization)** has defined a four-tier threat level hierarchy with respect to "situations" inside airliners. This enables controllers to immediately grasp the severity of any trouble without the need for a full explanation which may not be possible in some circumstances such as a hijacking.

While not every country follows it exactly, it is a useful guide:

1. Level 1: Disruptive behavior (verbal).
2. Level 2: Physically abusive behavior.
3. Level 3: Life-threatening behavior (or display of a weapon).
4. Level 4: Attempted or actual breach of the flight crew domain.

Threshold

Like the threshold to someone's house over which you must step, the threshold is the beginning of the runway with markings in the form of bars to indicate the threshold and the width of the runway.

See **Runways and Their Markings**.

A "displaced threshold" is the rare case where the threshold is not right at the beginning of the runway because of the proximity of buildings or a hill. Sometimes the threshold is displaced because the runway surface at the original threshold has become weakened due to the impact of repetitive landings. Though a displaced threshold means less runway is available, that only applies to landings, since the entire runway can usually be used for takeoffs.

Thrust Vectoring

Method to increase the maneuverability of fighter aircraft by diverting the thrust of the engine or engines to supplement control surface directional input. Found in fifth-generation fighters, such as the MiG-35 and the F-22 Raptor.

THS (Trimmable Horizontal Stabilizer)

THS-OVM (Trimmable Horizontal Stabilizer Override Module)

As its name implies the stabilizer keeps the aircraft stable. It is a large airfoil sticking up vertically next to the **rudder** or one horizontally in front of the **elevators**. In fact, they are like super rudders and elevators that can be moved slowly often by a jackscrew.

By adjusting them, in what is called trimming, the aircraft the aircraft can be made to without need for constant elevator or rudder input.

The two Boeing **737 MAX** crashes resulted from the horizontal stabilizers repeatably pushing the nose of the aircraft down.

Tibetan Plateau

No airline routes pass directly over the Tibetan Plateau because its elevation is 14,000 feet, which does not allow aircraft to descend to 10,000 feet in the event of a **Decompression (Loss of Cabin Pressure)**. Furthermore, there is a dearth of airports an airliner could divert to.

Tilt Rotor Helicopter

These are **Helicopters** with rotors that tilt through almost 90 degrees, combining, in theory, all the advantages of the traditional helicopter with those of a fixed-wing aircraft to allow much greater horizontal speed. The most well-known one is the **V-22 Osprey**.

Titanium

Metal with the best strength-to-weight ratio, increasingly used in aircraft after initially being mostly used for engine components. It maintains strength at moderate temperatures. However, due to its extensive use in the latest airliners such as the Boeing 787, it is in short supply, and its price has increased to the extent that alternatives are being sought where possible. Much of it is produced in Russia with the CIA obliged to set up dummy companies to purchase it for the **SR-71 Blackbird**.

TOGA (Takeoff/Go-Around)

A control that automatically puts the aircraft in takeoff configuration when, say, a landing is aborted under instructions from air traffic control because an aircraft or vehicle is still on the runway. See **Go-Around**.

Toilets

Toilets on airliners are flushed with a minimum of water and a vacuum to suck out the detritus and dispose of it in a holding tank where it stays until it is drained on landing at the next airport. Stories of blocks of "blue ice" falling to earth are mostly folklore.

A gap to allow the entry of air underneath the toilet seat is essential lest someone with a large posterior cover it entirely and risk being sucked downwards, as apparently happened at least once.

Effluent from the washbasin is discharged overboard, but with the aircraft traveling fast and usually at a great height it is safely dispersed. Emirates wanted to use it after being made safe following treatment for their showers in First Class on their Airbus A380s but could not get permission to do so.

Tombstone Technology, "Tombstone Agency"

The notion that largely due to cost, essential modifications and measures to prevent accidents only take place after major accidents and loss of life, and often under subsequent pressure from the public and legislators.

In the US, the **NTSB (National Transport Safety Board)** that investigates the causes of crashes can only recommend ways to avoid similar disasters. It is the FAA (Federal Aviation Administration) that decides whether to put them into effect.

The FAA has even been called the "Tombstone Agency" because it usually only mandates changes after crashes involving significant loss of life.

The crash* of a Turkish Airlines DC-10 taking off from Paris is a terrible example. Two years before, in what is called the Windsor incident, the cargo hold door blew out on a DC-10 with the difference in pressure causing the rear cabin floor to bend downwards damaging hydraulic lines and controls for the rear engine. Because the floor had been reinforced for a reception with a piano and extra people, it did not buckle as much as it might have, and there remained just enough control for the pilots to land safely. The NTSB recommended changes to the door mechanism which would really have needed an **Airworthiness Directive (AD)**, but a "gentleman's agreement" between the heads of the FAA and McDonnell Douglas meant nothing serious was done and the Paris crash resulted.

Following that and the court cases costing McDonnell Douglas an enormous amount of money, the cargo hold door mechanism was redesigned with the DC-10 becoming a very safe aircraft much loved by pilots. There was the subsequent fatal DC-10 crash where an engine fell off on takeoff but that was mainly due to improper maintenance using a forklift to remove engines for repair. See **"Faster May Be Safer!"**

Top of Descent

The aircraft may have been cruising for hours but finally it comes to a point when the pilots prepare its descent towards the destination. This point is called the top of descent.

They try to do this gradually and evenly from quite some distance away to save fuel.

Touch and Go

This is a maneuver usually done for training and practice where the pilot touches down on the runway, at the appropriate point, and immediately lifts off without applying the brakes to perhaps do a circuit and repeat the maneuver several times.

TRACON (Terminal Radar Approach Control)

Radar used for airspace from the airport up to a height of 18,000 feet in the US and a maximum of roughly 50 miles. It corresponds to radar used by approach control and departure control.

Track

The actual path of aircraft over the ground, which will differ from the heading due to the wind blowing from the side.

Transponder

Initially developed in World War II as an IFF (identification of friend or foe) device that would reply to show whether aircraft were friendly when interrogated by radar.

Nowadays, transponders squawk (transmit) an aircraft's height, flight number, and **Squawk** code indicating its situation—there are special squawk codes for an emergency, hijack, radio not working, and so on. The latest transponders, called S-mode transponders, can squirt (transmit) a whole stream of information, including actual position determined by GPS, and the maneuver—climbing, turning, and so on—being undertaken, thus making future collision avoidance theoretically much more effective.

Traffic Pattern

Aircraft fly very precise rectangular patterns around an airport prior to landing. Each side of the rectangle has a name. Depending on the terrain and nearby buildings, the pattern can be anticlockwise (i.e., to the left of the runway on which the aircraft is landing in which the leg just prior to landing is called "left base"); or clockwise (i.e., to the right of the runway on which the aircraft is landing in which the leg just prior to landing is called "right base").

The leg aircraft fly parallel to the runway but in the opposite direction to the direction of landing is the downwind leg as aircraft normally take off into the wind. As said, traffic patterns may now be flown with either right-hand or left-hand turns, though in the early days of aviation they were to the left, and that is why the captain traditionally has the left-hand seat—so they could see where they are going on the turns.

Transition Altitude (TA)

When aircraft are flying low their pilots need to know their altitude (**Height Above Mean Sea Level (MSL)**), which are the elevations shown for land, mountains, and the like on their charts and for the **Minimum Safe Altitudes**.

Higher up, hitting the ground is not of concern but hitting other aircraft is. For this reason, above a certain height, called the transition altitude (TA), aircraft are assigned **Flight Levels (FL)** by air traffic control. TA is the altitude above sea level at which aircraft change from the use of local barometer derived altitudes to the use of flight levels.

When operating at or below the TA, aircraft altimeters are usually set to show the altitude above sea level. Above the TA, the aircraft altimeter pressure setting is normally adjusted to the standard pressure setting of 1013.25 hectopascals (which is equivalent to millibars) or 29.92 inches of mercury. The altimeter will show the altitude in feet, but these will be flight level feet. Though not perfectly accurate they are good enough for managing the aircraft.

In the United States and Canada, the transition altitude is 18,000 ft (5,500 m) because they have high terrain. In Europe, the transition altitude varies and can be as low as 3,000 ft (910 m). There are discussions regarding the standardization of the transition altitude within the Eurocontrol area.

In the United Kingdom, different airports have different transition altitudes, varying between 3,000 and 6,000 feet.

Transport Aircraft [Military]

A lot of freight is carried in the holds of airliners flying scheduled routes. Also, while some airliners can have freighter versions from the start, those that are suitable can be converted later. This can increase their residual value.

Transport aircraft for military use often must meet special requirements such as being able to carry extremely heavy items like a tank. This can require extraordinarily strong floors that would be too heavy for civilian use where the extra weight for the reinforced floors would greatly increase the amount of fuel burned.

Military transport aircraft have often played an invaluable role in providing humanitarian aid in times of disaster since some are large and designed to land on poor fields.

Trim (Adjusting the Trim)

Supplementary (mini-) control surfaces integrated with rudder, ailerons, or elevators that can be adjusted so the aircraft flies naturally straight, with wings level, or even climbs or descends without pressure on the control column (or sidestick). Adjusting these to achieve the desired effect is called trimming.

Pilots taking over manual control from the autopilot have on occasions been disastrously caught out by the presence of extreme trim that the computer has been applying to correct what it considers the wrong control column input, say, by the other pilot.

The most important element in trimming the aircraft is the pitch trim, which received much attention related to the Boeing 737 MAX.

TSA (Transportation Security Administration)

Set up under the DHS (Department of Homeland Security) in the aftermath of 9/11. Though responsible for security in all transportation modes, it is in the aviation domain that it has the highest profile.

It is responsible for screening passengers for dangerous or forbidden items, and, in future, checking passengers against watch lists of individuals deemed a threat. This task used to be largely the responsibility of the airlines. The FBI's Terrorist Screening Center (TSC), with the help of other agencies, draws up those watch lists.

Federal air marshals—armed officers traveling incognito on aircraft in the US and abroad—come under its remit. They also have a program whereby some aircrew can be specially trained in the use of firearms and how to deal with onboard threats and act in an unpaid capacity as an extra line of defense.

Like its counterparts in other countries, the TSA has been criticized for excessive rules about what passengers can carry on board an aircraft and for not concentrating on people who could be dangerous. It is adopting a more thoughtful approach and no longer confiscating nail clippers.

See **Behavior Detection Officers**.

Trident (Hawker Siddeley Trident) 1964/117

British short- to medium-range airliner with three tail-mounted engines. Good aircraft but sold only in small numbers, because spec'd for British European Airways (BEA). Many found service in China—perhaps because of price and because the BEA specification met China's needs too.

Turbulence in General

In aviation, there are four clearly defined levels of turbulence:

1. Light, sometimes referred to as "light chop" when it feels like driving over a slightly bumpy road.

2. Moderate, worrying for many passengers, but not a threat.

3. Severe, with unstrapped passengers thrown about and the pilots momentarily losing control.

4. Extreme.

Turbulence is often what makes so many people feel nervous about flying. This may be largely due to its repetitiveness when it does occur, thus progressively reinforcing one's fear. One should think of it as a bonus roller-coaster ride, and even enjoy it, for there is virtually zero possibility of it being a threat to the aircraft since they are so strongly made. The sight of the wings flexing is an indication of how resilient they are, and if they could not sustain flexing, they might break off. One way for passengers to overcome their fear is for them to think of it as driving on a road with potholes.

Seats nearer the front and over the wings are less affected by turbulence than those right at the back. Also, night or early morning flights are subject to less turbulence often caused by hot air coming off the ground.

There is a website (https://turbli.com) that purports to predict turbulence for a particular flight worldwide which one can simply enter. It may be reassuring in warning of the likelihood of a patch of not too severe turbulence that may not last too long. However, it is unlikely to be able to predict those dangerous freak events and passengers should always keep their seatbelts attached.

The site has interesting worldwide maps showing storms, wind direction, and so on, with turbulence pilot reports (PIREPs) for the US. Interestingly, turbulence along the ITZ (Intertropical Convergence Zone) is demonstrated.

See **Aerophobia**.

Turbulence Resulting in Injuries to Passengers and Crew

Considering the number of flights worldwide, cases where turbulence results in injuries are extremely few. However, they do happen, and seem to be getting more frequent, due presumably to climate change.

An incident on Singapore Airlines Flight SQ321 from London bound for Singapore on 20 May 24 refocused attention on the dangers of sudden-onset turbulence. The Boeing 777-300ER plane had 211 passengers and 18 crew.

To be precise, according to the Transport Safety Investigation Bureau of Singapore (TSIB), the flight was normal prior to the turbulence event. At 07:49:21 (UTC) on 21 May 24, it was passing over the south of Myanmar at 37,000 feet and likely flying over an area of developing convective activity. The Gravitational force (G), recorded as vertical accelerations, fluctuated between positive (+) 0.44 G and positive (+) 1.57 G for a period of about 19 seconds. (This would have caused the flight to begin to experience slight vibration.)

Around the same time as the onset of the slight vibration, an uncommanded increase in aircraft altitude, reaching a peak of 37,362 ft, was recorded. In response to this uncommanded altitude increase, the autopilot pitched the aircraft downwards to descend back to the selected altitude of 37,000 feet. In addition, the pilots observed an uncommanded increase in airspeed which they arrested by extending the speed brakes. While managing the airspeed, at 07:49:32 hours, it was heard that a pilot called out that the fasten seat belt sign had been switched on.

The uncommanded increase in aircraft altitude and airspeed was most likely due to the aircraft being acted upon by an updraft. (The autopilot was engaged during this period.)

At 07:49:40 the aircraft experienced a rapid change in G, with the vertical acceleration decreasing from plus (+)1.35 G (with passengers pushed quite hard down in their seats) to negative (-)1.5 G, within 0.6 seconds, resulting in the occupants not belted up becoming airborne. At 07:49:41, the vertical acceleration changed from that negative (-)1.5 G to plus (+)1.5 G within 4 seconds. This made people who were airborne suddenly feel heavier and ultimately 1½ times their normal weight and fall back down hard.

Though the press may have suggested the aircraft dropped considerably, it was merely a dop of little more than 4.6 seconds' duration resulting in an altitude drop of 178 feet, from 37,362 feet to 37,184 feet over the 4.6 seconds. The pilots manually controlled the aircraft for 21 seconds and reengaged the autopilot at 07:50:05.

Subsequently, over the next 24 seconds, acceleration ranged from +0.9 G to + 1.1 G, while the aircraft returned to 37,000 feet at 07:50:23.

The flight diverted to Bangkok because of the injuries sustained, with one fatality, a 73-year-old British man who died of a suspected heart attack.

Forty-six passengers and two crew members, including citizens of the United Kingdom, Australia, Malaysia, and the Philippines, were being treated in hospital in Bangkok. Twenty of those were in intensive care with spinal cord, brain, and skull injuries.

In the light of this sad incident the airline and others reviewed their policies regarding turbulence. Although the airline is adopting a more cautious approach to turbulence, including not serving hot drinks or meals when the seatbelt sign is on, and having the crew return to their seats, this may be wise but not solve the problem of what is a freak event.

Of course, a lawyer is taking up the cases of the injured, saying the airline was at fault for "possible failure to take the normal precautions to avoid an obvious and large area of thunderstorms and the failure to alert passengers to fasten seat belts."

According to the TSIB one of the pilots had called out at 07:49:32 that the "fasten seat belt" sign had been switched on. Even were this true, passengers do not immediately react to that. Furthermore, a passenger hospitalized with bad spinal injuries said he had been asleep at the time and only woke up when he fell back down injuring himself. He would never have heard the "fasten seat belts "Ding," and anyway there was not really time.

Perhaps, just like for road vehicles there should be a law that passengers should be belted up—even just loosely—except when going to the toilet, or not granted compensation on being injured when seated without their seatbelts attached. However, proving it would be difficult, with people needing to go to the toilet from time to time.

All one can hope for is that cabin crew constantly go around trying to persuade passengers to keep belted up, but with such severe incidents so rare, strict policing by flight attendants might not be popular, though for people's own good.

Furthermore, **Clear Air Turbulence (CAT)** is impossible to predict because the weather radar can only detect the movement of air containing moisture.

Turn and Bank Indicator

An instrument found even in the smallest GA aircraft. Indicates how the aircraft is banking and turning and whether the operation is coordinated—in other words, the bank angle is correct for the speed and rate of turn, with passengers not feeling they are falling to one side, though perhaps pressed down slightly in their seat.

Two-Person Rule

This rule applies in situations not only where the launch of nuclear weapons is involved but also in chemical plants, biological research labs, and banks where enormous sums are involved. Though it is to normally to ensure one "crazy" person does not cause a major disaster, it is sometimes for safety should one person be incapacitated.

The two-person **Missileers** in ICBM silos in the US, for instance, have their own keys that must be turned simultaneously with the points of insertion too far apart for one person to do that.

Incidentally, in theory the president of the United States with the "Nuclear Football" can launch a strike on their own initiative with no one guaranteeing they are sane or not drunk (see Nixon drunk in **"Madman Theory" in International Relations**) because there would be little time to act should enemy missiles be seen incoming. Though everything involving nuclear strike procedures is shrouded in secrecy it seems the joint chiefs of staff can stop a "crazy" president's orders being acted on.

In the case of US submarines, some keys can be kept in safes with combination locks and the combination only revealed in conjunction with the launch order.

Two Persons in the Cockpit Rule

Following the Germanwings suicide crash (where the copilot purposely crashed the aircraft after shutting the captain out of the cockpit) many airlines introduced a rule whereby there always had to be two people in the cockpit with one of the cabin crew coming in when one pilot needed to absent himself. However, it was found to be impractical and, in most cases, abandoned and other precautions taken.

See **Mental Health of Pilots** and **Psychometric Evaluation of Pilots** where this complex question of a pilot's mental state is covered in considerable detail.

~ U ~ V ~

U-2 "Dragon Lady"

See **US MILITARY AIRCRAFT ~ Lockheed U-2**.

UAM (Urban Air Mobility), Air Taxis

Urban Air Mobility is expected to become a reality in Europe by 2026. New technologies such as electric propulsion and enhanced battery capacity, applied to vertical takeoff and landing systems, make this possible.

The first commercial operations are expected to be the delivery of goods by drones and the transport of passengers, initially with a pilot on board. Later, remote piloting or even autonomous services could follow.

Several exploratory projects are under way and some European manufacturers have already applied for certification, including for piloted vehicles for passenger transport. EASA is working with them on determining their airworthiness.

In the US, investors see a big future in UAM but do not expect it to really boom until the 1930s. With eVTOL air taxis already becoming viable, the expected advances in battery technology will surely take UAM to a new level.

For now, the airport shuttle—hitherto the domain of the helicopter—appears to be the be best place to start as there are the facilities for taking off and landing as well as a ready clientele. Electric charging facilities just have to be installed.

There are two companies in the US leading the field, Joby and Archer, with agreements with airlines with the aim to make reserving an air taxi seamless when booking the ticket for the flight.

Interestingly, Joby makes a point of how exceptionally quiet their craft is, with it difficult to distinguish its sound from background noise. It has a range of 100 miles on one charge and a maximum speed of 200 mph.

They are cooperating with Delta Airlines in the US, with service expected to begin in New York City where transport from

JFK Airport to Central Manhattan can take more than an hour by car, compared with just seven minutes by air taxi.

Joby expects to launch its commercial passenger service in 2025.

The picture below was chosen because it shows the four seats for the passengers and the space in front of them for the pilot, together with the six strongly built propellers. The redundancy they provide should be reassuring.

[Joby Aviation]

Arrangements are being made for Joby services to be offered in Japan, Dubai, and so on.

Archer, the other US UAM company mentioned, is treading a very similar path, and if anything, finding more contacts abroad. They have also been working with United Airlines with feelers from Southwest Airlines. Both companies have supplied the USAF and benefitted from subsidies from them.

Also, consider the UK-based company Vertical (https://vertical-aerospace.com) with their VX4 prototype carrying four passengers and one pilot eVTOL

The **Vertical Flight Society** based in the US has chapters all over the world and is a good place for following symposia and events related to eVTOL and other craft taking off vertically.

UAS (Unmanned Aerial Systems), the New IED

IEDs (Improvised Explosive Devices) on land are often deadly and a constant threat, no matter how sophisticated the troops. Now we have drones, often simple small UAVs, typically commercially available quadcopters outfitted with small explosives, able to not only kill and maim troops but also damage major installations, such as those that hit the Saudi Aramco ones in September 2019. US bases are also vulnerable, leaving the US feeling it no longer has air superiority in some respects.

UAV (Unmanned Aerial Vehicle)

The military are investing considerable funds in the development of usually very small pilotless aerial vehicles, such as the US Predator, which, thanks to modern technology, can fly under their own GPS guidance systems or under remote control to gather intelligence—SIGINT, photos, and so on—or even to attack small targets, such as vehicles carrying people to be "taken out."

The FAA has now made the registration of all privately owned drones obligatory, and while they are of great value for taking photos, they have been put to bad use as well.

See **Gender Inclusive Language, UAV = Uncrewed Aerial Vehicle**.

UCAV (Unmanned Combat Aerial Vehicle)

See **Collaborative Combat Aircraft (CCA), Loyal Wingman**.

UFO Now Called UAP (Unidentified Anomalous Phenomena)

Unidentified flying objects (UFO) are now being referred to as unidentified anomalous phenomena because they are not flying with wings, and some are believed to be able to move through water as well as air and space.

An interesting Aviation Week podcast hosted by Guy Morris discusses the question of UAPs with a former US Navy F/A-18F Super Hornet pilot and a combat veteran. He is also the first US active duty pilot to come forward publicly about regular sightings of UAPs or UFOs as they were once and sometimes better known.

Without going into detail, Ryan, the interviewee, describes how objects were sometimes picked up on their sophisticated radar and later also their weapons systems.

It seems UAPs are being taken more seriously these days.

Ullage

Ullage is the space above a liquid in a container, such as the air above the wine in a wine bottle.

A British Royal Air Force C130 Hercules was shot down by small-arms fire in Iraq in January 2005, killing all ten on board due to an explosive mixture forming in the ullage as the fuel was depleted. Had the aircraft's wing tanks been fitted with ESF (explosive suppressant foam) as recommended and had been the case for the US's Hercules since the 1960s, the disaster might well have been avoided.

Ultra-Elite

While those that can travel in first and business class do so mainly to make long-haul flights tolerable, another unspoken reason is that some premium classes can offer a degree of pampering and flattery that the beneficiary, although powerful, may not enjoy in the family setting at home! In the US, this is especially true if they are one of those very special passengers for whom the airline goes that extra mile, for even in business and first class, there is a distinction between the favored few and the rest. (*All animals are equal, but some are more equal than others*, as George Orwell famously wrote in *Animal Farm*.) The US majors have an unpublicized club for those special people, often ones who fly a lot and can determine how their staff fly.

In Australia there were complaints that Qantas had undue influence on the government, with politicians, officials, and even the prime minister's son being given privileged access to the exclusive Chairman's Lounge, like the ones for the "elite" in the US. This was cited in the context of a Senate inquiry examining the decisions behind a federal government ruling to block Qatar Airways' bid to double its flights to major Australian cities.

Ultralight Aviation (Microlight Aviation)

Because so many of us—including the retired with money and time but not able to pass a strict medical test—want to fly for fun, the chance to fly subject to minimal regulations is very appealing. In fact, in some countries including the United States, Canada, and the United Kingdom, ultralight aviation may represent as much 20 percent of the general aviation sector in terms of the number of craft.

Interestingly, advances in lightweight materials such as carbon reinforced plastics, a weight limit of say 300 kilograms for an ultralight means that with sufficient funds it is possible to purchase one able to fly a very considerable distance, and in some cases one carrying two people.

However, there are strict limitations regarding when (dawn to dusk), how high, and notably where they can fly, thus limiting their applications.

In the US, the FAA does not register ultralights meaning it is difficult to know exactly how many there are. Also, they lay the responsibility for their airworthiness and the aptitude and medical status of those flying them on the people concerned. Their main concern being that they are not flown in congested airspace.

Othe conditions are:

1. Under 5 US gal (19 L) fuel capacity.
2. Empty weight under 254 pounds (115 kg).
3. Top speed of 55 knots (63 mph; 102 km/h).
4. For aircraft with a maximum stall speed not exceeding 24 knots 28 mph; 44 km/h) the FAA has made provisions.

Though electrically powered ultralights are gaining attention, the FAA has not agreed to the "empty weight," treating the weight of the batteries as that of fuel.

In the UK, there is British Microlight (BMAA) providing information about all aspects of ultralight aviation, including training courses. While in the UK and Europe there may be more oversight than in the US, the weight limits are higher with twin occupancy even permitted in some cases.

Though not covered here, ultralights also include trikes, powered parachutes and paragliders, gyroplanes, hang gliders, balloons, and sailplanes (gliders).

"Unable!"

Though ATC has overall authority, the FAA, in a 2021 online video simulation aimed mainly at the general aviation sector, stressed that the PIC (Pilot in Command) is *in command* and the *final arbiter* and should reply "Unable" to an instruction from ATC that for one reason or another they do not feel they should follow.

Many will remember the dramatic exchanges between Sully and ATC in the *Miracle on The Hudson* incident, where the term unable featured:

> **Departure to Cactus:**
> "All right, Cactus 1549, it's going to be left traffic to Runway 31."
>
> **Cactus 1549 (Sully):**
> "Unable."

Unbundling

"Unbundling" can allow an airline to offer optional services for an added fee in even in business class, as low cos carriers do in their standard class.

Undercarriage/Landing Gear

For the complex mechanical system between the body of the aircraft and the ground, US pilots tend to favor the term "landing gear" as opposed to "undercarriage." They will say "Gear down" to confirm the undercarriage is down.

Most airliners have tricycle undercarriages, with almost all the weight carried by the main wheels near the center of gravity under the wings, and a relatively tiny nose wheel to provide equilibrium and steering.

Undercarriages are made to be able to withstand considerable mistreatment including hitting the runway with the wheels at a slight angle to compensate for a crosswind—at the last minute, the pilot will often dip the wing into the wind to reduce the angle required.

The B-52 bomber almost uniquely has an undercarriage that swivels. This is because the low-slung outboard engines and fuel tanks at the end of the downward sloping wings are so near the ground it cannot dip a wing into a crosswind.

Unintended Consequences of Guerilla Action

1. The **worst (multi-aircraft) disaster ever*** resulted from the closing of Las Palmas Airport in Spain's Canary Isles due to a bomb threat by Basque separatists.
   Aircraft, including two Boeing 747s, one from New York and one from Amsterdam, were diverted to nearby Tenerife Airport where there was not proper parking spaces for them all. When flights to Las Palmas resumed, the taxiway parallel to the runway was blocked by parked aircraft, resulting in the two 747s having to back-taxi down the runway itself, following each other with the Pan Am one behind meant to turn off well before the end. However, in the fog the KLM one in front turned round and began its takeoff run before they had done so. As a result, they collided with a total death toll of 583.

2. **Another major crash*** indirectly resulting from guerilla action occurred at Cali in Columbia where the air traffic controller at the airport could not see the position of the incoming American Airlines 757 flight because the radar had been sabotaged.
   To approach the airport, the aircraft had to fly down a valley with extremely high mountains on either side using waypoints (beacons) to confirm its position. Halfway down the valley, the captain entered the wrong

beacon into the flight management system causing them to veer more than 100 degrees off course and fly over the mountains into the next valley, which neither pilot noticed as they were rushing to prepare for the landing. Finally, when they turned back to return to the valley the ground proximity warning system (GPWS) sounded, and the pilots engaged full thrust and pulled back on the yoke to gain height.

The Boeing 757 is perhaps the airliner with the greatest amount of redundant power, and had they retracted the air brakes, it would have got them over the mountain. As it was, it struck the trees right at the top of a ridge and shattered.

With no radar the controller had no way of knowing where they were, and help only arrived the next day. Four people survived.

Upset Prevention and Recovery Training (UPRT)

US AIR FORCE (USAF) and Other Branches

~ "Special" Programs

1. Aurora
   SR-91 Son of Blackbird

2. Rods from God
   Kinetic Bombardment, or Kinetic Energy Projectile

3. Boeing X-37B Orbital Test Vehicle

4. RQ-180 Stealth Drone

5. NGAD (Next Generation Air Dominance)

~ Integration of Platforms, Sensors, and Troops

The US military has ambitious plans for aircraft, drones, and missiles, with some systems only to be realized around 2035. Furthermore, they envisage a war with sensors, aircraft, drones, soldiers, pilots, officers, and troops, all linked digitally, and working together. Despite AI making it (perhaps dangerously) quick to react, it can be slow to adapt.

The war in the Ukraine has shown the importance of adapting and the immense value of units reacting quickly on their own ideas, say, adapting drones with new software to counter jamming, and finding new ways to deploy them.

This has led US planners to consider a two-tier digital integration system with the lower tier with everyone and everything linked and a flexible upper tier.

~ Specialist Magazines with Info for Other Countries

We can only touch on this because there are so many aircraft, drones, missiles, and so on. However, the following three sites provide links to almost everything and enable one to keep up to date not only for the US but also for other countries.

Here are links to particularly useful pages:

1. AIR & SPACE FORCES Magazine
   Weapons and platforms
   (https://www.airandspaceforces.com/weapons-platforms/)

2. THE WARZONE
   Navies
   (https://www.twz.com/category/navies)

3. DEFENSE NEWS
   Air warfare (https://www.defensenews.com/air/)

~ Summary of Platforms and Missiles

We mention just some US military platforms and missiles together here as it is difficult for the reader to remember the name of the manufacturer, which sometimes changes.

Bombers:
B-1B Lancer
B-2 Spirit
B-21 Raider
B-52 Stratofortress

Fighters/Attack:
F-15 Eagle
F-15E Strike Eagle
F-15 EX Eagle II
F-16 Fighting Falcon
F-22 Raptor
F-35 Lightning II

Command Control and Communications (C3)
And Battlefield Management (BM):
E-4 NAOC
E-8 JSTARS
E-9A Widget
EA-37B Compass Call
EC-130H Compass Call

Air Lift:
C-130H Hercules
C-130J Super Hercules
C-17 Globemaster III
C-37 Gulfstream
C-5 Galaxy
C-5M Super Galaxy
VC-25 Air Force One

Intelligence Surveillance Reconnaissance (ISR) :
U-2S Dragon Lady

Special Ops:
CV-22 Osprey

Strategic (Non-nuclear and Nuclear):
AGM-183 ARRW
AGM-86 ALCM
B-61 Thermonuclear Bomb
LGM Minuteman III**
**** Work is being done at great expense on the next generation.**

AIRCRAFT in No Special Order

~ ~ B-2 Spirit, Stealth Bomber 1997/21

Designed by Northrop, now Northrop Grumman, as a low-observable heavy strategic bomber. With a crew of two it has a range of 6,000 nautical miles and 10,000 nautical miles with a single aerial refueling.

Though primarily envisaged to deliver nuclear weapons, it can carry a whole range of conventional weapons and even serve as a stealthy platform for large air-to-surface standoff weapons. Development began in the early eighties with a first flight at the end of the decade.

The overall per unit cost rose to US$1.4 billion (estimated cost of hull loss in accident on Guam described below) since only 21 were ordered rather than 132 envisaged. Congress had decided the cost not only of the aircraft was too much, but the great cost of operating and servicing had to be considered. They thought it was no longer needed.

Only 20 are in operation because one crashed on takeoff on Guam on February 23, 2008.

In that case, it was still on the runway and about to lift off when the nose suddenly pitched up 30 degrees making it impossible to fly or recover. Surprisingly, the two pilots ejected safely before it crashed back down and caught fire.

Guam is a hot, humid place, and moisture penetrating an air data sensor during calibration meant the computer was fed erroneous information. Note that stealth aircraft with no inherent stability depend on the computer and sensors.

With the B-2 departing from the US mainland for, say, operations in the Middle East with en route refueling, missions can be extremely long and for that reason they have food preparation and toilet facilities.

~ ~ B-21 Raider

Named "21" because it is to be the first US bomber for the twenty-first century.

The secretive sixth-generation bomber was unveiled with a rollout involving a lot of precautions to prevent detailed photographs on December 2, 2022, at Northrop Grumman's production facilities in Palmdale, California.

~ ~ Lockheed F-16 Fighting Falcon 1978/4,600+

A multirole single-engine fighter that first flew in 1974. It was originally developed and designed by the fighter division of General Dynamics, later bought by Lockheed Martin.

Besides playing a key role in the US Air Force for several decades, it has been exported to 25 countries or more.

The secret of its success is great performance combined with a relatively low price and simplicity making it cheap to maintain. Upgrades have made it useful even today, though the F-35 is taking its place.

Can operate at speeds of 1,500 mph and up to 60,000 feet, as a multirole fourth-generation fighter. Nevertheless, it is having a limited new lease of life with modern avionics, radar, and sensors due to problems over F-35 availability.

~ ~ Lockheed F-22 Raptor 2005/195

This is an exceptional but terribly expensive fifth-generation air superiority fighter. It was finally ready for service in 2005, with orders expected to be as many as 750 units only to find it was no longer needed following the collapse of the USSR.

Its great cost of around US$150 million per unit could not be even partially recouped through export orders as Congress had forbidden them for fear that the advanced technology would fall into enemy hands. Disappointed Australia and Japan were particularly interested, with Israel initially so, but finding it too expensive.

At the time of writing, it had only shot down a high-flying Chinese balloon.

Of late, its virtues are being appreciated and made use of.

~ ~ Lockheed SR-71 Blackbird 1966/32

Developed at the Lockheed **Skunk Works** in large part from the ultra-secret Lockheed A-12 Oxcart (1967/12) (first flight 1962) with much input from the talented Kelly Johnson, as a Mach 3 reconnaissance aircraft designed for the CIA to take the place of the U-2.

Because of cost problems, and the fact that satellites were able to take over much of the role of the U-2, the A-12 Oxcart was discontinued in favor of what was to become the SR-71.

Though it was said that leakage of fuel when cold was the reason only a little fuel would be loaded on the ground to allow it to take off, and the full fuel load added from an air refueling tanker up in the sky, this was only partly true.

Another reason would be to allow the airframe to heat up and expand so fuel could be loaded without leaking out.

There was also the risk of finding a fault after liftoff and having to land back immediately with a full load of fuel.

In fact, twelve were lost in accidents, but none to enemy action, though there was a close shave when an engine failed, and it had to drop down to a lower altitude with Russian fighters setting off to intercept it. However, Sweden sent up fighters to escort it to safety.

At the time of writing, it holds the record, established in 1976, for the fastest air-breathing manned aircraft.

Married Men Only!

Following the defection of a Russian pilot in his MiG to Japan with its secrets, the Americans decided it would be safer to have pilots of their ultra-secret ~ ~ **Lockheed SR-71** to be exclusively married men as they would be less likely to defect for a financial reward.

~ ~ Lockheed F-35 Lightning 2

Stealth multi-role fighter costing US$100,000,000, top speed of 1,200 mph and a range of approximately 1,350 miles.

The F-35 can operate in stealth configuration with weapons only in its internal armaments bay and very likely with other stealth aircraft and drones to knock out an enemy's detection systems in the first two or three days of combat with them acting as **Wild Weasels**.

Once there is no longer the need for stealth, it can switch to what Lockheed Martin calls "Beast Mode" with externally mounted weapons in addition to those in the internal armaments bay and hence a larger radar footprint. Much more armament can be carried, including weapons too long to fit in the bay.

~ ~ Lockheed SR-72 "Son of Blackbird"

"Son of Blackbird" is only a temporary name, for names are only properly attributed to aircraft on their entry into service depending on many considerations. A prototype is expected to fly in 2025 with entry into service around some years later.

In some respects, it will be like the SR-71 in that it will have an air-breathing engine and able to take off under its own turbine power without rocket assistance.

Flying faster than the SR-71, it will have a so-called turbine-based combined cycle (TBCC) engine using the same inlet nozzle and outlet but different paths between them with it able to work as a conventional turbine for takeoff (and landing) from zero up to Mach 1, and then switch to ramjet speeds up to Mach 3, and then scramjet speeds up to Mach 6, or potentially Mach 10.

Since the heat generated on the surfaces would be higher than for the SR-71, **Carbon-Carbon Composites** would have to be used instead of titanium. It would have to rely on its great speed rather than stealth because the high surface temperature generated would not only be obvious but impossible to coat with conventional **Radar Absorbing Material (RAM)**.

Most commentary suggests that it would be an unmanned vehicle and make great use of artificial intelligence. It would mainly be used for reconnaissance but also have a strike capability and represent a great leap in USAF capability.

~ ~ Lockheed C-130 Hercules Family 1956/2,700+

This includes the legacy and Super Hercules with 70 nations using them.

~ ~ Lockheed C-130J Super Hercules

This is the latest version with more powerful Rolls-Royce engines and six-bladed propellers. This changed the flying characteristics, and Lockheed initially had a hard task dealing with this. There was a problem over counterfeit computer chips used by a Lockheed subcontractor that could cause the instrument panel to fail. This was the subject of a US congressional investigation.

~ ~ Lockheed C-5 Galaxy 1969/131

Large US military transport aircraft able to carry heavy loads long distances. Teething and production problems at the beginning which coincided with the other financial losses Lockheed was making over the TriStar L-1011—a truly excellent airliner in advance of its time, but as mentioned above, not able to be sold initially due to Rolls-Royce being unable to supply the newly developed engines on time. The US government had to rescue Lockheed with a loan.

Besides being designed to be able to transport massive items, the C-5 Galaxy had a special undercarriage enabling it to land and take off from poor quality airfields near the front line. This has never really been used militarily though the ability of the undercarriage to *kneel* so trucks can drive straight in and out of the front is constantly taken advantage of.

Interestingly, its wings are dihedral to make it more maneuverable. At one stage the aircraft had to have its load restricted to 80 percent to avoid fatigue reducing its life. However, an updated M-version with better engines, improved avionics, and aluminum alloys not available in the 1970s mean its life should be extended to the 2040s.

It has not only been used in military conflicts but also quite frequently for bringing humanitarian aid following disasters.

~ ~ Lockheed U-2 (Lockheed Skunk Works) 1956/104

The original U-2, called the "Dragon Lady," first flew in August 1955 having been developed in Lockheed's **Skunk Works** in complete secrecy. It was made famous or infamous when one piloted by Gary Powers was shot down by a Soviet missile in 1960.

Many will have been surprised to have read in 2000 that it had completed flight testing and installation of the Collins Aerospace Senior Year Electro-Optical Reconnaissance System (SYERS-

2C). The multispectral camera can use its various sensors to see not only in darkness but also through haze, smoke, and cloud.

The US Air Force says the U-2 is also, of course, equipped with the usual spying gear, an optical bar camera, synthetic aperture radar, signals intelligence electronics, and communications gear. Communication and integration with other platforms such as the F-22 Raptor and F-35 is a matter of course.

The 33 U-2s flying today were built in the 1980s and further modified. It was thought that the U-2 would be superseded by the **Northrop Grumman RQ-4 Global Hawk** unmanned air vehicle, but the U-2 always seems to get a new lease of life. Though normally restricted to skirting sensitive countries in international airspace, the fact that it can fly at 70,000 feet means that it can peer deep into their territory.

Ever since the U-2's conception everything was done to make it as light as possible, for every pound saved meant it could fly that much higher. On takeoff, the wheels on stilts on either side supporting the wings are jettisoned leaving just a retractable bogie under the fuselage and a tiny wheel at the rear. These are used for landing with the aircraft tipping onto a wingtip when it slows and stops. Then the wheels on stilts are reattached allowing the U-2 to taxi to its hanger.

The U-2 is exceedingly difficult to land, being sensitive to crosswinds, and prone to float over the runway due to the ground effect associated with its high-lift wings. In fact, it must be fully stalled to land but with the undercarriage so feeble it cannot do this much more than two feet off the ground. Consequently, on landing the U-2 is followed down the runway by a chase car with another U-2 pilot briefing the U-2 pilot on his height and attitude and when it is possible to perform the stall.

The Soviet Union apparently finally managed to shoot down Gary Power's U-2 in 1960 by noting that it flew in a straight line and firing missiles—without arming the normal tracking systems that would show up on the U-2's systems—to where they predicted it would be.

~ ~ V-22 Osprey 2007/400+

A tilt-rotor helicopter developed by Bell Helicopter and Boeing Defense that would be able to take off and land like a helicopter and fly at relatively high speed like a fixed wing aircraft.

As said, it has been plagued by problems with accidents involving loss of life. In late 2023, just when the Japanese were about to receive a further batch, one operated there by the US marines crashed leading to the Japanese putting it on hold and other V-22 Ospreys in the US taken out of service for several months.

It is being improved with great modifications in progress or planned.

Despite its shortcomings, it is seen as an essential tool, particularly for certain special ops.

A new tilt-rotor craft is planned.

~ ~ Doomsday Plane

See **Doomsday Plane** and **Air Force One**.

~ DRONES

~ ~ MQ-1 Predator Drone 1995/360 (285 RQ-1, 75 MQ-1)

This historically important drone began as a "garage project" by an Israeli immigrant, Abraham Karem, that in 1983 became a long-endurance reconnaissance UAV prototype for **DARPA** (Defense Advanced Research Projects Agency) called the Albatross. Further work and designs, one of which was the Amber, culminated in the GNAT 750 at which point Karem's company, and the design, were acquired by General Atomics sponsored by DARPA.

The GNAT 750 was deployed by the CIA over Bosnia and Herzegovina in 1993 and 1994. Despite problems, its great potential for medium altitude reconnaissance was apparent and a larger improved version developed. First called the RQ-1 Predator, it later became the MQ-1 Predator able to carry weapons including **Hellfire Missiles** to take out individuals of "concern."

In RQ-1, R stood for Reconnaissance, Q for unmanned, and 1 as being the first in what might be a whole series. The change to M meaning “multi-role” was in response to its increasingly important role as a deliverer of offensive ordonnance. Even before that, Predators were equipped with lasers able to designate targets for piloted aircraft accompanying them, with them sometimes lighting up the target to automatically guide the missiles. All versions had sophisticated sensors, radar, cameras, and reconnaissance systems too complicated to describe here. The whole flight could be controlled by satellite from any command and control center staffed by two people. These control stations could be transported in a C-130 Hercules to take them nearer the action thus reducing the signal time lag.

It was retired by the US Air Force in 2018, with limited use by others continuing. Its place was taken by the MQ-9 Reaper.

~ ~ RQ-4 Global Hawk 2001USAF/40?

The R stands for reconnaissance and the Q for unmanned aerial vehicle. A high-altitude 60,000 feet, long endurance 32 hours, long-range 12,300 nautical miles or 22,780 km large UAV.

Made by **Northrop Grumman**.

Able to network ISR data (Intelligence Surveillance Reconnaissance data) to other platforms via ground-based command and control. Though still an asset, the Global Hawk is vulnerable in contested environments and after years of sterling service is being gradually withdrawn.

Developed into the **MQ-4C Triton**.

Made by **General Atomics**.

~ ~ MQ-4C Triton Drone 2018/65+?

Navalized Version of **RQ-4 Global Hawk**.

Range 24+ hours

Ceiling 50,000 feet

Speed 331 knots (max)

Length 14.5 meters

Reinforcements to the airframe and wing of the Global Hawk coupled with de-icing and lightning protection systems, enable it to descend through layers of cloud and get a close view of ships and other targets at sea.

Cooperation with Australia as regards their financing and deployment. Also deployed and redeployed to Guam but number confidential.

Made by **Northrup Grumman**.

~ ~ MQ-9 Reaper Drone 2007/310+

The MQ-9 Reaper is an upgrade of the General Atomics MQ-1 Predator that was withdrawn from US Air Force service in 2018. The Reapers are faster, have better sensors, increased payload of munitions, and greater range.

Range 1,100 km or 135 mph (endurance up 42 hours).

Made by **General Atomics**.

~ ~ Roadrunner

Drones can be made to do all sorts of things. An interesting example is the Roadrunner made by the US company Anduril:

Costing in the "low hundreds of thousands of dollars" and to get cheaper as quantities increase, with its cost justified by being so much cheaper than the patriot missile and the fact that it can loiter at the place of interest—say where drones are expected or tanks may be—and return to its launching place where it can land coming down tail first like the SpaceX Falcon 9 space launch rocket booster.

It can land, get refueled, and take off inside of a couple minutes. The system can allow for a single individual to launch, manage, and recover multiple vehicles by themself.

~ ~ MQ-9B "SeaGuardian"

This is a version specifically designed for maritime operations, with it being interesting to see how they and other versions work.

According to General Atomics publicity:

What the drone does is, in effect, serve as the distant eyes and ears for naval commanders. Its onboard sensors can see all through the visual and infrared spectrum, including—with its onboard multi-mode radar—through clouds, fog, mist, or smoke. Other onboard systems can hear throughout the radio frequency spectrum, collecting intelligence of all kinds that contributes to the most complete common operating picture possible.

Helicopters can drop sonobuoys with the drone following up to monitor them, and able to detect a submarine. This means other assets can deploy with the benefit of precise data about the target's location and course and then attack. Satellite operation means that MQ-9B's pilots and crews can be located anywhere.

SeaGuardian can be configured to conduct a number of maritime ISR operations, including but not limited to:

Humanitarian Assistance/Disaster Relief
Search and Rescue
Law Enforcement
Anti-Surface Warfare
Anti-Submarine Warfare
Airborne Mine Counter Measures
Long-Range Strategic ISR
Over-the-Horizon Targeting

While drones are militarily used all the time for multiple purposes, they do come to particular public notice, for instance when used by the US to "take out" most wanted terrorists.

See **Hellfire Missile, R9X: Knife Bomb, Flying Ginsu**.

~ MISSILES [Examples only]

~ ~ LGM-35A Sentinel

The five-decade-old Minuteman III will be replaced by the $100 billion LGM-35A Sentinel beginning in 2029, according to Defense News.

~ ~ Patriot Missiles

The MIM-104 Patriot missile system is the US principal system for defense against air platforms and missiles. It was first, in the 1960s, destined to be an antiaircraft system, but now can detect, track, and target ballistic, cruise missiles, munitions as well as aircraft.

USA (United Space Alliance)

In its internal functions, NASA uses "USA" to denote the United Space Alliance, a distinct entity based in Houston, with Boeing, Lockheed, and others working together, born out of NASA's desire to avoid dealing with too many individual contractors.

The "USA" marking on rockets and so on can be somewhat confusing as it does not necessarily stand for United States of America as one might suppose. The massive Saturn-5 rocket that sent astronauts to the moon had three rocket stages, each made by a different company: 1st stage Boeing; 2nd stage North American Aviation; and 3rd stage Douglas Aircraft Company.

UTC (Coordinated Universal Time/Zulu)

This odd but now commonly used abbreviation for Greenwich Mean Time was decided at the International Telecommunication Union. Many wanted CUT standing for *Coordinated Universal Time* in English, but those wanting TUC standing for *Temps Universel Coordonné* in French stood in their way. They ended up with this face-saving compromise pleasing no one.

V-Speeds

Airspeeds (velocities) of special significance—especially at takeoff and landing—calculated according to conditions like the total weight of the aircraft, air temperature, and flap settings.

There are dozens of V-speeds, referring to the speed at which the aircraft can or should do, or not do, such and such. However, though there are many including ones at which the flaps can be extended, and the undercarriage lowered, the four most significant ones are:

V_1, V_R, and V_2 for takeoff, and V_{REF} for landing.

See **Taking Off (Taxiing, Takeoff Roll, Liftoff)**.

Vacuum Bomb

See **Thermobaric Weapons.**

Variable Cycle Engine (VCE)

See **Adaptive Cycle Engine (ACE)**.

A jet engine that can vary the mode in which it functions to concord with the flying regime, say, subsonic, transonic, and supersonic, rather than being a compromise as generally the case now.

There is now the envisioned engine for the **Lockheed SR-72 Son of Blackbird** that will have a so-called turbine-based combined cycle (TBCC) engine using the same inlet nozzle and outlet but different paths between them with it able to work as a conventional turbine for takeoff (and landing) from zero up Mach 1, and then switch to ramjet speeds up to Mach 3, and then scramjet speeds up to Mach 6, or potentially Mach 10!

VASIS (Visual Approach Slope Indicator System)

Array of directional lights near the aiming touchdown point on the runway. They appear to change color according to whether the pilot is below, on, or above the glide path.

They are on the left of the runway and usually consist of two bars. If the aircraft is on the glide slope, the so-called upwind bar shows red and the downwind bar shows white. If they are too high both bars show white, and if too low both bars show red.

For fields handling large aircraft there may be a three-bar VASIS, giving the option of two visual glide paths.

~ PAPI (Precision Approach Path Indicator)

Similar but more precise and not always installed because it is more expensive to make and maintain.

VC10 (Vickers VC10) 1964/54

The VC10 was an excellent airliner, with four tail-mounted engines that passengers liked with the result that on transatlantic routes load factors were better than for the 707. Unfortunately, it had been designed with extra power to cope with the short runways and high-altitude airports—mainly in Africa—that lay on Britain's "Empire" routes to the Far East and Australia. This meant it used more fuel than the 707. Finally, with sales of the VC10 not taking off, those very airports began lengthening their runways for the "underpowered" 707.

The Russian Ilyushin-62 was similar and some countries that had been considering buying the VC10 when it was discontinued bought that aircraft.

Vents in Cabin Floors

When the cargo hold door of a DC-10 that had taken off from Paris blew out, the difference in air pressure between the hold and the passenger cabin above caused the cabin floor to buckle downwards and fracture the triple hydraulic lines attached underneath. This rendered the aircraft uncontrollable with the result that it plunged into the ground killing all remaining on board—six passengers had been blown out of the hole in the floor.

A similar cargo hold door failure had occurred previously on the DC-10 without tragic results because the floor of the passenger cabin had been reinforced to support the weight of a piano and extra people for a publicity event held on board the aircraft.

Vertical Flight Society

Formerly the America Helicopter Society (AHS).

Website: https://vtol.org.

The world's only international technical society for engineers, scientists and others working to advance vertical flight technology.
We bring together industry, academia and government to tackle the toughest challenges in vertical flight.

They have become international with chapters all over the world and a good place to find out about symposiums and conferences and the latest news.

Members receive their magazine *VERTIFLIGHT* (https://vtol.org/publications/vertiflite-magazine), published six times a year.

Vertical Speed, Vertical Speed Indicator (VSI)

The vertical speed is the rate of ascent or descent, and the VSI (vertical speed indicator) is one of the key instruments in all aircraft.

In modern aircraft the computer issues a verbal warning, "Sink rate … Sink rate …," if it finds the rate of descent is dangerously high for the circumstances. Some aircraft, such as the early Boeing 727s, could sink almost irretrievably fast without the pilots realizing the danger, for very considerable engine power was required just to stay level with full flap.

Vertical Stabilizer (Tailfin)

The normally fixed vertical fin on the tail which helps the aircraft fly straight without yawing. The **Rudder** attached to it enables the aircraft to turn left and right though most of the turning is done using the ailerons to make the aircraft to bank to the left and right and turn naturally.

Vertiport

We have airports and heliports, and now the vertiport for the anticipated **UAM (Urban Air Mobility),** air taxis and similar vertical takeoff platforms for **eVTOL (Electric Vertical Takeoff and Landing)** craft.

For instance, Joby Aviation of the US is planning air taxi services in Dubai and the United Arab Emirates, with four vertiports planned for the area and expected to be operational in 2026.

Archer has announced a new partnership with private aviation services company JetEx to integrate the Midnight eVTOL air taxi into the latter's network of several dozen FBOs.

VFR (Visual Flight Rules)—as Opposed to IFR

The pilot must be flying in good visibility conditions and is responsible for seeing and avoiding other aircraft, towers, mountains, and so on. A pilot may be flying under instrument flight rules (IFR) even in good visibility.

VHF (Very High Frequency)

A radio signal that gives good quality line-of-sight transmissions, which means the signal can be blocked by mountains.

Used for VOR and short-range radio communications.

Viscount (Vickers Viscount) 1950/445

A British four-engine turboprop airliner originally seating around 50 passengers, with a later version seating up to 71. It saw many years of service and was the most successful British airliner ever.

VOR (Very High-Frequency Omnidirectional Range)

A radio beacon on the ground whose VHF radio signal enables aircraft to determine both distance (range) and compass bearing (radial) from it. The basic principle is like that of a lighthouse with different-colored glass to indicate segments. In the VOR there are two signals (one omnidirectional pulse indicates the start of a rotating directional signal whose angle can be calculated by the time difference).

As they depend on VHF radio signals, VOR beacons have the disadvantage of not being visible over the horizon or behind obstacles, such as mountains. With the development of more accurate systems such as GPS many VOR beacons are being taken out of service, though some are being kept as backups.

VTOL (Vertical Takeoff and Landing), eVTOL

A VTOL aircraft or drone can take off vertically without the need for a runway and then switch to horizontal flight with the aid of wings to have better range and possibly great forward airspeed.

Traditionally, there were few such aircraft, notably the British Harrier "Jump Jet" and the US Bell Boeing V-22 Osprey tiltrotor military aircraft. However, with the advent of electrically powered aircraft likely to include short hops from places without runways, VTOL and eVTOL are now of great interest, with the FAA questioning how pilots should be certified as the delicate transition from vertical to horizontal flight and vice-versa is problematic, with qualification as a fixed wing pilot and helicopter pilot not covering it.

One version of the F-35, the F-35B, can take off virtually vertically.

eVTOL (Electric Vertical Take-Off and Landing) vehicles are a step toward Advanced Air Mobility (AAM), and/or **Urban Air Mobility (UAM)**.

~ W ~

WAAS (Wide Area Augmentation System) for GPS

See **GPS**.

Wake Surfing/Riding

Migrating birds ride the wake of those ahead to save energy. Airbus is thinking of applying this to airliners with a possible fuel-saving of 5 or 10 percent. See "**Fello'fly**."

Wake Turbulence

Like ships, airliners can produce an upsetting wake, but in their case, it is swirling vortices coming off the wingtips. Other aircraft caught up in them can be jolted and even destabilized, with small aircraft being particularly vulnerable.

Surprisingly, they are more persistent and hence dangerous in calm conditions for strong winds or gusts quickly break up the vortices.

Again, the vortices are especially powerful when emanating from the wingtips of large airliners laden with fuel just after takeoff and climbing in "clean configuration," that is, with the slats and flaps retracted. For this reason, air traffic controllers refer to large airliners as "heavy," thus alerting other aircraft to the possible danger.

As described in our book *Air Crashes and Miracle Landings*, in 2001 an Airbus A300 taking off slightly prematurely from New York's JFK airport flew into the wake of a Japan Airlines 747 that had taken off ahead of them. Climbing out in clean configuration, and heavily laden with fuel for the 14-hour flight to Tokyo, the 747 was generating a powerful wake. Instead of giving the aircraft its head, the pilot flying the Airbus overreacted and swished the rudder back and forth from stop to stop, putting so much force on the tail that it broke off and fell into the bay with the aircraft flying on to crash onto houses.

Air traffic controllers determine the separation required on landing and takeoff according to the weight of the airliner. It seems the 757 has a notoriously bad wake despite not having the takeoff weight to normally be classed as a "heavy." The superjumbo Airbus 380 is proving better than first feared in this regard, though there was an incident where an A380 upset a light aircraft flying below in the Middle East.

Washington, DC

Capital city of the United States, not to be confused with Washington State, situated in the Pacific Northwest. The letters DC stand for District of Columbia, a special federal district.

Weapons of Mass Destruction (WMD)

The term has been used ever since reference by the Archbishop of Canterbury to the bombing of Guernica in 1937. Over the years with ever more lethal weapons, biological and atomic, the meaning has been expanded. Now it sums up chemical, biological, radiological, and nuclear weapons.

However, terrorists, for example, can set off a primitive dirty bomb with it being seen as a weapon of mass destruction. The dangers of chemical and biological weapons should not be dismissed, with the fear that systems using artificial intelligence set them off without constraint.

Weather Radar

Weather radar on airliners has greatly improved, enabling pilots to circumvent dangerous storms. However, it cannot detect **Clear Air Turbulence (CAT)** because the air in question is dry and does not show up.

Weighing Passengers

In February 2024, Finnair announced it would weigh passengers, with their consent, not to charge them extra but to "refine weight estimates for planes before takeoff." This is instead of calculating a passenger's weight based on an arbitrary average that might not be applicable for Finnish passengers who in winter tend to wear very heavy coats in view of their cold climate. The airline said only the member of staff at the desk would see the weight.

As one might expect, there was an instant reaction on the Internet with overweight people saying it was humiliating, adding that extra seats or a special section on the aircraft should be provided for hard-done-by extra-large people and that anyway they should never be required to pay extra.

On the other hand, slimmer passengers were saying it was unfair that they were charged for slightly overweight luggage when someone weighing vastly more did not have to pay extra.

The weight of a passenger is a factor in:

1. Total weight of the aircraft and the fuel required,
2. Center of gravity (see **Croc' in Carryon Luggage**),
3. Unfairness of ticket pricing,
4. Encroachment on space of adjoining passenger(s).

Airlines use assumed averages for passenger weight, often adjusting them according to the time of year. For long-haul flights, airlines in the US assume at least 100 kg for adult males, with the FAA advising airlines to "visually" estimate and consider the weight of obviously very heavy passengers.

The EASA updated its passenger weight data in 2022, finding that the average weight of passengers was 75.6 kg, while the average carry-on baggage weight was 7.6 kg. Passenger weight had risen by 1.1 kg and carry-on luggage by 1.5 kg since 2009.

The now defunct airline, Samoa Air, made news in 2013 as the first to charge passengers according to their overall weight including baggage. This was because Samoa has one of the highest rates of obesity in the world.

Though it did not charge according to weight, Hawaiian Airlines with flights from Samoa's capital, Pago Pago, also started weighing passengers boarding there so it could distribute them accordingly throughout the cabin to ensure the appropriate center of gravity. Incidentally, the airline had noticed the fuel burn on those flights was consistently much higher than predicted using standard weights!

As was the case with Samoa, the assumed average passenger standard weights can be wildly wrong, with mistakes being made especially where children are concerned, there being great variations with age, not forgetting the fact that nowadays many of them may be obese, too.

No one mentions that airlines operating in the Far East such as Japan have the advantage that their passengers will, on average, weigh very much less than those elsewhere and not require so much legroom.

To sum up, ticket pricing is very unfair as regards passengers' weight, with light ones paying extra for luggage that is a little overweight when someone who may be double their weight overall pays nothing extra.

See **Plus-Size Passengers**.

~ Impact of Passenger Weight on Ultra-Long-Haul Flights

The weight of individual passengers on ultra-long-haul flights could perhaps be a starting point for factoring passenger weight into the cost of the ticket. To make it more politically palatable, very light people could perhaps be given a discount.

Incidentally, Qantas was happy to discover its 787 nonstop flights from Perth to London are very popular with women with babies and tots who presumably would happen to not weigh much. What will Qantas do when it starts its Sunrise Project flights in 2026 from Sydney to London and New York lasting 20 hours or longer? Already it is charging more for those non-stop flights from Perth to London, Paris, and Rome.

Whittle, Frank, Gloster Meteor Fighter in World War II

A brilliant British engineer born in Coventry in 1907 who joined the Royal Air Force as an apprentice, became an officer and a pilot, and was even sent to Peterhouse College, Cambridge University, where he attained a first-class degree.

He patented a form of turbojet engine, and had he been supported consistently, Britain might have had an operational jet-engine fighter by the middle of World War II. When in 1935 the German physicist and engineer, Hans von Ohain, was about to apply for a turbojet patent his lawyer showed him one filed by Whittle, making him modify his application.

It seems that the two were working in parallel and not copying each other. Von Ohain with support of Heinkel created the first turbojet engine to actually power an aircraft, the He-178, on August 27,1939. However, the Heinkel jets fell by the wayside and it was to be the Messerschmitt Me-262 jet fighter that finally entered service in April 1944. It could fly 93 mph (150 km/h) faster than any Allied aircraft and could achieve a 1:4 kill ratio. However, as mentioned in **! >>> World War II ~ Strategic Bombing (On Germany)** lack of fuel meant many Me-262s could not be deployed. Also the shortage of materials for the alloys meant the aircraft could not fly for long, not to mention the lack of trained pilots due to lack of fuel.

Working with Whittle and his company Power Jets Ltd., the Gloster Aircraft company in England produced the Gloster Meteor twinjet which first flew in 1943. It did not have the firepower of the Me-262 nor its sweptback wings, and did not engage in actual combat because of fears the Germans would learn its secrets if shot down over enemy territory. However, its speed meant it could be used to enable it to successfully tackle the German's V-1 "buzz bomb" cruise missiles on their way to targets in England.

As many as 3,947 Gloster Meteors were produced up until 1955 with it seeing service not only in the British RAF but also in the Australian, Belgium and Argentine air forces.

Whittle was a remarkable man, overcoming many obstacles and at times suffering from nervous breakdowns. He entered the RAF as an apprentice and left on medical grounds as an air commodore, to be showered with honors and honorary degrees.

Whiteout

A situation found in polar regions where snow on the ground and the absence of shadows and the presence of cloud make the pilot unable to identify the horizon or even a looming mountain.

A notorious case was the Air New Zealand DC-10 on a sightseeing flight in the Antarctic in 1979*. Without telling the pilots, airline staff had mistakenly reprogrammed the aircraft's computer with a flight plan taking it over Mount Erebus. Flying under visual flight rules the pilots (but not the flight engineer) thought they were safe for they could see out, and the flight plan took them over water. However, the whiteout prevented them noticing the volcano, Mount Erebus, ahead. They flew right into it in accordance with the flight plan they had been given. There were no survivors, and the scandal was worsened by the attempted cover-up that followed*.

Whole Aircraft Recovery Parachute System

Very light aircraft are sometimes fitted with parachutes that can be deployed should they run into trouble such as engine failure, spin, or stall. These lower the rate of descent and can allow the aircraft to come down with less or even no harm to the occupants.

Unfortunately, they are not effective when the aircraft is too low or travelling too fast. The Cirrus SF-50 Vision Jet is one such example.

Why Are Most Airliners Painted White?

There are several very good reasons:

1. **Cost**
   The cost of the paint alone can be as much as $200,000 to $350,000. White paint is more resistant to fading than dark colors and has to be redone less often with the aircraft out of service.

2. **White paint reflects sunlight**
   This keeps the cabin cooler, and this is especially important when under burning sun with the air-conditioning not running.

3. **Facilitates detection of leaks**
   Most important fluids such as fuel, oils, and hydraulic fluid are relatively dark in color, and leaks are more easily identified.
4. **Leased aircraft more easily rebranded**
   Around half of commercial aircraft in the world are leased. Leasing involves an airline hiring the aircraft (and sometimes the crew) for a pre-determined number of years. As such, when aircraft are regularly changing between operators, a predominantly white livery is easier and cheaper to rebrand.

Branding using colors is mostly done on the tail with it sometimes extending to the rear fuselage below.

In the event of a crash airlines may try to paint out their conspicuous tail logo!

Wicks

Looking like pieces of rigid string, these are attached to the trailing edges of the wings and other airfoils to dissipate static electricity, which might affect the avionics or cause other trouble. They are not designed to handle lightning and are, in fact, liable to be damaged by it and need replacement after a strike.

Even with the wicks dissipating most of the static electricity, there can still be some remaining meaning the aircraft must always be earthed when being refueled.

"Wild Weasels" Preemptively Suppressing Enemy Air Defenses [US]

US Air Force code name for platforms of any type carrying missiles able to sense an opponent's radar and other emissions and destroy the emitting sites, thus enabling other craft to follow relatively safely.

Confusingly, these are referred to as "antiradiation missiles," suggesting, to the uninitiated, that they can resist nuclear radiation, which is not the case.

Wind Tunnel

Wind tunnels that push air through them to see how airfoils behave are nothing new, with the first rudimentary ones being tried out in the eighteenth century. In World War II the Germans had a number including some supersonic ones, and at the end of the war the head of Boeing's technical staff was a member of a team visiting Germany to see what they had achieved, and he learnt about the value of a swept wing for high speeds from tests the Germans had done using supersonic wind tunnels. He cabled the company to tell it to stop the work it was doing on a conventional wing. (See **BOEING** Postwar.)

The Chinese are alleged to have a wind tunnel for speeds 25 times the speed of sound.

Windows of Airliner

Airliner windows normally have three layers:

1. A thin easily replaceable inner panel made of plastic about 5 mm thick called the "scratch panel" because that is what passengers are liable to scratch.

2. Another 5 mm thick panel with a little hole at the bottom, slightly smaller in area than the scratch panel because it is "gasketed." The tiny hole, called a bleed hole, serves to ensure the pressure between the inner and outer panels is balanced and to get rid of condensation.

3. A structurally supported high-strength outer panel made of strong plastic about 10–12 mm thick.

Windscreen/Windshield

Aircraft windscreens must be very strong and very high tech.

Even so, windscreens do very occasionally crack. In one well-publicized case in the UK in 1990, a windscreen panel of a BA-11 twin-jet blew out because maintenance staff had used the wrong bolts. It was on the captain's side, and he was sucked halfway out and only prevented from disappearing completely by the copilot grasping him. A flight attendant took over, grabbing and holding onto the legs of the captain, hanging halfway out. The aircraft landed successfully, with the captain having sustained various injuries, including some frostbite, but subsequently being able to return to duty. The flight attendant was not so lucky and suffered frostbite and eye damage.

Windsock

A sleeve that is open at one end and closed at the other (like a sock), with the open end attached to the top of a pole that indicates the direction of the wind and to some extent its velocity. Usually placed at airports and airfields but sometimes at chemical plants and windswept bridges and so on.

It is sometimes designed to rise (bloated) when the wind reaches 15 knots, with 5 other flexing points showing indicating speeds in 3-knot steps.

Wing

Though there used to be simple theories taken as gospel regarding the way wings provide lift, nowadays designers realize there are phenomena that are not fully understood, and the designing of them is something of an art. Much of the work designing a wing depends on hundreds or thousands of tests in **Wind Tunnels**.

Some wings can change their configuration according to the speed of the aircraft to enable them to take off and land at slow speed with the wings outstretched, and fly with them swept back for supersonic flight.

Winglet and Sharklet

"Mini wings" inclined at various angles and attached to the wingtips. They reduce drag rather than lessen wake turbulence. More importantly, they make the wing work as if it were longer.

On business jets where the wing can simply be extended to achieve the same or rather better effect, they are not really needed. According to Mentour Pilot—see **Internet Material on Piloting and**

Aviation—a prospective purchaser of a business jet allegedly complained to the manufacturer that their aircraft did not look sexy without winglets!

Where handling space at airports may limit wingspan, winglets can be a means whereby carrying capacity/range may be increased. While not being quite what one might consider winglets to be, the raked wingtips that the A350 and most Boeing 787s use are wingtip-type devices giving more lift for the same span. Sometimes these are called sharklets.

Wiring and Wi-Fi (Special Wave Band)

There are miles of wire on an airliner, posing problems of weight, possible short circuits and fire, and complexity.

See **IFE (In-Flight Entertainment) and Kapton**.

WORLD WAR 2 (Selected Facts)

Germany—Post-War Interview with Albert Speer

In May 1945, at the end of the war in Europe, there was a race to gain access to Albert Speer, Nazi Germany's Armaments Minister who knew better than anyone how Germany had survived for so long despite the relentless bombing.

The race to gain access to him was won by the American economist Paul Nitze, who together with John Kenneth Galbraith and George Ball, managed to interrogate him for seven days before the United Nations War Crimes Commission took him in hand.

Nitze, a former investment banker, was vice chairman of the United States Strategic Bombing Survey, and he thought information from Speer could be invaluable for dealing with Japan.

Their most significant finding was that Speer believed it was US bombing of Germany's synthetic oil production facilities that was the critical factor and **not** the bombing of the ball bearing plants and other industrial sites.

In fact, Speer regarded the USAAF attack on the ball bearing plants at Schweinfurt on October 14, 1943 as a net win for the Germans.

> The Americans had lost 60 B-17s, with another 17 no longer airworthy, and an additional 121 suffering minor damage. Worse, the numerous crew on each B-17 meant the number of aircrew killed, wounded, or missing in action was more than 600, totaling almost 20 percent of the men sortied, with it having virtually no effect on the German war effort for they had built up considerable stocks of ball bearings, had managed to simplify many bearings to need fewer or none at all, and anyway could get supplies via Switzerland and other countries.

In the interview, Speer said the day he realized the war lost was May 12, 1944 for it was then that 935 American bombers of the Eighth Air Force successfully attacked the German synthetic oil production facilities in central and Eastern Germany, with the Leuna synthetic oil production plant at Merseburg again the main target.

Consequently, though Germany still had plenty of aircraft, they could only fly a very limited number of missions, and worse still fuel could not be allocated for training new pilots, meaning they were almost immediately shot down. Also, the first ever operational jet fighter, the ME-62, could not be used to advantage, with it impossible to properly train pilots on such a gas-guzzler.

According to Speer, the terrible allied bombing raids on German cities did little, as in the case of the Blitz in London, to lower morale. It seems it was only the most devastating allied raids such as those on Hamburgh, Cologne, and dare one say Dresden, that really impacted morale.

Germany—Was Strategic Bombing Worthwhile?

Towards the end of the war in Europe the famous Jimmy Doolittle took command of some US bombing raids at a time where fighters such as the P-51 Mustang were finally able to accompany the bombers right into Germany. He changed tactics, saying the fighters no longer had to stay with the bombers but could go ahead and take on the German fighters as they took off from their bases to attack the bombers, and then take them on again as they returned to base.

The bombers akin to bait, served to make possible the depletion of the Luftwaffe before the Normandy landings.

Germany—Allied Casualty Rates

US and UK bomber crews initially faced terrible casualties.

For the UK's Bomber Command, the fatality rate was 44.4% (55.573 out of 125,000 aircrew). Morale was at risk and fearing crew would refuse to go on missions, the top brass came up with the pejorative term "LMF" (Lacking Moral Fiber) to trickily stigmatize aircrew finding themselves unable to continue going on missions.

Anyone classed as "LMF," would be removed from their unit, and, if officers, busted to the ranks and given menial tasks such as peeling potatoes. If pilots, their pilot's license would be revoked to prevent them getting a flying job after the war.

Though not widely publicized, the policy was applied more leniently later on for a number of those deemed LMFs had, prior to cracking up, heroically completed a good number of missions.

The US Army Air Force also had morale problems due to their at times horrendous losses but managed to turn things around by giving its fliers some hope by saying they would only have to do 25 missions. Also when Curtis Lemay took over, he declared anyone refusing to fly would be court-martialed, with this reducing the number.

That said, towards the end of the war with fighters such as the P-51 Mustang able to go all the way with the bombers to their targets and the Luftwaffe depleted, losses were far less anyway.

Nevertheless, before that, in 1943, it was said:

> *Only about 25% of US Eighth Air Force bomber crewmen completed their 25-mission tours—the other 75% were killed, severely wounded, or captured. Despite knowing the poor odds of finishing their tours, bomber crews courageously pressed their attacks mission after mission.*

Germany—Conclusion

The strategic bombing campaigns weakened Germany, for as Speer much later said, it took troops, guns, and munitions firstly away from the Eastern front, and later also from the Western. However, it was not the hammer blow winning the war as "Bomber Harris" in charge of the UK's Bomber Command envisaged, with, as said, the precision raids against the synthetic fuel production facilities much more significant.

Japan—The Doolittle Raid

Due to the vast distances bombers would have to fly, for two years the only air raid on Japan itself was the famous "Doolittle Raid" on April 18, 1942, where sixteen B-25 U.S. Army Air Force bombers led by Lieut. Col. James H. Doolittle taking off from the U.S. Navy aircraft carrier Hornet bombed Tokyo and other cities.

> To save weight, wooden broomsticks took the place of the guns sticking out of the back of the bombers. As the Hornet neared Japan, it encountered a Japanese patrol boat that managed to get off a report to Japan. Fearing this could lead to the loss of his aircraft carrier its captain thought it too dangerous to continue, forcing the bombers to take off prematurely often without enough fuel to reach their intended landing sites in China.

Though the damage the bombers inflicted on Japan was minimal, it shocked the Japanese, who felt they had lost face with the bombs falling near the Emperor. However, it did much to raise morale back in the United States—unfortunately at the expense of thousands of Chinese killed in reprisal for having collaborated.

Japan—The B-29

The US developed the B-29 bomber at a cost one and a half that of the Manhattan Project creating the atomic bombs. While being a technological marvel, the B-29 had many problems and initially aircrew were more than twice as likely to be lost from accidents than enemy fire.

Magnesium was used in the engines to save weight, but had the disadvantage that when the engines overheated as they so often did, they would suddenly ignite at such high temperatures that the adjacent wing spar would immediately soften and fail.

In 1944, with US airbases still too far from Japan for even the B-29, a scheme was hatched to ship B-29s to India and fly them "over the hump" (the high Himalayan mountains) to Chengdu in China together with fuel, supplies and maintenance personnel.

The first raid on Japan from China was on 15 June 1944 and consisted of 68 B-29s and the last on 29 March 1945. Not much was achieved at tremendous cost, and operations were shifted to the Mariana Islands as soon as these were taken over.

To begin with, the B-29s would fly to Japan from the Marinas at very high altitude and drop their conventional bombs very inaccurately with them blown even further off target by the jet stream.

Then a new commander, Curtis LeMay, took over and ordered them to come in low and drop incendiary napalm with terrible consequences due to the houses and buildings being made mostly of wood.

Japan—Deadliest Air Raid in History

On March 9,1945, more than 300 U.S. Army Air Force B-29s took off from those island bases on Saipan, Tinian, and Guam, taking some seven hours to fly the 1,500 miles to firebomb Tokyo. They first dropped small bombs as markers followed by continuous drops of a total of 500,000 M-69 napalm bomblets, terminating at 3:00 a.m.

100,000 or more, people—mostly civilians—were killed outright and an area of some 15.8 square miles turned to rubble. The immediate death toll almost exceeded that for **both** atomic bombs dropped some five months later.

By the time the two atomic bombs were dropped (on August 6 and 9, 1945) Curtis LeMay had already burned down some 75% of Japan's cities using B-29s with them having to fly those difficult 3,000+-mile round trips.

With runways on Okinawa (captured on July 2, 1945) close-by, not to mention more B-29s and hundreds of bombers no longer needed in Europe, Le May said the atom bombs were not needed.

Japan—Russia Joins The Fray and The Atom Bombs

The atomic bomb was developed primarily out of fear Hitler might develop one first, and not with Japan in mind.

To get an understanding of Japan's surrender it is useful to bear in mind the following dates:

August 6	1st atom bomb dropped on Hiroshima
August 8	Soviet Union declares war on Japan
August 9	2nd atom bomb dropped on Nagasaki
August 8-15	Russia decimates Japanese Forces in China
August 15	Formal surrender & Emperor's public address

> Even before the two atomic bombs were dropped, Japan's Supreme War Council consisting of six top officials had been deliberating on what terms surrender would be acceptable, with them all the while aware the war was essentially lost.
>
> They could not come to a decision for there was the 3-man peace faction wanting to surrender very much in line with the Potsdam Declaration facing the 3-man hardline faction wanting to pursue the war.
>
> The latter thought that preparations (involving the whole population and Kamikaze attacks) being made would make the invasion of Japan so costly it would force the US to grant Japan much better peace terms.

Finally, with Council deliberations being pursued in the presence of the Japanese Emperor in an underground chamber at the Palace, in an unprecedented move, they asked the Emperor to decide, with him proclaiming the war must end on the terms offered.

Though the atomic bombs had not militarily forced the hand of the council, they had certainly preyed on the mind of the Emperor, with him even citing them in his speech broadcast to the nation.

The Japanese had been all along hoping that the Russians could continue to act as intermediaries with the US but their declaration of war against Japan between the dropping of the two atomic bombs had put an end to that. Furthermore, the Russians were defeating the enfeebled Japanese troops in China and were about to arrive in Japan proper.

Many think it was the threat of the Russians rather than the atom bombs that was militarily critical in bringing about the end of the war. However, the existence of a "terrible new weapon" cited in the Emperor's speech to the nation was probably what made the him decide the war must end, and certainly made it easier for him to justify doing so, though the word "surrender" was not mentioned.

> Had the war continued according to the desires of the hardline faction the whole nation would have endured terrible suffering.
>
> Le May would have been able to pursue his desecration of Japan's cities and towns even more terrifyingly with many more aircraft flying much shorter distances.
>
> Those not killed, would have been starving and without fuel because of the blockade of the country by submarines and warships.

It can be argued that the atom bombs, though not militarily decisive, saved many Japanese lives and prevented much suffering.

"Killing Machines" had to be Defeated

US historian Victor David Hanson points out that the allies were faced with two implacable killing machines. Having started out trying to limit bombing to military-related targets, they finally had to resort to dreadful means to avoid the slaughter of millions of their own young men.

SOME OTHER WORLD WAR 2 FACTS

B-47—Boeing Learnt from Germany

In May 1945, the US Army Air Force sent a technical mission to occupied Germany led by von Kármán that visited the secret German aeronautics laboratory near Braunschweig.

George S. Schairer, head of Boeing's technical staff, was part of the mission team, and on seeing the work the Germans had done realized their wind tunnel tests confirmed the controversial thesis proposed by R. T. Jones that a swept wing is best for high-speed jet aircraft. Schairer immediately cabled his people back at

Boeing telling them to halt work on the B-47 bomber and its wing in view of the information he had garnered.

Incidentally, it was Boeing's adoption of the swept wing in its jet aircraft designs that gave it an advantage over Douglas's DC-8.

Norden Bombsight Myth

The US Norden Company bombsight as used by the US Army Air Force in Europe in World War 2 was purported to be such a technical marvel that its secret had to be protected at all costs. Pre-mission, bombardiers (bomb aimers) were not allowed to take notes and had to memorize everything and were asked to be willing to give up their lives to protect its secrets.

After their bombers landed back at base in England, the "brains" of the bombsight had to be removed and kept in a bag in a secure room at the airfield

All this, when the Germans allegedly knew all about it even before US entered the war thanks to a spy at Norden in the US, and had deemed it was not worth copying! Admittedly, the Germans preferred dive bombing to bombing from great heights.

Whatever the truth, the Norden bombsight never came near matching the hype surrounding it. For instance, it was claimed it could drop bombs in a pickle barrel from 10,000 feet. This was ridiculous as it would be impossible to even see the barrel obscured by the crosshairs, but the policy was pursued and the Norden bombsight promoted because it suited the various parties.

In reality, the accuracy of the Norden bombsight largely depended on the ability of the bombardier and the conditions in which the bomb run was being made. Though not as bad as some say, it was no way near meeting the exaggerated expectations.

V-1—The First Cruise Missile

This was the world's first cruise missile, using a pulse jet engine that was made by the Germans but lacked an accurate guidance system such as we have today. However, it was cheap, and cost-effective compared with the V-2.

It could do a lot of damage to property and unnerve civilian populations due to the sound of its pulsating engine, which would stop just before it plunged earthward and detonated. It was just slow enough to be shot down by antiaircraft fire and some fighters, including the Gloster Meteor (See **Whittle, Frank, Gloster Meteor**).

The British had to devote considerable resources to deal with it, and even evacuate children including the author of this book from London who as a child in London remembers hearing a V-1 approaching, and its pulse jet engine stopping (meaning it would be coming down), and, after a pause it detonating with the windows of the sweet shop he was in blowing out.

It was only when the launch sites in France were overrun following the Normandy landings that V-1 strikes on England ceased.

Some 7,573 V-1 missiles (representing 73 percent of those launched) were unsuccessful, of which 1,847 were downed by fighters, 1,878 destroyed by antiaircraft fire very often over the coast, 232 blocked by barrage balloons , and others landing in open country.

In total, 2,419 V-1 missiles (23 percent) did succeed in reaching the general London area, killing and injuring civilians.

V-2—First Operational Ballistic Missile

Wernher von Braun (see **Operation Paperclip**), led the teams designing the V-2 rocket, the world's first proper ballistic missile.

It took only five minutes from launch for a V-2 to reach London, with it coming down supersonically after reaching a height of 50 miles en route, thus making it unstoppable.

A total of 1,402 V-2s were launched at England, of which 1,358 reached London. They killed 2,754 people and injured 6,523.

It came down randomly, again with civilians suffering most.

However, Germans spies sent to England were mostly turned by the British and forced to falsely report where the V-1 and V-2 missiles were landing. This resulted in the Germans mistakenly setting them to land short or off to the side and saved many lives.

The absence of accurate final guidance systems meant the V-2s in particular were a great waste of resources.

Kamikaze (A Human Guidance System)

While the German V-1 and V-2 weapons were let down by the lack of an accurate guidance system, the Japanese had a guidance system in human form—the Kamikaze suicide pilots. These were more of a threat and trouble than the US public were allowed to know at the time.

The effectiveness of the Kamikaze was limited by the fact that the velocity and mass of the aircraft used was not usually enough to cause significant damage.

END OF WORLD WAR 2 ENTRY

WSI Weather Information for Pilots

WSI provides weather information worldwide for diverse industries and organizations and in particular has an app for pilots enabling them to see predicted weather over their route and which they can consult en route. Their route can be automatically entered to show on charts various conditions such as wind direction and velocity, temperature, cloud, and, in particular, turbulence.

WSPR (Weak Signal Propagation Report) [MH370]

With it proving impossible to find the wreckage of Malaysian Airlines **MH370**, which disappeared on a flight from Kuala Lumpur to Beijing, it has been claimed that WSPR (pronounced whisper) has been successfully used to track it and even determine where it came down.

WSPR radio signals are designed to test the strength of amateur radio frequencies with global transmitters sending thousands of low-power radio pulses around the world, and across oceans, every two minutes.

When an aircraft crosses any radio signal it will, according to believers in this tracking method, visibly disturb it.

Anyway, a widely publicized analysis was done with it purportedly showing MH370 going into a 20-minute holding pattern before turning south towards the South Pole and having used up fuel coming down further down than supposed, in fact, due West of Perth. This site coincided with the one worked out by drift analysis of wreckage such as the **Flaperon** found on Reunion.

With top academics ridiculing it, how much credence should be given to this determination by WSPR is doubtful.

See **Victor Iannello, Radiant Physics [MH370]** for scientific appraisal.

~ X~Y ~ Z ~

X-, Y-, Z-Axes

When describing how an aircraft behaves, it is sometimes useful to cite the axis about which it rotates:

1. X-axis: Rolling or banking. One wing goes up and the other down.

2. Y-axis: Pitching. The aircraft nose goes up and down.

3. Z-axis: Turning movement to left or right, sometimes referred to as the *compass direction.*

Yaw

Term used to express the turning or swerve of the aircraft to the left or right (about the z-axis).

Your Life Is Worth How Much?

In the context of the regulatory framework, such as the FAA's, cost-benefit comes to the fore in deciding whether a new safety feature should be made mandatory. Various formulae are used to calculate the value of a life, not only in terms of lost earnings but also of the loss to society and family. In February 2008 the FAA raised to $5.8 million the value of a statistical life to be used by analysts in the Department of Transportation when assessing the benefit of preventing fatalities. Note this is quite different from the sums the courts will pay in compensation after an accident, which vary enormously from country to country and consider the earning power and the family responsibilities of the deceased, and whether the airline or manufacturer was at fault.

The phrase *tombstone mentality* is not officially used but sums up the recorded fact that safety improvements are very often only introduced after a serious accident has caused legislators, the media, and the public to focus on the problem and demand immediate action.

Zell (Zero Length Launch) System

In the 1950s and 1960s, the US and NATO worried that the USSR could take out airbases, experimented with mobile fighter launch systems that could be deployed anywhere after an attack on airfields.

The aircraft would be mounted on a rocket-powered booster and launched at the push of a button. Though it worked, it was found to be costly and impractical, and was abandoned.

Even so, the recent **ACE (Agile Combat Employment) with Dispersed "Austere" Airbases** is something similar and facilitated by modern high-powered fighters not requiring the extremely long runways they did in the fifties.

Zeppelin

The Germans developed sophisticated airships before World War I, with them being used largely for sight-seeing trips for the rich.

When they were used at night to bomb London, it was initially not easy to see them and shoot at them. Later in the war, when powerful searchlights were able to make them visible, artillery proved relatively ineffective. Even when planes were sent up with a weapon firing incendiary bullets, they did not have much success because a flaming bullet in pure hydrogen with no oxygen did nothing.

Finally, the life of the Zeppelin as a bomber ended with England's development of a bullet with nitroglycerine that ejected particles shattering the sides of the gas-filled balloons to allow ingress of oxygen.

See **Lady Grace Drummond-Hay, "Around the World by Zeppelin."**

Zulu (GMT/UTC)

In the 1950s, "Z" was used to denote Greenwich Mean Time, and as Zulu stands for Z in the NATO phonetic alphabet, it is the word used to denote GMT or rather **UTC (Coordinated Universal Time**). An example would be: 10:45 Zulu.

~ 0123456789 ~

1934 Air Mail Act [US]

This act mandated that aircraft manufacturers be separate from airlines resulting in the breakup of the United Aircraft & Transport Corporation (UTAC) made up of United Airlines, Pratt & Whitney, and Boeing.

2030s

Many * projects, whether they be military or civil, seem only to be destined to be realized in the next decade.

On one level airframers such as Boeing seem to be waiting for new engines to be perfected before embarking on the development of a new airliner.

As mentioned elsewhere in this book, the **blended wing** may come into its own enabling small upstarts to erode the Boeing–Airbus duopoly based on experience with wing-and-tube designs.

5G Impact of Cellular Phone on Radio Altimeters [US]

When the 5G mobile phone service was rolled out in the US at the beginning of 2022, there was great concern on the part of the FAA with it warning that 5G could make it impossible to use autoland systems that depend largely on radio-altimeters working on frequencies close to the 5G ones. This was surprising as they had had two years to prepare for it.

5G had been installed in other countries without presenting problems for aircraft. The problem in the US was that the frequency allotted to 5G was closer to that which radio-altimeters use, making interference more likely. Possibly, the size of the country meant their 5G signals were more powerful.

After certain flights were cancelled, US 5G providers agreed to constrain signals at airports, and the problem seemed to have gone away. It is possible, that with many US airports being so geographically large, it became possible to lower the impact in the vicinity of runways.

However, a proper solution must be found because airports are one of the places where thousands of people will most want to avail themselves of the great facilities 5G can offer.

9/11: Two Unarmed F-16s Scrambled to Down 4th Airliner

Much has been written about the intentional crashing of the two hijacked airliners into the World Trade Center in New York and the third into the Pentagon near Washington, DC, on September 11, 2001.

However, not so much is said about the measures taken to deal with the fourth hijacked airliner, United Airlines Flight 93, that had been delayed and still posed a threat.

Aloft in **Air Force One**, US President Bush was asked by Vice President Cheney for shootdown authority for United 93 which was over the Ohio valley and had turned towards Washington, DC. The president duly granted him that authority should it fail to comply with commands from the fighters.

Col. Tillman commanding Air Force One at the time said that because so many people were on their cellphones, the President was having difficulty contacting officials, and notably the Secretary of Defense who was outside the Pentagon observing the damage caused by the third hijacked aircraft.

There was some confusion over this because the only people able to authorize the shooting down of a civilian airliner were the President or the Secretary of Defense who could not, as said, be contacted because he was outside the Pentagon.

The only fighters near enough were at Andrews Air Force Base near Washington DC, but with no time to arm them, the task of downing United 93—by ramming their F-16s into it—fell to Col. Marc Sasseville accompanied by a female Lieutenant, Heather Penney, nicknamed "Lucky."

"I'm going to go for the cockpit," declared Sasseville.

"I'll take the tail," replied Lucky.

Scrambling with no time for the usual checklists, Sasseville and Penney headed northwest over the Pentagon from where fumes were still rising from the impact on it by the third hijacked airliner.

Uncertain what would happen when they rammed United 93, Penney feared she might eject prematurely and miss altogether.

Not being given any information regarding United 93's whereabouts, and not knowing whether there might be another rogue airliner posing a threat, they had to fly low, doing sweeps to East and West, looking out for ones on their radar and liaising with air traffic control who were having to switch from their normal task of keeping aircraft away from each other to a conflict management mode, having to confirm potential bogies and eliminate friendly flights such as medivac ones and even helicopters, showing up on the F-16's radar.

This continued for almost an hour and a half, with them landing back at Andrews still not knowing the passengers on United 93 had confronted the hijackers, with the aircraft crashing in open country.

As a matter of fact, had United 93 not crashed, it would very likely have reached Washington 15 minutes before Sasseville and Penney even lifted off. Nevertheless, if there had been a fifth hijacked aircraft they could well have found it during their sweeps and been obliged to ram it. So, it was not in vain and could have meant many lives saved.

In a recent interview, USAF major Penney, now retired, said:

We took off knowing that if we were successful in our mission we would not be coming back.

Sasseville was years later asked how he about the possible suicide mission with no weapons, he said graduating in the Air Force and later over the years learning about the heroic things servicemen had done not only in the air but also on the ground say in Vietnam one just did not give it a second thought.

One should stress that, unlike today, the fighters at Andrews were not integrated with any rapid response network such as NORAD, more concerned with nuclear threats from the North. Much of what the two did was on their own initiative, with their unit working in collaboration with the White House and the Secret Service with which Andrews had a close relationship on a daily basis cooperating about the use of Air Force One and the need to halt other flights in an and out.

On hearing that Flight 93 had crashed, the President and his staff on board Air Force One at first assumed it had been duly shot down and were relieved to subsequently learn that was not the case. The passengers on United 93 were the real heroes.

After returning to Andrews after a couple of hours, Sasseville's and Penney's aircraft were refueled and loaded with live weapons.

On taking off again, they found a Norad AWACS aircraft had taken over command with its sophisticated equipment giving a much better view of the scene than could civilian air traffic control radar. However, their F-16s were at the time not equipped with the secure communications equipment required for communicating with NORAD AWACS so messages between them had to be whispered vocally.

Finally, they were told the president was returning to the White House and they were to provide a fighter escort for Air Force One. On their arrival at Andrews Penney made up and down maneuvers to ensure Air Force One was not shot down by mistake.

Interestingly, there were two other F-16s from Andrews aloft patrolling Washington DC and NYC airspace that day with both pilots later to become generals. One was Dan Caine.

TABLE OF CONTENTS

ACKNOWLEDGEMENTS

So many have contributed in one way or another to this book that it is impossible to thank them individually. Also, key ones cannot be cited without running the risk of them automatically being assumed to be the source of some of our more contentious material or that we are quoting their views.

The sources listed here under the entry **Sources of Aviation Information** were invaluable.

Finally, many thanks to Ting Baker for doing the copyediting, with me wondering how she picked up on some obscure points, and to Nikita Wood for her most helpful check.

Christopher Bartlett

APPENDIX

Air Crashes and Miracle Landings

FYI

https://chrisbart.com

www.ingramcontent.com/pod-product-compliance
Lightning Source LLC
LaVergne TN
LVHW021739071126
851857LV00039B/540

* 9 7 8 0 9 5 6 0 7 2 3 9 9 *